In the Kingdom of Coal

An American Family
and the Rock That Changed the World

Dan Rottenberg

ROUTLEDGE
NEW YORK AND LONDON

Published in 2003 by
Routledge
29 West 35th Street
New York, NY 10001
www.routledge-ny.com

Published in Great Britain by
Routledge
11 New Fetter Lane
London EC4P 4EE
www.routledge.co.uk

Routledge is an imprint of the Taylor & Francis Group.
Printed in the United States of America on acid-free paper.

10 9 8 7 6 5 4 3 2 1

Library of Congress Cataloging-in-Publication Data

Rottenberg, Dan.
    In the kingdom of coal : an American family and the rock that changed
the world / by Dan Rottenberg.
        p. cm.
Includes bibliographical references and index.
    ISBN 0-415-93522-9 (hardback : alk. paper)
1. Leisenring family. 2. Businessmen—United States—Biography. 3.
Givens family. 4. Coal miners—United States—Biography. 5. Coal
trade—United States—History. 6. Coal mines and mining—United
States—History. I. Title.
    HD9550.L4R68 2003
    338.7'622334'092273—dc21
                                                    2003009386

Contents

PART III: BIG STONE GAP

PART IV: TO THE POWDER RIVER

Introduction
The Message in the Necho Allen Hotel

When I was a boy in the late 1940s my maternal grandfather's business transferred him from New York City, where he had always lived, to the town of Pottsville in eastern Pennsylvania. As an eight-year-old New Yorker I naturally wondered why my grandfather would forsake the world's most sophisticated metropolis for a provincial burg of twenty-four thousand souls stuck in the middle of nowhere. My confusion was further compounded by my subsequent discovery that nothing in my experience seemed quite so bizarre and exotic as a visit to my grandfather's new hometown.

Pottsville in 1950 was famous for its venerable Yuengling brewery (America's oldest), for the forty-two murders committed in the 1870s allegedly by the legendary Molly Maguires, and for its equally legendary native son John O'Hara, who set four of his novels there. But I knew none of this. All I knew was that visiting Pottsville felt like walking onto the set of a movie—one of those Hollywood small-town musicals in which even the lowliest extras radiate a mystical sense of glamour, confidence, and prosperity.

Downtown Pottsville in those days bustled with the energy and wealth of cities many times its size. Within perhaps a dozen square blocks were crammed five department stores, seven jewelry shops, four furriers, nine shoe stores, eleven furniture stores, and thirty-seven clothing shops. The town supported two daily newspapers and dozens of restaurants and taverns, as well as several nightclubs and three movie theaters, the largest of which, the Capitol, seated twenty-seven hundred patrons. Then there were the showplace Victorian mansions of the anthracite coal barons ascending the hills of Mahantongo Street, better known as Pottsville's "Millionaire's Row." And there was Pottsville's ingenious system of underground steam pipes, the heat from which kept the streets clear of ice and snow in all but the worst winter storms.

No fewer than nine hotels catered to the streams of coal brokers, entertainers, and tourists who constantly passed through Pottsville. The grandest hotel, which endures in my memory to this day, was the red-brick, nine-storey Necho Allen, named for the hunter who discovered anthracite coal on a nearby mountain in 1790. This hotel's subterranean Coal Mine Tap Room, decorated with anthracite walls and ceilings supported by mine timbers, attracted visitors from across the country who would never dare to set foot in an actual coal mine. In the Necho's ballroom, big bands entertained standing-room-only crowds. The Necho's elegant lobby-level dining room, with its ornate chandeliers and heavy white table linens, was dominated by an enormous three-piece oil-on-canvas painting of Sistine Chapel pretensions—but the subject of this monumental triptych was not God awakening Adam, but Necho Allen arousing succeeding generations of coal miners, coal towns, and coal machines to their heroic destiny.

On the Necho's seventh floor the big local coal companies maintained suites to entertain coal brokers and other visiting businessmen, and bellhops delivering sandwiches and drinks to the men playing cards there sometimes came away with tips of $5 or $10—more money than many Americans then earned for an entire day's work. Because the Necho's guest rooms were usually fully booked, you could always find people standing in the lobby, waiting for a cancellation. And the centerpiece of this lobby, enclosed behind a glass display case, was a giant and massive chunk of anthracite coal, taller than anyone who walked past it. In much of the rest of the world this ugly rock symbolized dirt, danger, and pollution, not to mention man's exploitation of his fellow man; but in my grandfather's magical town virtually everyone venerated it as a precious jewel and the source of all their blessings.

Like most eight-year-old New Yorkers (and many much older New Yorkers), I assumed that electricity came from a wall socket, that heat came from a radiator, and that coal was a blight on humanity. Now I found myself wondering: Was it possible that the bourgeois burghers of Pottsville knew something that we erudite New Yorkers did not?

In time my grandfather retired and returned to New York, coal and Pottsville receded from my memory, and I became a journalist grappling with the eternal challenge of my trade: Why do the world's most pressing problems seem to steadfastly resist solution? The answer, I concluded after nearly four decades as a reporter and commentator, is that the most important things in life are often the least attractive, and consequently the world's attention wanders elsewhere—to the Super Bowl, say, or *Sex and the City*. The development of the common law, the workings of financial markets, the raising and processing of the world's food supply, the evolution of indoor plumbing and modern sewage systems, the complexities of bookkeeping, the blessings of life insurance and property insurance—these and similar milestones in the history of human progress rarely inspire best-selling novels or feature films.

Certainly coal belonged high on such a list. Here was the critical force behind yesterday's industrial revolution and today's electricity supply—which is to say the critical force driving the modern world for the past two centuries. Yet with the exception of a few books on mine disasters, a few biographies of John L. Lewis, and a few movies set in coal towns, the titanic human struggle to extract coal from the ground and bring it to a relentlessly capricious market had never been chronicled for a popular audience. California gold miners and Texas oil men may have seized the public's imagination; but outside of a few places like Pottsville, coal has been dismissed, as the historian Nathaniel Burt put it, as "a dirty business in every sense of the word."

So how, precisely, does a writer dramatize such a subject in accessible human terms? I despaired of answering that question for many years until, serendipitously, I first encountered E. B. (Ted) Leisenring Jr. in 1997. Ted Leisenring had recently retired as chief executive of Westmoreland Coal Company, America's oldest surviving coal concern. His own family traced its coal lineage back even further than Westmoreland's birth in 1854—all the way back, in fact, to the very dawn of Pennsylvania's anthracite coal industry in the early nineteenth century. Ted represented the fifth generation of a dynasty that had built and operated coal mines not only in the hard-coal anthracite regions of eastern Pennsylvania, but also in the soft bituminous coal mines of western Pennsylvania, the hills of southwestern Virginia and, finally, in the surface mines of Montana, Colorado, and Wyoming. In the course of two centuries the Leisenrings had tangled personally with, and outlasted, such legendary predators as Henry Clay Frick, Andrew Carnegie, John L. Lewis, and Tony Boyle.

In a fractious industry comprising literally thousands of companies, no single family could be said to personify the history of coal, but the Leisenrings came closer to that description than anyone else. And virtually all of the family and company papers had been preserved and catalogued at the Hagley Library in Wilmington, Delaware, just waiting for some author like me to delve into them. Much of this trove consisted of correspondence between the company's executives back East and their mining superintendents in remote places like Connellsville, Pennsylvania, and Big Stone Gap, Virginia—the sort of day-to-day communication that at most companies would be spoken and consequently lost to posterity, but in this company's case had been put in writing and consequently preserved.

Ted Leisenring himself was a Yale graduate who possessed a curious mind, a deferential manner, a pleasantly manageable ego, a bachelor's degree in English, and an interest in words and language that he had never put to practical use. But as I came to know him in the ensuing years I discerned something else beneath his stoic and self-deprecating Republican businessman's exterior: the heart of a romantic. He had entered the coal business less for fortune or duty than for the sheer masculine adventure of it; and at a critical point when the future of his company was at stake he had followed his heart rather than his head. Although his company and his family's legacy lay seemingly on the brink of extinction when I first met him, Ted was eager to explore the history of coal and his family's role in it—not for self-aggrandizement or self-justification, but simply as a way, in his seventies, to extend the adventure. Thus was conceived my plan to use the Leisenring family saga as a microcosm for the story of coal and, by extension, the story of the modern industrial world.

It was left to my editor, Karen Wolny of Routledge, to suggest an even more dramatic possibility. Instead of examining coal solely through the prism of entrepreneurs like the Leisenrings, she suggested, why not enhance my drama and balance my picture with a coal miner's perspective as well?

In due course I traveled to Big Stone Gap, Virginia, to meet a few retired miners who had worked underground with Ted Leisenring in his earliest days on the job, in 1949. I hoped to find at least one whose family might provide a compelling counterweight to the Leisenrings. My prayers were answered beyond my expectations when I stumbled upon the Givens family, whose men had worked in the Leisenring family's mines for four generations.

In some respects the Givens family story was even more remarkable than that of the Leisenrings: At least two dozen Givenses and relatives had worked in the mines, many for thirty or forty years; two were killed there, and one lost an arm; yet the survivors seemed to have attained an ennobling level of serenity that I have rarely encountered. My subsequent discovery that one of the Givens descendants became a famous professional athlete merely provided the icing on this story's cake.

Like two trains running along parallel and frequently intersecting tracks, the Leisenring and Givens families had pursued their separate destinies—one above ground, the other below—for generations while locked together, for better or worse, in a common quest. This was no mere upstairs/downstairs fable of servants and masters; it was the story of two very different families captivated by the same demanding mistress, whose enticing potential rewards were rivaled only by the sacrifices she exacted in return.

Yet of course all of us who dwell in the modern industrial world have reached a similar accommodation with coal, albeit unwittingly. Ultimately the dominant character in this story is neither a Leisenring nor a Givens nor any of the gifted players they encountered along their journey. It's that massive chunk of black rock that I first encountered, in 1950, behind a glass showcase in the lobby of the Necho Allen Hotel.

—Philadelphia, January 2003

Leisenring and Givens
Family Trees

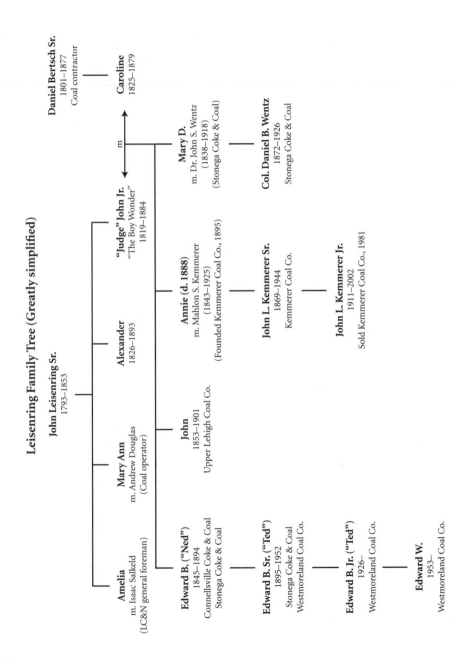

Leisenring Family Tree (Greatly simplified)

Daniel Bertsch Sr.
1801–1877
Coal contractor

Caroline
1825–1879

m

"Judge" John Jr.
"The Boy Wonder"
1819–1884

Mary D.
m. Dr. John S. Wentz
(1838–1918)
(Stonega Coke & Coal)

Col. Daniel B. Wentz
1872–1926
Stonega Coke & Coal

John Leisenring Sr.
1793–1853

Alexander
1826–1893

Annie (d. 1888)
m. Mahlon S. Kemmerer
(1843–1925)
(Founded Kemmerer Coal Co., 1895)

John L. Kemmerer Sr.
1869–1944
Kemmerer Coal Co.

John L. Kemmerer Jr.
1911–2002
Sold Kemmerer Coal Co., 1981

Mary Ann
m. Andrew Douglas
(Coal operator)

John
1853–1901
Upper Lehigh Coal Co.

Amelia
m. Isaac Salkeld
(LC&N general foreman)

Edward B. ("Ned")
1845–1894
Connellsville Coke & Coal
Stonega Coke & Coal

Edward B. Sr. ("Ted")
1895–1952
Stonega Coke & Coal
Westmoreland Coal Co.

Edward B. Jr. ("Ted")
1926–
Westmoreland Coal Co.

Edward W.
1953–
Westmoreland Coal Co.

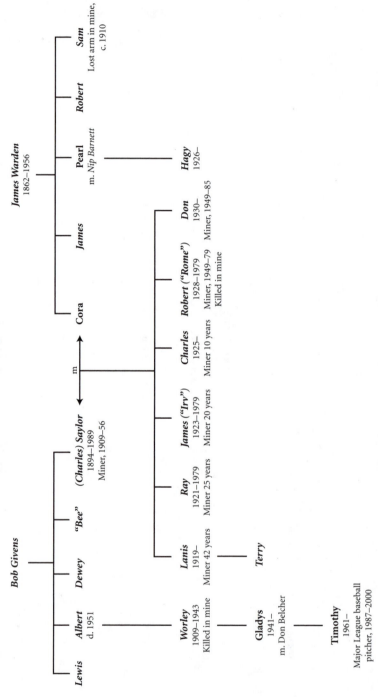

Givens Family Tree (Greatly simplified) (Miners' names in italic)

James Warden
1862–1956

Bob Givens

Lewis **Albert** **Dewey** **"Bee"** **(Charles) Saylor** **Cora** **James** **Pearl** **Robert** **Sam**
d. 1951 1894–1989 m. *Nip Barnett* Lost arm in mine,
Miner, 1909–56 c. 1910

m

Worley *Lanis* *Ray* *James ("Irv")* *Charles* *Robert ("Rome")* *Don* **Hagy**
1909–1943 1919– 1921–1979 1923–1979 1925– 1928–1979 1930– 1926–
Killed in mine Miner 42 years Miner 25 years Miner 20 years Miner 10 years Miner, 1949–79 Miner, 1949–85
Killed in mine

Gladys **Terry**
1941–
m. Don Belcher

Timothy
1961–
Major League baseball
pitcher, 1987–2000

x

Chronology

1792—Lehigh Coal Mining Co. created to mine and transport anthracite coal in Pennsylvania's Lehigh Valley.

1801—First horse-drawn railroads operate in England.

1813—Workman at Josiah White rolling mill in Philadelphia discovers long-lasting burning qualities of anthracite coal, solving need for fuel during British embargo of 1812.

1818—White and partner Erskine Hazard create Lehigh Navigation Co. as subsidiary of Lehigh Coal Co., begin to develop town of Mauch Chunk (now Jim Thorpe) as center of its operations.

1819—White and Hazard sell their mill at the Falls of Schuylkill to focus their attention on anthracite coal at Mauch Chunk.

1823—Josiah White sends 9,000 tons of anthracite downriver to Philadelphia, launching nineteenth-century coal boom.

1825—Opening of Erie Canal links New York City with Buffalo and Great Lakes.

1827—LC&N's gravity-based "switchback" railroad begins hauling coal from mines to Mauch Chunk.

1828—John Leisenring moves to Mauch Chunk from Lehighton, becomes proprietor of Mansion House, the LC&N Co.'s hotel.

1830—Peter Cooper builds the "Tom Thumb," powerful steam railroad locomotive.

1832—Josiah White retires from active direction of LC&N; John Leisenring succeeds him as postmaster of Mauch Chunk for next 16 years; LC&N sells Mansion House hotel to Leisenring.

1833—Reading Railroad chartered.

1835—John Leisenring gives up hotel, becomes general merchant; remains a central figure in Mauch Chunk/LC&N community.

1835–38—LC&N extends Lehigh Canal northward from Mauch Chunk to White Haven.

1835—Philadelphia firm of Eastwick & Harrison develops locomotive capable of burning anthracite coal.

1836—John Leisenring's son, also named John, starts work with LC&N engineer corps at age 17, working on canal extension.

1837—LC&N decides to construct Lehigh & Susquehanna Railroad between White Haven and Wilkes-Barre; young John Leisenring is placed in charge of eastern division.

1838—Hazard brings the Welsh iron master David Thomas to America.

1839—Reading Railroad carries first traffic between Philadelphia and Reading.

1840—David Thomas fires America's first anthracite-fueled hot blast iron furnace at Catasauqua, near Allentown.

1842—Reading Railroad reaches coal country at Mount Carbon, first step toward its absorption of the rival Schuylkill Canal to dominate anthracite transport. First collective miners' protest, Minersville, Pa.

1843—LC&N begins extending its mine railroads west of Mauch Chunk; young John Leisenring moves to Ashton (now Lansford) to be near the project.

1846—Pennsylvania Railroad incorporated; first iron suspension bridge spans Monongahela River at Pittsburgh.

1848—Bates Union, first miners' union, founded and collapses. Gold discovered in California.

1849—Henry Clay Frick born in southwestern Pennsylvania.

1852—Pennsylvania Railroad links Philadelphia with Pittsburgh.

1853—Death of John Leisenring Sr., July 9.

1854—John Leisenring Jr. forms new partnership of Sharpe, Leisenring & Co. to do contract mine operating on leased land, first of a series of anthracite mining companies around Mauch Chunk for the "Boy wonder of the anthracite." His partnership also creates the Pennsylvania mining town of Eckley.

1854—Westmoreland Coal Company founded by Philadelphia investors.

1855—Bessemer steelmaking process developed.

1859—Oil discovered in western Pennsylvania.

1860—LC&N makes John Leisenring chief engineer and superintendent, forcing him to resign from his partnership. He moves back to Mauch Chunk.

1861—Civil War begins. American Miners' Association, first national miners' union, launched in St. Louis.

1862—John Leisenring heads rebuilding project after flood destroys Lehigh Canal from White Haven to Mauch Chunk.

1863—Robert E. Lee invades Pennsylvania, hoping to cut North's coal supply.

1864–67—John Leisenring expands his private operations through development of Upper Lehigh Coal Co., north of Eckley, the first mining venture in which he holds the major interest.

1868—John Leisenring resigns from LC&N to devote full time to his own companies and acquires 3,000 acres of coal land southwest of Wilkes-Barre.

1869—Knights of Labor formed in Philadelphia.

1871—John Leisenring elected to five-year term as associate judge of Pennsylvania District Court; Henry Clay Frick makes first purchase of coking coal lands in western Pennsylvania.

1872–73—John Leisenring, as director of Central Railroad of New Jersey, helps it assemble a mining subsidiary, Lehigh & Wilkes-Barre Coal Co.

1873—Andrew Carnegie opens Braddock Steel Works in western Pennsylvania, declares iron passé; Frick expands his coal holdings during Panic of 1873.

1875—John Leisenring's son-in-law Dr. John S. Wentz becomes a partner in John Leisenring & Co. and superintendent of its mine at Eckley.

1876—First murder trial of members of Molly Maguires in Pottsville, Pa.

1879—John D. Imboden promotes mineral deposits in southwestern Virginia.

1880—John Leisenring acquires coking coal lands around Connellsville, Pa. General John D. Imboden and Tinstman brothers form Tinsalia Coal & Iron Co. in Wise County, Va.

1880—John L. Lewis born in Lucas County, Iowa.

1881—Leisenring group's Connellsville Coke & Iron Co. makes coke for the first time in April 1881.

1882—John Leisenring group acquires Tinsalia Coal & Iron Co., which becomes Virginia Coal & Iron Co.

1882—Andrew Carnegie becomes silent partner (and biggest customer) of Henry Clay Frick, principal operator in Connellsville coal field.

1883—Frick becomes general manager of all of Carnegie's properties; organizes Coke Syndicate to control prices in southwestern Pennsylvania.

1884—Gas explosion at Leisenrings' Connellsville shaft claims 19 lives. John Leisenring dies of Bright's disease, age 65, succeeded by son Edward ("Ned"), who is assisted by brother John and brothers-in-law Dr. John S. Wentz and Mahlon S. Kemmerer.

1886—First labor strike in Connellsville coke regions; total union victory.

1889—Henry Clay Frick becomes head of all Carnegie operations, forces Leisenrings to sell him their Connellsville operations at cost. With proceeds, Leisenrings turn to develop their untapped deposits of coking coal in southwestern Virginia.

1890—United Mine Workers of America founded in Columbus, Ohio.

1892—After suppressing Homestead strike, Frick survives assassination attempt.

1894—Edward Leisenring dies (Sept. 20). Leadership of Leisenring group passes to his brother-in-law Dr. John S. Wentz; operations become known as "the Wentz interests."

1896—Virginia Coal & Iron opens its first mines and coke ovens at Stonega, Va. Superintendent John K. Taggart is killed in an explosion.

1902—Working of family's Virginia mines leased to newly created Stonega Coke & Coal Co., with Dr. John S. Wentz as president and largest stockholder.

1904—Dr. Wentz resigns presidency of family's bituminous coal companies, succeeded by his oldest son, Daniel, but retains presidency of family's anthracite companies.

1908—John L. Lewis, age 28, moves to Panama, Ill.

1909—John L. Lewis appointed legislative agent for District 12 (Illinois) of the UMW. Saylor Givens, age 14, begins work as miner for Stonega.

1910—Stonega absorbs Keokee Consolidated Coke Co.

1911—John L. Lewis appointed representative of American Federation of Labor.

1914—Ludlow Massacre: Colorado National Guardsmen open fire on camp of striking UMW miners.

1915—Stonega begins decade of rapid expansion.

1916–17—Dr. John Wentz acquires large block of stock (3,583 shares) of Westmoreland Coal Co., the oldest surviving bituminous coal company (founded 1854).

1917—National Coal Association founded by operators, including Daniel Wentz of Stonega. U.S. declares war on Germany.

1918—John L. Lewis appointed statistician and then vice president of UMW.

1918—John Wentz dies (July 1); wealth estimated at nearly $5 million. Succeeded by son, Colonel Daniel Wentz.

1919—John L. Lewis assumes duties of UMW president.

1920—John L. Lewis elected president of UMW.

1920–21—Daniel Wentz, a founder of the National Coal Association, elected its president.

1922—Leisenring family's various Pennsylvania anthracite companies merged into Hazle Brook Coal Co.; anthracite and bituminous strikes.

1924—Jacksonville Agreement promises labor peace.

1926—Colonel Daniel Wentz dies suddenly (Feb. 8), age 53. Succeeded by longtime Virginia associate Otis Mouser. E. B. (Ted) Leisenring, at age 31, becomes vice president of most family companies.

1929—Stonega's output peaks at 3.25 million tons, not to be exceeded until 1964.

1929—Stock market crashes. Westmoreland Coal Co. president S. Pemberton Hutchinson dies suddenly; Westmoreland's management turned over to Stonega's officers, doubling size of bituminous properties under the Leisenring group's management. When Mouser also dies, E. B. (Ted) Leisenring succeeds him as president of Virginia Coal & Iron, Westmoreland, and Stonega.

1933—Amid Depression, Congress passes National Industrial Recovery Act, giving impetus to massive UMW organizing drive.

1935—John L. Lewis withdraws UMW from AFL, organizes Congress of Industrial Organizations.

1939—World War II begins in Europe.

1941–42—UMW agrees to "No Strike" pledge in wake of Japan's attack on Pearl Harbor, then rescinds it.

1943—Worley Givens, age 34, killed in mine accident, Amonate, W.Va.

1945—World War II ends; Westmoreland and Stonega undergo extensive modernization program. Production nearly reaches pre-Depression high.

1946—UMW Welfare and Retirement Fund established.

1946–47—In face of nationwide UMW strike, federal government operates bituminous mines for 13 months.

1947—Mine disaster in Centralia, Ill., kills 111 men.

1948—John L. Lewis names W. A. (Tony) Boyle his assistant.

1949—E. B. (Ted) Leisenring Jr., at 23, goes to work for Stonega Coke & Coal in Big Stone Gap, Wise County, Va.

1951—E. B. Leisenring Sr. resigns his presidencies due to ill health; succeeded by Daniel Wentz's protégé, Ralph H. Knode.

1952—E. B. Leisenring Sr. dies, age 57.

1955—AFL and CIO merge.

1956—Saylor Givens, 62, retires after 47 years in mines.

1959—E. B. Leisenring Jr. becomes president of Stonega Coke & Coal.

1960—John L. Lewis retires as UMW president after 40 years.

1961—E. B. Leisenring Jr. becomes president also of Virginia Coal & Iron and of Westmoreland Coal Co.

1962—Rachel Carson's *Silent Spring* launches environmental movement.

1963—W. A. (Tony) Boyle elected UMW president.

1964—Westmoreland and Stonega consolidated into single operating company, christened Westmoreland.

1967—Virginia Coal & Iron Co. changes name to Penn Virginia Co.

1968—Westmoreland acquires West Virginia coal properties owned by Sprague family of Massachusetts for $13.59 million plus 300,000 Westmoreland shares, becomes nation's largest coal exporter.

1968—Strict new mine safety law enacted in wake of disaster at Morgantown, W.Va.

1969—"Black lung strike" prompts passage of federal Coal Mine Health & Safety Act. UMW dissident leader Jock Yablonski murdered, apparently under orders of union president Tony Boyle. John L. Lewis dies.

1970–71—E. B. Leisenring Jr. serves as president of National Coal Association.

1972—After failures to merge Westmoreland and Penn Virginia, the two companies are more completely separated.

1972—Reformer Arnold Miller defeats Tony Boyle for UMW presidency.

1973—OPEC oil embargo and energy crisis trigger new demand for coal. Coal prices, profits, and wages soar.

1973–76—Wildcat strikes by UMW increase, becoming critical problem.

1974—Former UMW President Tony Boyle convicted of 1969 slayings of union rival Joseph Yablonski and Yablonski's wife and daughter.

1974—Westmoreland acquires access to 1 billion tons of coal in Montana, joins movement to develop low-sulfur, non-unionized coal resources in western states.

1974—Federal government removes price controls from coal; companies not locked into restrictive long-term contracts begin to generate huge profits.

1975—Westmoreland announces plans to open two new western mines at cost of $20 million and $35 million, total cost equal to all of its 1974 earnings.

1975–76—Coal industry achieves record employment and earnings in 1975 and 1976. Westmoreland records highest return on investment of any U.S. company.

1976—Westmoreland begins mining in Colorado.

1976–78—Ted Leisenring serves as chairman of Bituminous Coal Operators Association, industry's negotiating arm.

1977–78—UMW strikes for 110 days.

1977—Congress passes Surface Mining Control and Reclamation Act. Also, Clean Air Act amendments require all new plants to be equipped with scrubbers, even if sulfur emissions could be limited by less expensive means. Over next 2.5 years, 40 percent of surface mining in Southwest Virginia shut down because of new law's requirements; Westmoreland Coal lays off 1,500 workers, or 30 percent of its employees.

1978—Decline of U.S. steel industry in face of foreign competition hurts metallurgical coal market. Westmoreland earnings drop to 20 cents a share in 1978 and 26 cents in 1979, down from record $8.82 a share in 1976.

1978—Meltdown at Three-Mile Island, Pa., undermines appeal of nuclear power.

1979—Robert "Rome" Givens killed in mine accident, Pine Branch, Va.

1979–83—Nationwide recession; contraction in U.S. steel industry hurts mines producing metallurgical coal, especially in West Virginia.

1979—Arnold Miller resigns as UMW president because of ill health.

1980—Westmoreland ranks number 13 among U.S. coal producers, with 12.7 million tons (vs. 59 million for industry leader Peabody Coal Co.).

1980—Westmoreland suspends dividends.

1988—Westmoreland president Pemberton Hutchinson becomes CEO; E. B. Leisenring Jr. remains chairman.

1993—Westmoreland suffers $92.3 million loss in fourth quarter, wiping out more than two-thirds of its previous $136.2 million net worth. Loses $94 million for year.

1994—Auditor KPMG Peat Marwick raises doubts about Westmoreland's ability to survive; Westmoreland seeks Chapter XI bankruptcy protection from demands of UMW Pension and Benefit Funds.

1995—Westmoreland moves headquarters to Colorado Springs, Colo.

1999—Westmoreland discharged from Chapter XI bankruptcy.

Lots of chocolate for me to eat,
Lots of coal making lots of heat;
Warm hands, warm hair, warm feet—
Now, wouldn't it be loverly?
—Lerner and Lowe, *My Fair Lady,* 1955

The conditions of coal mining are and always have been a disgrace to a civilized people.
—Henry Ford's *Dearborn Independent,* 1921

Christopher White crossed Atlantic 1676

Joshua White (great grandson)) 1728 ' 1785

son

& Joshua White II 1781

Prologue
"There Will Come a Time"

Josiah White grew up an ambitious and industrious young man in a society where concepts such as ambition and industry were scarcely comprehended, much less accepted. When he was born in 1781 the industrial revolution had already arrived in England, but the best and brightest Americans were still preoccupied with agriculture, war, and politics. Washington's victory over Cornwallis at Yorktown was still seven months in the future, and America's independence from Britain was hardly a foregone conclusion. The most productive Americans worked on farms; cities and towns were places where, as Benjamin Franklin observed in his autobiography, it was unusual for a man to be seen as busy with his business the entire day, much less actually *be* busy with it. The typical merchant or craftsman conducted his business not as a means to wealth or achievement but as an end in itself: His shop was the focal point of a leisurely daily routine spent communing with his friends and neighbors, punctuated by long breaks for meals and trips to the tavern.

Like Franklin's before him, White's restless and inquisitive mind responded to a different stimulus. From his ancestors he had inherited a willingness to sacrifice serenity for progress, so it was inevitable that he would disrupt the complacent patterns of his time and place. Josiah's great-great-great-grandfather Christopher White, a carpenter, had crossed the Atlantic with a contingent of two hundred Quakers to settle the colony of West Jersey in 1676, five years before William Penn founded Philadelphia across the Delaware River. Christopher had chosen his lot, sight unseen, from a map drawn in London before he departed for the New World. Over the next three generations the Whites transformed this wilderness into a productive wheat and timber estate whose wealth and ingenuity was symbol-

1

ized by the family's brick mansion, shaped like a tall cross so that sunlight streamed into every room.

But Christopher's great-grandson—the first Josiah White—had another interest beside farming. To alleviate flooding on his land, he began tinkering with dams along the Alloway River, a tributary of the Delaware that flowed through his property. The notion of harnessing or diverting the natural course of a river was unknown in the New World at that time, but in 1728, when the first Josiah White was only twenty-three, he approached his neighbors with an audacious proposal: He would build a permanent earthen dam along the Alloway and advance all the costs for materials and labor; his skeptical neighbors need not reimburse him for their share unless the dam lasted for more than a year. His dam did indeed stand firm for more than eleven months, but a few days short of its first anniversary it was maliciously wrecked—either by neighbors hoping to avoid payment for his work, or by others who envied his status as a young inheritor.

In the wake of this disappointment the first Josiah White subsequently sold his estate and moved to the village of Mount Holly, where he opened a small textile and dyeing mill along the Rancocas River. In effect he was launching a new life in a new business and a new community, relying on his own imagination, sociability, and courage rather than his landed inheritance. This was the very essence of a quality that Americans centuries later would praise as entrepreneurship. But such a mindset was virtually unknown among colonial Americans. Benjamin Franklin stood out in such a society because he was the exception, and the same was true of White.

The mill was still being operated by Josiah's son and daughter-in-law when the most entrepreneurial of the Whites, Josiah's namesake, was born to them. This blond-haired second Josiah White proved a precocious youngster from an early age, which is to say he was self-willed, impulsive, and difficult to lead. Adults considered him too playful for a Quaker child, yet his "play" merely reflected a curious and intelligent mind: Much of young Josiah's time was spent creating and fooling around with contraptions for no useful purpose other than to see how they would work. "I want to see if it will go," was his customary explanation. His elders found these activities pointless, but educators much later (Maria Montessori, for example) would recognize them as the foundation of learning.

Josiah's father died in 1785, when Josiah was only four, leaving the mill to his two eldest sons. As the third son, Josiah inherited some land and goods but would clearly need to learn a trade in order to survive. His mother, Rebecca, left to run a business and a home while raising four small boys, packed Josiah across the Delaware River to Philadelphia as soon as he was old enough to go off on his own.

When Josiah arrived in Philadelphia in 1796 he was a fifteen-year-old apprentice to a hardware merchant. His adopted city was a community of 40,000 basking in its reputation as America's financial and intellectual center and, indeed, the new nation's capital and its largest and most cosmopolitan city. On Philadelphia's neatly paved streets this provincial teenager found elegant women in furs and velvets and curving plumes, and refined men in satin breeches and powdered wigs, speaking half a dozen languages. To Thomas Jefferson, freshly returned from abroad, Philadelphia was "more impressive than London, or Paris"; to the great portrait artist Gilbert Stuart, Philadelphia was "the Athens of America." But to Josiah White's observant mind it was something much less and potentially much more.

The Philadelphia he perceived was a city top-heavy with shopkeepers, mariners, and mechanics but virtually devoid of producers. "People here earned a living by mercantile work," he later wrote disparagingly, "but at home"—that is, in New Jersey—"by farming." The three emblems on Philadelphia's coat of arms, adopted just seven years earlier—a plow, sheaves of wheat, and a ship under full sail—reflected the city's priorities. Philadelphia's skyline was dominated not by factories or bridges but by church spires and the masts of sailing ships. Its trading offices, coffeehouses, shipbuilders, wheelwrights, weavers, and glassblowers hugged the shoreline of the Delaware River because the city drew its lifeblood from the river and the sea trade. Beef, flour, hides, and timber arrived daily from the hinterlands to be loaded onto seagoing ships, while more than ten ships arrived and departed daily from Europe, China, South America, the West and East Indies, and the coasts of North America, bearing the goods of the world to be loaded and unloaded on Philadelphia's Front Street wharfs.

What little manufacturing existed—rope works, cooperages, boatyards—functioned to support this oceangoing trade. Goods were manufactured only in response to demand, not to create demand. Philadelphia's first iron foundry, the Mars Works, had been set up in 1792 and produced a high-pressure steam engine by 1803. But the concept of an industrial district, in Philadelphia or elsewhere, had yet to occur to anyone. Cabinetmakers and hatters and mechanics worked in tiny two- or three-man shops next door to doctors and stables and Georgian red-brick family homes trimmed in white. The city's tree-lined and pebblestone-paved streets, laid out in a checkerboard utopian vision by William Penn 120 years earlier, were now strolled with alacrity not only by men and women but by cows and pigs as well. The air was filled not with the diesel fumes or sulfur of industry but with the agricultural odors of wood, charcoal, manure, and beer being brewed.

So abundant were the city's ocean trading opportunities that few Philadelphians noticed that they possessed other natural resources as well. Unlike every Atlantic port other than Baltimore, Philadelphia was blessed with powerful internal streams capable of supplying water power to factories, if only someone would build them. A vast network of streams (many of which later dried up or were paved over by city streets) already powered waterwheels that drove flour mills as well as the sawmills that cut the marble front steps and door frames of Philadelphia's brick houses. In 1807 Pennsylvania's state government sold a Philadelphia tavernkeeper named Robert Kennedy the rights to the water power at the Falls of Schuylkill, west of the city, on condition that he build locks around the falls to ease the passage of freight boats coming down the Schuylkill River from Reading, some fifty miles to the northwest. Kennedy built a chain bridge there but did nothing to tame the river; nor did he attempt to harness the river by building any sort of mill. Presumably he reasoned that the rewards of industrial ventures didn't justify the potential risk.

In this environment Josiah White initially gravitated where the opportunities lay. Liberated from his apprenticeship on his twenty-first birthday in 1802, this short and somewhat stocky young man successfully established himself as an ironmonger—that is, a hardware dealer, like his former master. But his heart wasn't in the work. His mother, he later wrote, "had inculcated in her children a dislike to store-keeping, as too much encouraging pride and idleness, and rather tending to a cunning craftiness"; White himself preferred "a mechanical trade, a joiner or carpenter, as I was fond of tools." The previous year he had been fascinated by a tiny item from London published in the *Philadelphia Aurora*: "Railroads are strips of oak, laid upon a level road, about eight feet apart, with wheels to fit the rails thus laid. . . . The horse goes in the middle of the track. . . . Three times the weight can be moved than on a common road. . . . Railroads may sometimes supersede canals."

White's local heroes in those days were not his fellow bourgeoisie but eccentric visionaries like the local inventor Oliver Evans, who regularly amused Philadelphians with such public declarations as, "There will come a time when wagons will move on land with steam-engines, at twenty miles an hour, a man breakfasting in New York, dining in Philadelphia, and supping in Baltimore, all the same day." In his journal Josiah remarked, "There are not a half-dozen men in the state of Pennsylvania who know how to make a screw water-tight; machines are unknown." He resolved to save $40,000 by the age of thirty—an awesome goal, equivalent to millions in the twenty-first century. But that was the figure Josiah calculated would liberate him to spend the rest of his life tinkering with the technological innovations that most fascinated him but seemed then to lack financial reward. (Even Josiah didn't then perceive of technology as a source of wealth rather than a hobby.) However much he may have chafed at the life of a hardware merchant, Josiah wrote, "I will make it a strong point to endeavor to like whatever I judge necessary to this business. . . . Then I'll quit in toto."

White's plan, like Philadelphia's entire economy, suffered from a critical flaw. Philadelphia's dependence on commerce left its merchants entirely at the mercy of outside forces. Britain's navy, the world's most powerful, had begun stopping and searching foreign vessels, in theory to search for deserters but in practice often seizing American sailors and goods. In 1808, from the nation's new capital city of Washington, President Thomas Jefferson reacted by placing an embargo on all foreign commerce, causing Philadelphia's maritime activity to grind to a halt. Ships were tied up in port, shipbuilding stopped, and mobs of sailors besieged Philadelphia's City Hall, shouting, "How shall we live?" Merchants like Josiah White were effectively cut off from their sources of supply. An enterprising Philadelphian had no choice but to seek new opportunities.

Josiah White spotted his chance in 1810 when a newspaper ad announced that the water rights at the Falls of Schuylkill, having been neglected by Robert Kennedy, were once again for sale to anyone willing to build locks for navigation there. The "Falls" on the Schuylkill were actually swift rapids requiring immense locks, and it was then generally believed that locks couldn't be constructed on any major river. White responded to this challenge much as his grandfather had in New Jersey. He and his younger partner, Erskine Hazard, proceeded to build a dam, one hundred feet long, and two locks, each eighty feet by seventeen feet—the first time a major U.S. river had been tamed by human ingenuity. The lock opened the river to barge traffic and enabled the young partners to make money by charging a toll of 50 cents per barge— at least until the following spring, when a flood washed the locks away.

Threatened now with bankruptcy, White and Hazard thought up another way to profit from the Falls of Schuylkill. White designed and patented a machine for "rolling and molding" iron, and on their dam site the partners opened the Fairmount Pennsylvania Nail and Wire Works, powered by the flow of the river. This new mill's purpose was to produce some of the essential tools, like wrought-iron nails, that the embargo on foreign trade had choked off.

The Falls provided a cheap and effective source of basic power for White's mill, but there remained the challenge of heating his furnace for the necessary smelting of iron products. Wood was still plentiful in nearby forests, but wood couldn't satisfy the voracious appetite, not to mention the higher-temperature heat requirements, of an iron-producing blast furnace. At first White and Hazard used charcoal, a carbon residue distilled from wood. But the rising price of charcoal made it economically unfeasible to use in large quantities. So White and Hazard's fuel of choice was bituminous coal from Virginia. This soft coal, first mined in Virginia in 1750, was unpopular in towns because of the dark smoke and foul smell it produced when burned. But White's factory stood more than a mile from the heart of town; no one would sniff the despised bituminous coal there.

To be sure, a harder, cleaner, more potent, slower-burning type of coal, called *anthracite,* had been already discovered only a hundred miles north of Philadelphia on the banks of the Susquehanna River in 1762. As early as 1769 a blacksmith at Wilkes-Barre had actually used it to heat iron at his forge. The blacksmith had found to his amazement that this so-called "stone coal" or "black diamond" coal burned slowly, with a blue flame, producing little soot or smoke. Anthracite was the purest coal the world had ever seen—extremely high in the carbon content that generated heat but virtually devoid of smoky sulfur—and it existed in abundance in northeastern Pennsylvania and seemingly nowhere else. A small circle of Philadelphia businessmen, excited by its possibilities, had already created the Lehigh Coal Mining Company in 1792 to buy up land there and float the anthracite by barge down the Lehigh and Delaware Rivers to Philadelphia. By 1803 they had actually delivered a shipment to Philadelphia.

But the very hardness and purity of anthracite also made it difficult to ignite. Just as a huge log in a fireplace usually needs kindling to ignite but will burn for hours once it catches, anthracite coal lacked an igniting agent like sulfur. Whereas bituminous coal with its high sulfur content blazed up the moment it was thrown on a fire, anthracite seemed to take forever, if indeed it caught fire at all, and ignited only under heat so intense that few existing furnace grates could withstand it.

Thus that first 1803 shipment of anthracite coal proved impossible for Philadelphia homeowners and shopkeepers to burn in their small conventional wood stoves. Having been pronounced unburnable, the coal was broken up for use as gravel on the city's walkways. In 1812 another anthracite entrepreneur, one Colonel George Shoemaker of Pottsville, loaded several wagons at his mine and brought them to Philadelphia for sale, only to flee the city when his angry customers obtained a writ for his arrest as an impostor and a swindler. Even the state senator from the anthracite region declared that there was no coal there—only a "black stone" that was called coal but would not burn. In some places the sale of anthracite was declared a fraud punishable by law, and in short order all but one of the principal backers of the Lehigh Coal Mining Co. went bankrupt; two were sent to debtors prison.

Their travails at first seemed no concern to Josiah White, whose mill at the Falls of Schuylkill prospered as long as bituminous coal was available from Virginia. But the mill had been open only a few months when the War of 1812 began and the British navy blockaded America's major ports, choking off their fuel supplies. Unable to get bituminous coal by boat from Virginia, White again faced the prospect of bankruptcy. Desperately, he searched for an alternative.

The same war that threatened Josiah White's mill also brought another ambitious young man to Philadelphia. John Leisenring was the grandson of Johann Conrad Leisenring, who had emigrated from the German duchy of Saxony in 1748. He subsequently bought a 150-acre acre farm on the Lehigh River, in the midst of the vast—and by 1812 presumed worthless—recently discovered anthracite coal fields. Here Johann Conrad Leisenring built a stone house, where his grandson John was born in 1793.

Young John went to work as a currier—that is, a leather tanner—but at the age of eighteen his career was interrupted when he was called to serve briefly as a cavalryman during the War of 1812. In the army he met a Captain Charles Steadman from Philadelphia, who persuaded young Leisenring to ply his trade in the City of Brotherly Love. There John Leisenring boarded with Steadman's brother Alexander, a silversmith and goldsmith, and in 1814 married Alexander's daughter Anna. As a metalsmith, Alexander Steadman was probably acquainted with Josiah White and his wire mill at the Falls of Schuylkill. John Leisenring met with only middling success as a currier. Much like Josiah White, he began casting about for alternatives.

In the darkest moment of Josiah White's search for fuel he bought a quantity of the supposedly unburnable anthracite coal. Since the Lehigh River wasn't navigable, White had the coal carted down from the Lehigh Valley by teams of mules and horses. He spent some $300 on experiments with these "black stones," but his millhands couldn't make them ignite.

During one trial, after stoking and stirring the furnace for hours into the night without success, the disgusted workmen threw the anthracite chunks into the furnace, shut the furnace doors and went home. Through one of those serendipitous accidents of history, one of the departing workers left his jacket at the mill. When he returned to retrieve it about a half-hour later, he found the whole place in flames—or so he thought. On closer inspection, the blaze was entirely contained within the furnace, but it was an unwavering, smokeless sheet of pale flame generating a heat and intensity that no mortal had previously witnessed.

Anthracite, it turned out, was so potent that it had no need to be stirred. All it needed was time: It became red hot if it was simply left alone. Unlike bituminous coal, the more anthracite was scratched and poked, the less it responded. In the process of quitting their experiment altogether, White's

Slow!

workers had inadvertently found the secret to igniting anthracite: Use it in a closed furnace with a carefully controlled bottom draft, and leave it alone.

The millhand, dazzled by the blaze, immediately summoned his fellow workers and, according to Erskine Hazard's subsequent account, "Four separate parcels of iron were heated and rolled by the same fire before it required renewing." They had accidentally discovered how to harness for industrial use the most efficient fuel yet known.

Josiah White was quick to grasp the meaning of this discovery. Anthracite coal's superior ability to power steam engines held the potential to liberate manufacturers from their geographical dependence on water power: With anthracite coal, an industrialist could locate a factory not just along a river or waterfall but anywhere he chose. Access to coal could transform cities like Philadelphia from mercantile centers dependent on foreign trade into self-sufficient industrial powerhouses. Textile mills in New England could switch from water power to more potent steam power in order to exploit advanced machines like the new English power loom. Huge space-consuming open fireplaces could be replaced by compact stoves and eventually by central heat. Blacksmith shops dotting the countryside could be replaced by larger and more efficient forges and mills. Coal power could vastly expand the production of iron and, consequently, iron products, from bathtubs to printing presses to the iron cables needed to support suspension bridges.

If anyone in 1813 understood that a great global industrial revolution had just begun, Josiah White was that man. Almost immediately he began experimenting with various kinds of cast-iron grates capable of burning anthracite coal not only in factories but in homes. He soon sold off his mill and his water rights at the Falls of Schuylkill and headed north to the anthracite country to revive the moribund Lehigh Coal Mining Company. Coal might change the world, but first someone must organize companies to extract this black rock from dark, narrow, and dangerous caverns beneath the earth's surface. And someone must promote its use to skeptical consumers, build the canals and roads and bridges to transport it, hire and house miners, and raise the capital for all these endeavors. The opportunity was there, but only for men of imagination and nerve.

At just the time Josiah White was moving to the Lehigh Valley to seek his fortune in anthracite, John Leisenring's failure in the morocco dressing business in Philadelphia was driving him in a similar direction. Around 1816, when he was twenty-three, John Leisenring returned to Lehigh County to operate a small tannery with his cousin, Daniel Leisenring. Some four years later, when John was a new father, the two cousins bought a larger tannery at Lehighton, just three miles downstream from the spot where Josiah White was busy creating his headquarters town of Mauch Chunk (an Indian translation for "Bear Mountain").

Josiah White and John Leisenring hadn't known each other in Philadelphia, but now the magnetic pull of the awesome prospects of coal were about to throw them together. Like gold, like oil, this natural resource had remained submerged beneath the earth for thousands of years for lack of a compelling practical reason to extract it. Now Josiah White had discovered such a reason, and the race to find coal and sell it had begun. Over the next two centuries, new men would find entirely new uses for new types of coal: steam engines, railroads, steel mills, electricity—all the basic building blocks of the modern industrial (and post-industrial) world. Over that time, coal operators would make and lose fortunes, coal towns would flourish and die, and miners and mine owners would battle the earth, the atmosphere, and each other in their quest to satisfy the world's appetite for coal. Eventually even the pluckiest players, and even Josiah White himself, would burn out from the intensity of the struggle.

The Leisenring family alone would prove the exception. John Leisenring was about to launch his descendants on an adventure that would transport them into the bowels of the earth, across oceans, and into the halls of government. By the time they were through, this family would define much of the history of the American coal industry, and coal would define much of the world. But of course John Leisenring had no inkling of what lay ahead on the day he went to work for Josiah White. He was simply looking for a way to improve his station in life; he had cast his lot with the iconoclastic Josiah White, and White had cast his lot with coal.

PART I
Mauch Chunk

CHAPTER 1
"A Rock That Burns"

In his film *2001: A Space Odyssey,* the director Stanley Kubrick conjured a world in which the sudden and inexplicable appearance of a supernatural black slab on Earth or in space periodically causes momentous changes — transforming apelike cave dwellers, for example, into rational beings capable of organizing communities and working together cooperatively. Yet true human history is stranger and certainly more complex than Kubrick's 1968 flight of fancy suggested. The resources available to sophisticated humans of the twenty-first century are no different from those that were available to the first Neanderthals. No external intervention was ever needed to change the world—only the human mind's ability to find those natural resources and devise uses for them. "The thing called man had once been a shrew on a forest branch," the anthropologist Loren Eiseley has remarked. "Now it manipulates abstract symbols in its brain from which skyscrapers rise, bridges span the horizon, disease is conquered, the Moon is visited."

As early as the 1930s the Soviet geochemist Vladimir Vernadsky made a more startling observation: People—by virtue of their technology as well as their sheer numbers—were themselves becoming a geological force, shaping the planet and its atmosphere just as rivers and earthquakes had shaped its past. Over the ensuing seventy years many scientists and people of conscience routinely held their species responsible for overheating our planet, polluting the atmosphere, threatening the ozone, and otherwise tampering with the balance of nature. Yet by any commonly accepted definition of disorder and change, the turmoil produced by humans since their arrival on the planet remains tiny compared to the anarchy that reigned when natural forces were left to their own devices, unchecked by human interference.

The Earth of some 225 million to 350 million years ago, long before the age of dinosaurs, was a far warmer and more humid place than today—so warm that much of the land in what is now considered the Earth's "temperate zones" consisted of large, muddy swamps where tropical plants, trees, ferns, and mosses grew and multiplied with reckless abandon. As these plants and trees inevitably invaded each other's turf and collided, their accumulated leaves, twigs, branches, and trunks broke off and fell into the swamp bottoms or matted over the water's surface in thick floating masses. Younger plants lived on these mats of dead plant life and eventually died as well, further swelling the accumulated plant debris.

Stagnant moisture and bacteria preserved this dead plant material and converted it into peat, a spongy substance whose primary elements—carbon, hydrogen, and oxygen—are also the primary elements of coal. As these peat bogs sank beneath the surface over the course of millions of years, the heat from the Earth's core, combined with pressure from above, compressed and hardened this peat into a sedimentary rock that differed from other rocks in one basic respect: Because of its carbon content, when dry it was capable of igniting and burning freely. Through this haphazard process, the force of nature first created coal, which is simply another name for "a rock that burns"—just as, say, a diamond is simply another name for "a chunk of coal made valuable under pressure." (In theory, at least, if untapped coal reserves were left untouched for hundreds of millions more years, the earth's pressure would eventually convert all of it into diamonds.)

From the moment of coal's first appearance on Earth, more than two hundred million years would pass before the landmark moment, some five hundred thousand years ago, when humans first mastered the ability to control and reproduce fire. Civilization essentially began with this critical tool for illuminating and warming caves, for cooking, for clearing brush and trees (and consequently for farming and creating villages), for killing insects and ultimately for making pottery, bronze, and iron. The entire history of technology, from the first forges to the steam engine to electricity and even atomic power, is to a large extent the story of humankind's continuing ability to generate greater quantities of energy by fueling and harnessing ever-larger and more powerful kinds of fires. And of all the fuels for burning fire, none could rival coal for its intensity of heat, its economy, its versatility, and its sophistication. Thus in some ironic joke that only the ancient gods could appreciate, two of nature's most dangerous and unpleasant elements—fire and coal—provided the essential tools with which humans imposed order on the chaos that nature had bequeathed to them.

Because of the haphazard process by which coal was formed during the Carboniferous Age, there never existed such a commodity as "standard" coal. The older the coal, the deeper it lay in the earth, and the more it was subjected to pressure from the ground above, the more its moisture and gases were driven off and therefore the higher its potency. What appeared to laymen as a dark and dirty stone was actually an entity of almost infinite variety: Anthracite—hard, glossy, and almost pure carbon—was the queen of coals for its ability to burn at high temperatures over long periods, but it was also buried deepest in the ground and thus the most difficult to reach. Bituminous coal, with anywhere from 47 to 65 percent carbon content, was softer, dirtier, and less potent but useful for other purposes (such as making coke for the production of steel). Lower grades such as sub-bituminous coal eventually proved useful for heating the steam that generates electricity. The lowest-grade coals, such as lignite, were only marginally more effective than wood or charcoal.

The coal seams themselves varied drastically as to their thickness (from less than an inch to more than five hundred feet), their area (from a few acres to thousands of square miles), and their depth (from a few feet below the surface to hundreds of feet underground). Most coal seams are horizontal, but some are vertical, inclined, folded, or fractured, depending on how they were formed by geological forces. Thus there was no standard process for extracting coal from the ground, just as there was no standard use for the coal once it was extracted.

Given the abundance of coal deposits throughout the world's vegetative climates and coal's unquestioned status as a miracle fuel, historians sometimes profess bafflement that early humans were so slow to take advantage of it. But for thousands of years most of the world's coal supply remained beyond man's sight and reach: Miners used picks and shovels to remove coal at or just beneath the surface. By contrast, wood from trees was plentiful, visible, and accessible everywhere. Throughout those millennia it was simply easier to break off or chop down visible trees than to locate and extract coal under the ground.

Iron, the basic component of human tools, was widely found in ore and even meteorites, but it needed to be smelted—that is, cooked—to separate it from other substances and to alloy it with another necessary element: carbon. The making of iron and even steel is mentioned in the works of Homer and in the Hebrew Scriptures; an iron blade was buried in an Egyptian pyramid three thousand years before the birth of Jesus.

For thousands of years the fuel of choice for smelting iron was neither wood nor coal, but charcoal. This impure form of carbon could be made easily by burning wood while artificially limiting the air supply, usually by partially covering the fire with earth. (The word "coal" derives from the Anglo-Saxon *col,* which originally referred to charcoal. But coal has no connection to charcoal—just as, say, the Native Americans, who were designated "Indians" by Columbus, had nothing to do with India.) Coal was far more effective than charcoal as a smelting fuel but was rarely used in early smelting—not only because it was difficult to extract, but because of the smoke and foul odors it gave off during burning.

Coal was also used sporadically for heat and funeral pyres in ancient China, Syria, and Greece (it is mentioned in the writings of Aristotle). Evidence also exists that during the Bronze Age, some thousand to two thousand years before the Common Era, people in southeastern Wales used coal for funeral pyres to burn their dead. The Chinese probably used coal as early as 1100 B.C.E. King Solomon was probably familiar with the coal deposits of Syria, since coal is mentioned in the Book of Proverbs as well as several later books of the Bible. But many more centuries passed before coal had any lasting effect on civilization.

For much of that time it was considered more of a danger than an asset. In medieval England coal was thought to fill the air with deadly poisons (correctly, as miners diagnosed with black lung disease discovered many centuries later). In 1306 King Edward I declared the use of coal punishable by death, and at least one man was executed for violating the king's decree. Many people doubted that coal could ever be used successfully. But in fact the problem lay not so much with coal per se as with the inefficient methods of burning it, which utterly failed to filter out its smoke or bad odors.

Thus when the first iron furnaces appeared in the Rhine Valley early in the fourteenth century, charcoal was the principal fuel they used. The production capacity of those German iron furnaces held the potential to transform the world. But the furnaces' voracious appetite for charcoal—and consequently for wood—threatened to deplete forests across Europe and especially in Britain. By the early eighteenth century charcoal was becoming increasingly costly, and English smelters were actively searching for a substitute fuel.

In 1735 an English Quaker named Abraham Darby first successfully smelted iron using coke, a derivative of coal with a heating power superior to that of charcoal or coal but which cost far less than charcoal. Thanks to this innovation, Darby was able to produce and sell cast-iron pots, kettles, and other small iron products at reasonable prices and still earn a handsome profit. In this manner Darby and his fellow countrymen became the first people to grasp coal's commercial value and to use it widely. That concept expanded arithmetically when James Watt's steam engine, invented in 1765, was driven by coal-powered heat. Thanks to the combination of coal and Watt's engine, England's great industrial revolution was under way.

Yet coal was more than an industrial tool. As forests were cleared and wood supplies were depleted, coal became a desperately needed (not to mention a much warmer) source of heat. "Wood, our common fuel, which within these hundred years might be had at every man's door, must now be fetched near one hundred miles to some towns," Benjamin Franklin noted in 1744. Since "so much of the comfort and conveniency of our lives, for so great a part of the year, depends on the article of *fire,*" Franklin around that time invented a new stove designed to conserve wood. But the Franklin stove was merely a holding action against the day when a better fuel could be found and harnessed.

Even while generations of chilled European princes and nobles continued to heat themselves with feeble wood cut from rapidly vanishing forests, reports trickled back to them from North America of simple farmers warming their feet before fires fit for a king. Of the world's various land masses where coal deposits developed, nowhere were they so abundant as in the area of North America now occupied by the United States. Here 1.7 trillion tons of coal were buried beneath 458,600 square miles, or 13 percent of America's land area.

The first North Americans to use coal were probably the Pueblo Indians of the Southwest, who burned it to make pottery long before its discovery by white men. French settlers found bituminous coal on Cape Breton Island, off the Atlantic Coast of Canada, in the 1670s and mined it there on a small scale. In 1679, the Jesuit missionary Father Louis Hennepin, traveling with a French exploratory party led by Robert de la Salle, observed the black mineral along the Illinois River near what is now Ottawa, Illinois, about eighty miles southwest of Chicago.

After Father Hennepin's discovery, coal was eventually found on surfaces throughout all of England's thirteen American colonies. A settlement of Huguenots, near what is now Richmond, Virginia, found a few pieces of bituminous coal in 1700. Fifty years later, a boy hunting crawfish nearby discovered a rich bed of bituminous coal. About the same time, trappers and hunters in southwestern Pennsylvania began noticing outcroppings of bituminous coal on the ground or in riverbeds—evidence of rich coal deposits below. These two locations—along the James River north of present-day Richmond, Virginia, and in the Monongahela Valley south of present-day Pittsburgh—provided most of America's coal mining activity until about 1800.

The extent of the early demand for Virginia coal is suggested in the fact that Virginia miners shipped six hundred bushels of coal to England in 1758. Before the American Revolution, coal was exported from the Richmond area to Philadelphia, New York, and as far north as Boston. One Samuel Davis advertised coal for sale at Richmond for 12 pence per bushel in the *Virginia Gazette* of July 1776. Pennsylvania's General Assembly authorized coal purchases from Virginia in 1776. During the American Revolution, an iron furnace located at Westham on the James River used coal to produce shot and shell for the colonial cause until the furnace was destroyed by the turncoat General Benedict Arnold in 1783.

Meanwhile, in Pennsylvania's Monongahela Valley, commercial coal mining began in 1759. Between 1768 and 1784 the family of William Penn acquired all of western Pennsylvania's bituminous coal fields from the chiefs of the Six Nations at a total cost of $10,000, or less than a cent an acre—a transaction that ranks alongside Peter Minuit's purchase of Manhattan Island from the Algonkin Indians for trinkets worth $24 in 1626 as one of the great bargains of history.

With the arrival of the first steam engine in Pittsburgh in 1809, cheap and abundant coal became the principal fuel used in expanding the surrounding area's glass- and salt-making industries. Previously, salt had been transported to Pittsburgh from the East by packhorse over the Allegheny Mountains, or it had been painstakingly separated from brine pumped from wells by horsepower. Now, with coal-powered heat separating salt from brine, Pittsburghers could produce enough salt for one million people from brine deposits in their own region. In and around Pittsburgh, many settlers concluded that the warmth and technological benefits of coal outweighed the foulness of its odors.

"The air is always smoky," wrote a visitor to Pittsburgh in 1817. But he conceded that coal "makes the best fire I ever saw, equal to the best walnut wood."

Thus as settlements grew along the Atlantic coast and the Eastern forests were depleted, settlers' resistance to coal as an alternative fuel source gradually wore down. Still, coal was a bulky commodity that could be transported only on navigable rivers. As long as the population of the colonies remained sparse, as long as transportation remained primitive and as wood remained plentiful, coal provided less than one-tenth of the new land's energy needs.

Only a tiny fragment of America's coal supply—less than 1 percent of the nation's original coal reserves—consisted of high-grade anthracite coal. Yet this minute 1 percent constituted most of the entire world's supply of anthracite. And for reasons never explained by the caprices of nature, virtually all of America's anthracite—which is to say most of the world's anthracite—lay concentrated within a 1,400-square-mile area of what ultimately became northeastern Pennsylvania.

Anthracite was discovered later than bituminous coal because it lay deeper in the earth. The first American traces were found in Rhode Island and Massachusetts about 1760. Two years later, a map prepared by John

Jenkins Sr. noted "stone coal" in two locations in eastern Pennsylvania. But even after Josiah White's successful experiment with anthracite in Philadelphia in 1813, no systematic attempt was made to mine anthracite in any significant quantity. The anthracite coal fields were simply too far and inaccessible from port cities like Philadelphia and New York. In the early nineteenth century it cost almost as much to ship goods one hundred miles along U.S. roads as it cost to ship the same goods across the Atlantic from America to Europe. So it remained easier and cheaper to import bituminous coal by ship from England and Virginia, until Josiah White took the first steps toward solving that problem.

CHAPTER 2

A Passage from the Mines

George Frederick Augustus Hauto was a German expatriate of distinguished bearing, the nephew and putative heir of a baron, and a man who claimed other moneyed connections as well as a reputation as an authority on coal. Whether he was any of these things remained an open question among some Philadelphians—there was no quick way to check his references—but he professed fascination with White's work and ideas, and he proved his sincerity by investing in a wire bridge that White built across the Falls of Schuylkill in 1816. The following year he and White rode sixty miles north on horseback to visit the moribund Lehigh Coal Company's original anthracite mine, at Summit Hill on top of Mount Pisgah. Soon Hauto was raising funds to help White and Hazard acquire the old company and its anthracite coal lands.

Lacking the equipment and the inclination to dig shafts and send miners deep into the earth—as was already being done in the soft-coal mines of England, Wales, and Virginia—White and Erskine decided to simply cut a horizontal pit from a spot on the side of the mountain where outcroppings of coal were exposed on the hillside. This did not require much technical skill or capital, since they were tunneling not into rock, but into coal. For miners, they could simply hire farmers from the nearby countryside, who would work with pick and shovel not in darkness but in an open quarry in broad daylight. The farmers' children could be hired to separate slate from genuine coal—work which, White assured himself, would be no more strenuous than helping their parents on the farm.

The coal mine, Hauto wrote in 1820,

> lies at the top of a mountain and seems to extend over some hundreds of acres
> of land, covered by about twelve inches of loose black dirt, resembling moist

gunpowder, which can be moved by cattle, with a scraper, and thrown into the Valley below.

We have since uncovered about two acres of land, removing all dirt, earth, slate, leaving a surface of what is nothing but the purest coal, containing millions of bushels. . . . We have cut a passage through the rocks, so that teams can drive through the mine, to load. The coal quarries very easy. We have worked the stratum about thirty feet, and how much deeper it is, we do not know.

But this was a relatively minor challenge. The great obstacle with coal, as White well appreciated, was not mining it but transporting it. So primitive were American roads of the early nineteenth century that transportation to a population center might account for as much as nine-tenths the cost of coal, so that coal selling for $4 a ton at the mouth of the mine might sell for $28 by the time it reached Philadelphia. In many cases it was easier and cheaper to ship coal (or anything else) across the Atlantic from Britain than to transport it from America's hinterlands.

For that matter, transportation had been the overriding obstacle for every important commodity since the dawn of time. A vital substance like salt, which the body must replace daily, is taken for granted in modern societies, but it was so highly prized in medieval communities far removed from the sea or from salt mines that anyone who spilled salt was presumed to have been influenced by the devil. The spices of Asia, the teas of India and the cocoa beans of the West Indies and Africa—all critical elements in the development of healthy diets, and all items of minuscule cost today— were prized as delicacies in Europe precisely because of the immense distance and difficulty involved in transporting them. For centuries before the industrial revolution the principal function of financiers was neither raising nor lending capital but the more basic challenge of making capital portable, so that merchants and traders could move gold (or themselves) safely and conveniently from one place to another without risking the loss of their lives or their entire fortunes to highway robbers (which occurred throughout medieval times with remarkable frequency).

Necessity, the often-praised mother of invention, devised new methods of producing salt from brine, drove Marco Polo and Columbus to seek new trade routes for spices, and fostered the development of bank notes and insurance. Yet for all their importance to the advancement of civilization, neither salt nor spices nor portable capital provoked human ingenuity sufficiently to inspire people to develop canals, railroads, bridges, and steam engines to carry them. Coal did. And the transportation revolution it spawned began along the Lehigh River, where the Lehigh Company established its headquarters community of Mauch Chunk.

The company's mine at Summit Hill stood nine miles west of the river and the town. And the rugged and unpredictable Lehigh River itself suffered from serious limitations as a shipping route. Under the best of conditions, it required great skill to steer the barges on the river and keep them off of rocks and out of shallows. Getting a barge to Philadelphia was often a matter of luck, and many unsuccessful barges became stranded or ripped apart. With the annual coming of the spring freshets—surges of water driven by melting snow—the river became too swift to barge coal, and the power of the water would damage the locks, sometimes totally destroying them. The river's power also made any return trip impossible. To the extent that the Lehigh was navigable at all, it was strictly a one-way river flowing south toward Philadelphia—a condition that created a constant demand for new barges at Mauch Chunk, which in turn exacerbated the destruction of the remaining forests. And in the summer, the Lehigh became too shallow for barges to use in either direction.

A canal was the obvious solution, and White and Hazard and their partners set out to quickly construct what they called "the mightiest navigation system in the world." White designed a system through which the stone river bed of the Lehigh itself created the necessary force not only to move the boats when pulled by horses or mules, but to create water deep enough for a boat loaded with up to 150 tons of coal to pass on its own power while traveling in the opposite direction.

Unfortunately, this ingenious work was wiped out in a massive flood that struck the Lehigh Valley on January 8, 1818. There was "nothing left, not even discernible," Josiah wrote home. The Lehigh River, wrote a newspaper a few days later, "is a clear path, from source to mouth again." To compound the disaster, a financial panic struck the nation a week later, on January 15. Undeterred, White and Hazard mortgaged most of their worldly possessions and returned to Philadelphia to drum up additional financial support to reconstruct their canal.

At first they found no takers. A canal on the Lehigh was an even harder concept to grasp than anthracite coal or the notion of a business corporation (in that age of partnerships and sole proprietorships, the Lehigh Coal Company had been one of America's very first corporations). The Philadelphia banker and merchant Stephen Girard, the wealthiest man in America, turned White and Hazard down. So did Napoleon's brother Joseph Bonaparte, who following the Battle of Waterloo had settled first in Philadelphia and then at Bordentown, New Jersey. White himself subsequently summarized the general reaction over the next few months:

> Joshua Longstreth appointed an evening for us to explain the subject to him. In the evening we called to see him but he was gone next door to a party to have some fun (magnanimous interest, thought I, and worthy the man).

Benjamin Stillé was polite enough to allow of some general remarks, but said he was unable to appreciate them.

John Stillé said politely, "How do, sires? How do you come on with the Lehigh?" and before we could give him an answer he proceeded to read his newspaper and was so engaged at reading the chitchat of the day, he no longer had time to bid us good night.

At last they found a few investors. One was a man named Shoemaker—possibly the same Pottsville coal entrepreneur who had been chased out of Philadelphia in 1812 for peddling anthracite to the locals. Since he perceived the potential value of anthracite, Shoemaker promised to raise $50,000 in exchange for twenty shares of Lehigh stock. Another investor was Condy Raguet, a merchant and banker who had recently founded the Philadelphia Saving Fund Society, the first savings bank in the United States. The third was their old friend and investor George Hauto. With more than $100,000 in hand, in August 1818 White and Hazard were able to create the Lehigh Navigation Co., which they merged with the coal company in 1822 to form the Lehigh Coal & Navigation Co. The following April White and Hazard sold their water rights and their mill at the Falls of Schuylkill to Philadelphia's Watering Commission for $160,000, of which $90,000 remained to them after paying off their debts.

Now that they could focus completely on their coal enterprise, the new company's partners approached the Pennsylvania legislature with a proposal to develop the Lehigh River for shipping. In the context of that time this was somewhat like proposing to slather tanning lotion over the sun. "Gentlemen," the legislative committee's chairman is said to have told them, "You have our permission—to ruin yourselves!" But the legislature had nothing to lose by giving its assent. For the price of one ear of corn per year, the legislature leased 10,000 acres around Mauch Chunk to the new LC&N. By the end of the year, the LC&N employed 500 workers at 75 cents a day.

White's goal was to create, by 1824, a one-way descending navigation system, capable of hauling coal from Mauch Chunk downstream to the Delaware River and then to Philadelphia. The more challenging upriver return passage, by his figuring, would take twenty years. By that time—1838—White would be fifty-seven years old and the Lehigh River, as he envisioned it, would be the heart of a "royal web of waterways," leading to New York, Ohio, and the South. "When sales warrant the expense," his prospectus added, "a *railroad* is to be made, on excellent ground."

But this vision was lost on White's readers: Although horse-drawn iron railroads had operated in England since 1801, no such contrivance existed in North America in 1818, and only a handful of people on the continent even understood what the term meant. White was virtually alone in recognizing that wagons drawn by horses or mules along dirt roads made little economic sense for hauling heavy bulk freight like coal. River transportation made much more sense, but rivers didn't necessarily flow where people wanted their goods to go. Canals—which connected one river to another—represented one solution. But even as White envisioned his futuristic canal system he was looking beyond them to the age of railroads.

The great flood of January 1818, it now developed, was something of a blessing in disguise. A drought that summer dropped the Lehigh River to such low levels as to make White's original canal system unworkable. Now he set out to design a new system that would withstand floods and droughts alike. The result was the bear trap lock system, named for Bear Mountain. Previously, all canals had operated by opening up and shutting gates on the locks to let the ships through, a long and cumbersome effort that required whole crews of men. White's invention would enable a single man to open the gates within a few moments. In its day the bear trap lock would become to rivers what the Internet subsequently became to communication: It meant that any river could be navigated at any level of water.

To build these locks, according to a contemporary account,

> Immense stones were dragged from the mountains for lock-walls. A trough was constructed on the riverbed. The walls were completed, the trough done. Water was let in from a reservoir under a construction which seemed like overlapping cellar doors. A pool gathered. The lock-tender opened a sluice beneath a gabled platform—and water pressure in the pool, against the dam created by the gable, pushed the platform flat. The downcoming boat could swoop forward on a long wooden chute thus made. On the high breast of water, the craft floated safely and smoothly over the rapids.

At first the canal permitted traffic to float downriver only, since, there was no practical way to return the boats up river. Upon arrival in Philadelphia, the coal-bearing barges were broken up so their wood could be sold as lumber; their metal fillings were returned by horse and mule to Mauch Chunk for reuse in new barges. In this way the new company-owned town of Mauch Chunk, carved out of a wilderness in 1818, became more than just a loading dock. To attract workers, the LC&N provided virtually all of Mauch Chunk's infrastructure: mills, machine shops, a store, a school, an inn, and a doctor. White's German partner George Hauto described a town "with forty buildings now, gristmill, etc."

By 1820 White had tamed the rapids of the Lehigh River just as he had done on the Schuylkill years earlier. His Lehigh Canal stretched forty-six miles from Mauch Chunk to Easton along the Delaware River, where the coal could be floated downriver to Philadelphia. When their first delivery of 365 tons of anthracite arrived in Philadelphia in June of 1820, the commercial anthracite industry was under way. (Between 1824 and 1829 the LC&N extended its canal an additional twenty-six miles westward to the new village of White Haven, named for White himself. Here White built the greatest of his new bear trap locks, stretching a mile in length.)

But White's work wasn't finished. Mauch Chunk stood nine miles east of Summit Hill, where the first coal was mined atop Pisgah Mountain. The ground between Summit Hill and Mauch Chunk, as Hauto described it, "can hardly be compared to any more unfavorable, for production of a good road." Over this terrain White laid out, using scientific instruments, a road that was graded so as to descend the entire nine miles from Summit Hill to Mauch Chunk—a downhill road to accompany White's downriver canal.

"We have constructed a good road in three months," Hauto noted proudly. "And most of it in winter. On it, a horse can draw four tons with ease. And on this road we have sufficient number of teams to haul several thousands of bushels daily." The horses could move such heavy loads because, of course, they were pulling downhill all the way.

Now White confronted the commercial challenge of making his enterprise a paying operation. There was so little industry in America that factories alone couldn't constitute an adequate market for anthracite coal. For his venture to succeed, coal had to win acceptance from homeowners as a heating fuel. Yet Philadelphia's consumers had already sampled and bitterly rejected anthracite a generation earlier, and now their minds were set against it. The fact that, in the interim, White had discovered the secret to igniting anthracite made no difference to most of them. They were not unlike American consumers of the 1950s who refused to believe that frozen food could be eaten, or that Japanese automakers could build better cars than American companies.

But the prospect of changing their minds seems to have struck White as no more difficult than taming the Lehigh River. In newspaper articles and advertisements he sought to seduce the city's homemakers. Stone coal, he noted in one piece, "is free of all smoke or black or yellow dust in burning. The residue in a parlor grate is a clean white and consequently the chimney never gets foul and wants no cleaning. . . . And its beauty in an open grate exceeds any other fuel known."

Ultimately Philadelphians' resistance was broken down by Josiah White's wife, Elizabeth. Overcoming her Quaker reticence, this otherwise shy woman launched a sales campaign among Philadelphia ladies by setting up a burning grate in her house that was visible from the street. "Betsey" White, as she was known, also placed newspaper advertisements announcing that "orders will be taken for Lehigh coal at 172 Mulberry St. It may be seen burning at the above address." Gradually local ironmakers designed ornamental grates suitable for use in the parlor; other women began showing off the anthracite fires in their new grates; and soon every woman wanted one. By 1823 the power of these women had won over the city. When White risked sending nine thousand tons of anthracite downriver in 1823, the entire cargo was sold. By 1826, Philadelphians had bought thirty-one thousand tons from White's company. The age of wood fireplaces was over; the age of the coal grate was under way. It would last for fifty years, until central heating finally replaced it.

With this demand, White's ingenious downhill road from Summit Hill to Mauch Chunk soon became obsolete. Its capacity was limited, and in bad weather it could be washed out altogether. The situation demanded one last flash of ingenuity from White.

The first railroads, designed for use by horse-drawn cars and subsequently by steam-powered engines, were just being introduced in Britain. But they hadn't yet appeared in America, which lacked the technology as well as the capital to build rails and trestles capable of supporting England's heavy steam locomotives. Nevertheless, in January of 1827 the LC&N began installing wooden rails strapped with iron along the entire nine-mile road between Summit Hill and Mauch Chunk; by May it was completed. Like the LC&N's canal system, this "switchback railroad," as it came to be known, relied entirely on the force of gravity to move cars loaded with coal along its downhill path. A safety car equipped with a brake attached to a cable—and manned only by a single courageous "runner" to operate the brake lever— was attached in front to slow down and stop the train. The nine-mile descent took twenty minutes, with cars reaching speeds of up to fifty miles per hour. And unlike the barges on the canal system, the loaded coal cars were not broken up once they reached their destination. Instead they were dragged back uphill from Mauch Chunk, empty, by teams of mules.

The logistics of White's rail system are impressive even today. Each downhill train bound for Mauch Chunk consisted of seven coal wagons, each loaded with as much as four tons of coal. For every six trains (that is, forty-two wagons bearing a total of 168 tons of coal), a separate train was sent down bearing seven mule wagons, each wagon carrying four mules— twenty-eight mules in all. Upon their arrival and unloading at Mauch Chunk, the forty-two empty wagons were reassembled as three trains of fourteen wagons each, each train pulled back up to Summit Hill by a team of eight mules. The four remaining mules returned the seven empty mule wagons. When it came to imposing order out of chaos, it seemed, the natural world was a piker compared to Josiah White.

Buoyed by his success and his new prosperity, White built himself a mansion in Mauch Chunk, surrounded by parks, pleasances, and even peacocks. Curious visitors flocked to Mauch Chunk in the summers to see the switchback railroad and the canals and to sample Mauch Chunk's mountain air and picturesque hilly streets. Within a few years the switchback railroad had become a tourist attraction, hauling coal only in the mornings so that it could carry passengers in the afternoons. "We went at FIVE MILES THE HOUR," wrote one excited visitor, "swift as the wind!" Eventually the switchback became America's second-leading tourist destination—second only to Niagara Falls—attracting seventy-five thousand visitors annually, who paid $1 for the ride.

To serve these visitors, White expanded the serviceable Mauch Chunk Inn into a grand hotel called the "Mansion House," and Mauch Chunk became a spa to rival other early summer resorts like Saratoga or the Catskills. "Possibly in the whole United States there is no place so secluded, wild, romantic," rhapsodized one newspaper article. "We went to Mauch Chunk's hotel, large and elegant; there were fashionable and genteel people, and beautiful females. The Lehigh River is at the foot of this hotel. The mountains are sheer in the background; beauty and grace are there. . . . The piano and books are indoors, utter solitude outdoors if desired. . . . The meals are wonderful."

Here, at the Mansion House, John Leisenring's connection with Josiah White and the coal industry began.

CHAPTER **3**
Holy Trinity

The Lehighton tannery that John Leisenring and his cousin Daniel Leisenring acquired around 1820 stood just three miles downstream from Mauch Chunk, the coal and tourist town created and entirely owned by Josiah White's Lehigh Coal & Navigation Company. It could not have escaped them that Mauch Chunk was the de facto capital of what was poised to become the Silicon Valley of its day. In the surrounding countryside rival operators were already opening their own mines and canals wherever coal and rivers to transport it could be found. The Schuylkill Navigation Company, for example, floated seven thousand tons of anthracite (a fraction of the LC&N's haul, to be sure) from the Schuylkill River Valley to Philadelphia in the late 1820s. Another new company, the Beaver Meadow, took an option on a coal mine bordering the Schuylkill. Along the Susquehanna River, Maurice Wurtz and his two brothers joined forces in a hasty plan to build a canal from the Lackawanna River to city markets.

An ambitious man with a growing family would be foolish to linger around Lehighton when so many opportunities beckoned in Mauch Chunk. For one thing, the LC&N was in the process of expanding Mauch Chunk's inn to accommodate its tourist trade. In 1827 or 1828, when John Leisenring was about thirty-five, his father-in-law, Alexander Steadman, proposed a business partnership with his old Philadelphia acquaintance, Josiah White: Steadman and his son-in-law John Leisenring would take charge of the inn, now renamed the Mansion House. John would operate the hotel while Steadman ran the hotel's metal and tanning concessions. The deal was struck, and John sold his half of the Lehighton tannery to his cousin Daniel and moved to Mauch Chunk.

The Mansion House he took charge of was no mere hotel. It was Mauch Chunk's central gathering place and a major civic institution, and consequently its proprietor automatically became a leading civic figure. The building rested near the river and was artfully constructed against the side of a mountain so as to shade it from the afternoon sun. During John's tenure there it housed a dining room that could seat four hundred people, a livery stable, a fresh water spring, and a large front porch where visitors and local businessmen could exchange news of the outside world. As part of the building's expansion program, its old dining room was converted into an anthracite coal exchange, in which John may have had a financial interest. Here at the Mansion House John hired a cousin, Edward Leisenring, as the hotel's clerk. And here John and his wife, Anna Maria, produced eleven children, seven of whom grew to adulthood.

This large brood of Leisenrings was uniquely well positioned, for things were happening very fast in Mauch Chunk. For all the excitement of building a boom town, Mauch Chunk was too raw and rustic for Josiah White's wife, Betsey, so in January of 1832, when White was fifty years old, he retired from an active role with the Lehigh Valley Coal & Navigation Co. and returned to Philadelphia. "It is now the twenty-second year since I commenced operations in the work of internal improvements at the falls of Schuylkill," he explained to the stockholders, "in which time I have been absent from that kind of service but very few days. It is also the fourteenth year since I began with my colleague, Erskine Hazard, our labors at Mauch Chunk and on the Lehigh. The whole work is now done."

At that point John Leisenring assumed White's previous post of Mauch Chunk postmaster while continuing to manage the hotel. The postmastership was more than a clerical job: It was Mauch Chunk's sole federal position and the office through which residents of Mauch Chunk communicated with both the outside world and the U.S. government. In exchange for his services, John received a franking privilege—that is, free delivery of his letters—a valuable consideration roughly akin to free Internet service today.

About the same time, in response to anti-monopoly pressures from Andrew Jackson's federal government in Washington, the LC&N opened the company town of Mauch Chunk to individual businessmen and leased its mines to independent contractors. In practice, many of the company's former employees simply continued to perform their old functions, but as private contractors. In 1832 the LC&N divested itself of its store and hotel as well, and John Leisenring, no longer a company employee, became proprietor of the Mansion House in his own right at the age of thirty-nine.

Andrew Jackson may have lacked economic credentials (he explained his opposition to the Second Bank of the United States by remarking, "Ever since I read the history of the South Sea bubble, I have been afraid of banks"), but the LC&N's loss of its monopoly seems to have produced precisely the effect that Jackson had intended: It stoked the company's competitive fires. For the LC&N to flourish, new uses for anthracite would need to be found. From his new home in Philadelphia, Josiah White (who, to be sure, had never needed competitive prodding) turned to a new technological challenge: using anthracite coal to produce iron.

Iron was commonly smelted in a "blast" furnace, powered by a sudden blast of air (hot or cold, depending on the season). Anthracite coal, as White and Hazard had learned earlier at the Falls of Schuylkill, was a smooth, long-burning fuel, decidedly unreceptive to sudden blasts of air. To solve the problem, in 1838 Hazard brought to America the Welsh ironmaster David Thomas, who had already developed a process of smelting iron with anthracite. In July of 1840 Thomas fired America's first anthracite-fueled hot blast iron furnace at Catasauqua, near Allentown. The following month, when Thomas sent the first boatload of iron made with American anthracite down the Delaware to Philadelphia, a group of jubilant Pennsylvania ironmasters toasted the unique resources of their state: "Plenty of coal to warm her friends, plenty of iron to cool her enemies!"

These were heady and even pleasant times for anyone involved with coal. The first coal seams tended to lie close to the surface, so that a pair of skilled miners, or even a single miner assisted by a handful of unskilled workers, could dig several tons a day by burrowing into a hillside and extracting the coal with a pickaxe or shovel, after which the coal could simply be rolled downhill in a handcart to a waiting river barge. Since little capital was required, most operators were small entrepreneurs. In Josiah White's first mines the work was largely a family activity, in which fathers plied their picks in the sunshine while their rosy-cheeked (albeit filthy) children gathered up the nuggets. Before 1840, most mines drew their workers from neighboring farms and villages. Most of them were farmers and day laborers who worked the mines in winter when there was little demand for farm labor.

All that was about to change. Surface mines couldn't long satisfy Americans' growing appetite for anthracite coal. The best and largest coal deposits lay hundreds of feet beneath the surface. Reaching it came to involve start-up costs beyond the assets of any individual. Large tracts of land had to be acquired and cleared, deep shafts had to be dug, water had to be pumped out of them, armies of workers had to be hired and coordinated. Small entrepreneurs would have to give way to partnerships and then corporations capable of providing the necessary capital and organization.

The growing national appetite for coal drove a nationwide obsession to develop transportation systems capable of hauling coal from mines to cities, first along canals and then, as Josiah White had foreseen, by railroads. New York State's Erie Canal, at its opening in 1825, offered barges a single smooth, continuous water passage from Buffalo and Lake Erie at its western terminus to Albany, 363 miles to the east (and, by extension, with New York City, another 150 miles south of Albany on the Hudson River). Virtually overnight this fabled canal enabled New York City to displace Philadelphia as the nation's commercial center, and the Erie quickly spawned an era of canal building throughout the United States.

But canals suffered their limitations as well. They were expensive to build; their routes were limited; they could be shut down by winter freezes; and they provided no motive power. By contrast, railroads were cheaper to build, could be routed almost anywhere, could operate year-round and moved faster than either roads or canals—if a way could be found to power them safely and efficiently. Josiah White had foreseen this all along; the New York City boosters who staked their future on the Erie Canal did not.

In 1830, only five years after the Erie Canal opened, the industrialist Peter Cooper built the "Tom Thumb," a steam locomotive engine capable of negotiating the curves and steep grades along the right-of-way of America's first railroad, the Baltimore & Ohio. Its arrival did for steam locomotives what Josiah White's experiments at the Falls of Schuylkill had done for anthracite coal, and by 1835 more than two hundred railroad charters had been granted in eleven states. By 1845 at least a dozen railroads were built in the Pennsylvania anthracite region to carry coal from the mines to the Lehigh and Delaware Rivers and their various canals.

Now railroads threatened to make canals obsolete, and Philadelphia's jealous business leaders perceived a chance to recapture their city's commercial primacy from New York. Seizing this opportunity, Pennsylvania's General Assembly granted incorporation papers to a group of thirteen Philadelphia investors who formed the Pennsylvania Railroad in April of 1846. The first president was Samuel Vaughn Merrick, a dynamic Yankee and Unitarian who manufactured hand engines for volunteer fire companies. The new railroad's ambitious purpose was to link Philadelphia as well as the anthracite coal fields with the bituminous coal fields of Pittsburgh, three hundred miles across the Allegheny Mountains at the western end of the state. Construction soon began on a railroad extending existing lines from their western terminus at Harrisburg to Pittsburgh, 249 miles farther west. When this railroad entered Pittsburgh on December 10, 1852, Philadelphia enjoyed a commercial link to the west that was superior to anything New York could provide.

Virtually overnight America's railroads grew exponentially, in much the same way, say, as television grew in the 1950s and '60s and the Internet in the 1990s. In 1832 only 229 miles of railroad operated in the entire country; by 1840 America's railroads operated some 2,300 miles of track; by 1850 that number had increased to 30,000; it rose yet again to 53,000 in 1870 and 94,000 in 1880.

Once the railroads were developed and capable of hauling coal inexpensively, textile mills in New England were able to switch from water power to steam to take advantage of advanced machines such as the English power loom. That meant that spindles that had once run no faster than fifty revolutions per minute could now speed up to more than ten thousand. As one consequence, the output of New England mills increased tenfold between 1820 and 1840. This greater productivity per textile worker enabled U.S. mills to compete effectively against lower-paying mills in Europe, which lacked the firepower of anthracite. (This textile prosperity also produced an unintended political consequence: Northern mills relied on Southern cotton picked by slaves, rendering both North and South alike dependent on the insidious evil of slavery.)

Locomotives built to haul coal were powered at first not by coal but by wood, because coal was too powerful for the first primitive locomotive engines. Coal, unlike wood, generated fires so hot that they burned through the locomotive's firebox, created voluminous smoke, and generated sparks that flew through the smokestack and sometimes set nearby farm buildings aflame. But wood was growing scarce and, consequently, more expensive. Like Josiah White at the Falls of Schuylkill a generation earlier, railroad operators began searching for alternative fuels. In 1835 the firm of Eastwick & Harrison of Philadelphia developed a locomotive strong enough to withstand anthracite coal. Once put into use, the cleaner-burning anthracite coal also made railroads more attractive to travelers, so that soon railroads were moving passengers as well as freight. By the eve of the Civil War, America's principal railroads—the Reading, the Pennsylvania, and the Baltimore & Ohio—had all converted to coal-burning locomotives. In this way the railroads, originally built to haul coal, themselves became major consumers of coal. (As late as the 1930s, nearly one-fourth of all mined coal was used by railroads.)

The growth of railroads fed the coal industry in another way: by spurring the demand for iron. Until the railroads arrived, ironmakers functioned primarily to provide agricultural tools for farmers and grates and stoves for homes. Now railroads were demanding huge quantities of iron to make locomotives, cars, and rails. And as heavier iron trains replaced wooden ones, the old wooden strap rails would no longer suffice. A new industry of iron manufacturers specializing in rail production sprang up to meet the need. America's production of iron rails quickly rose from 25,000 net tons in 1849 to 365,923 in 1865. In 1849 iron rails accounted for only 4.8 percent of America's iron output; by 1860 that figure was up to

31.8 percent; by 1880, 48.6 percent. No wood-based smelting fuel, such as charcoal, could satisfy the railroads' constant demand for greater iron production at lower cost. By the mid-nineteenth century anthracite had become the principal fuel used in iron smelting, just as it had become the principal fuel for home heating.

Coal, iron, and railroads—these three interconnected industries became the holy trinity of prosperity for Philadelphia as William Penn's "greene country towne" metamorphosed into the world's greatest industrial city of the nineteenth century. Each leg of the trinity fed the other: Coal produced iron; iron produced railroads; and railroads delivered coal even as they simultaneously consumed both coal and iron.

And their mutual growth generated unexpected by-products. As America's production of pig iron expanded from barely fifty thousand gross tons in 1810 to six hundred thousand by 1850, Americans for the first time acquired all sorts of iron products, from bathtubs to kitchen utensils, from printing presses to sugar mills. Once people had learned how to burn anthracite coal in iron stoves, anthracite's popularity stimulated the rapid growth of the cooking-range industry, so that by 1860 Americans were making and buying three hundred thousand iron stoves per year. The development of iron cables enabled the construction of the first suspension bridges, one across the Monongahela River at Pittsburgh in 1846, and an eight-hundred-foot bridge crossing the Niagara Gorge in 1850. Transatlantic travel by steamship swelled between 1848 and 1860, thanks to the happy confluence of four technical advances: iron, steam engines, screw propulsion, and boilers fired by anthracite coal.

The Pennsylvania Railroad, built to ship Philadelphia products to the West, soon discovered the benefits of shipping Pittsburgh coal and iron to the East. Gas for heating homes and illuminating streets, churches, and auditoriums could now be extracted from bituminous coal mined in large quantities around Pittsburgh. A huge market for this coal gas existed in the newly created municipal gas works of major cities like New York, Philadelphia, and Baltimore, but until the Pennsylvania Railroad opened these cities had needed to import their coal for gasification from Virginia or even Wales. In 1854 the newly created Westmoreland Coal Company (founded by Philadelphia, Pittsburgh, and Baltimore businessmen to mine bituminous coal in western Pennsylvania) appealed to the Pennsylvania Railroad's state chauvinism to reduce the company's freight rates in order to make Pennsylvania's gas coal competitive.

"If you will grant us a freight rate that will permit us to enter into a contract with the [Philadelphia] city gas works to supply it with coal," one of Westmoreland's owners, General William Larimer Jr., reasoned in a note to the Pennsylvania Railroad's directors, "your company will enjoy the benefits of earning on that tonnage and we shall be able to reap benefits for our miners and shareholders. All of us would be benefited and none injured in the slightest degree. Other sales would assuredly follow."

The railroad accepted his logic in 1855 and reduced its rates to less than $6 a ton to haul coal east. The following year Westmoreland acquired twenty-four railroad cars and began shipping coal—the first time coal had been shipped by rail eastward across the Alleghenies. Westmoreland was on its way to becoming one of the nation's largest mining companies, and the Pennsylvania Railroad was on its way to becoming, by the end of the nineteenth century, the world's largest corporation twice over.

This monumental growth was still to come in 1835 when John Leisenring inexplicably backed away from it at the age of forty-two, giving up his hotel in Mauch Chunk to open a general store and leather shop. The reason remains unclear. It may have been related to the death of his father-in-law and partner, Alexander Steadman, or John's own declining health may have prompted him to seek less stressful work. By 1840 he was spending increasing amounts of time trying to restore his health in the ocean waters of Cape May, New Jersey.

Merely reaching Cape May was an achievement: As John described the journey in a letter to his wife, Maria, he left Mauch Chunk by stagecoach on August 5, 1840, and traveled only as far as Pottsville by that evening. The next day an 8 A.M. stagecoach took five hours to deliver him to Reading. After "an excellent dinner for the trifling sum of fifty cents," he boarded a 2:45 P.M. train which arrived in Philadelphia at 7 that evening. John spent the next day attending to business in Philadelphia, then the next day took an 8 o'clock steamboat that brought him, finally, to Cape May at 3 P.M.—three days' travel, in all. But John evidently considered the trip worth the trouble: He wrote Maria of walking on the beach with a friend after dinner on his first night there, "not being able to withstand the temptation any longer, Mr. Nowlin and meself stripped and plunged ourselves in the roling surf of the beautyfull ocean. Although somewhat cool it created a most delitefull sensation. I had an uncommon good nights rest. Got up this morning at 5 o'clock. Felt most like a new born. . . . I think I shall derive much benefit from being here a while."

In any case John Leisenring retained his Mauch Chunk postmaster's job and had built up a comfortable fortune. He had worked at the periphery of the coal business but never actually in it. Now he would turn his influence to securing a future for his children in America's fastest-growing industry.

CHAPTER **4**

Boy Wonder of the Anthracite

When John Leisenring's eldest son, John, a tall and rangy youth of seventeen, went to work for the LC&N in 1836, few people could boast a better coal lineage. John Leisenring Jr. had been nine years old when his parents moved to Mauch Chunk in 1828. As a boy he was on hand when Josiah White and Erskine Hazard built their first canal from Mauch Chunk to White Haven and then the gravity railroad linking White Haven to the anthracite fields at Summit Hill. Instead of serving an apprenticeship under his father or another local merchant, when he was seventeen his father utilized his nascent web of family connections to find a better opportunity. Of John's seven grown children, three married children of Lehigh Coal & Navigation Co. officers, including the general foreman and the chief bookkeeper. These blood relationships created the nucleus of what became one of the two major business factions in Mauch Chunk (the other revolved around Asa Packer, who founded the Lehigh Valley Railroad).

From young John's perspective, the most important match was his sister Mary Ann's marriage to the coal operator Andrew Douglas, whose older brother Edwin A. Douglas was the LC&N's chief engineer. Edwin Douglas, who came from upstate New York, had begun his career some years earlier under the great canal engineer Canvass White (no relation to Josiah), who beginning in 1816 had worked on the seminal projects of the canal age: the Erie, the Union, the Delaware and Raritan and ultimately Josiah White's own Lehigh Canal in 1827. Canvass White had studied canals in England and patented hydraulic cement. By 1834 he was dead at the age of forty-four, but his legacy was being carried on by disciples like Edwin Douglas who had learned from him on the job. Now, like most early-nineteenth-century engineers, young John Leisenring would learn from Douglas in the same manner that Douglas had learned from the master Canvass White.

John Jr. joined the LC&N at the moment when Douglas was extending the Lehigh Canal northward from Mauch Chunk to White Haven. As a sub-assistant engineer on one of the sections, John worked in the survey crew and also monitored the contractors' work. His pay was a relatively comfortable $2.50 per day. His problem, at least at first, was that he himself required monitoring. In November of 1837, after he had been on the job for about a year, his father wrote to him from Wilkes-Barre about a conversation he had had with Douglas.

> I commenced it by asking him whether you was [sic] improving any as respects being careful in your business (the complaint has been that you are entirely too careless). He says that he finds that you are trying to be more careful than you had been but that you had suffered yourself to run into carelessness so deep that it will take a good deal of your particular attention to divest yourself of your careless ways. He says if he can impress on your mind the absolute necessity of being more careful than you had been, there would be no anger but you do well enough. I asked him whether you would be kept on the line this winter. He tells me you would. He tells me that he intends keeping you and Baldwin on all together. I asked him whether you was [sic] capable of having the charge of one half of that line. He says you are if you choose to be careful enough.

The letter does not end there. After the Lehigh navigation project is finished, the senior Leisenring reports, Douglas would like to make use of young John and his colleague Baldwin on future projects if he can "have an eye on you both himself," but otherwise Douglas was inclined to hire someone else as his principal assistant.

"I asked him which end [of the next project] you would be on," the father writes. "He said he thought the lower end, but then said he did not know, for that he would leave to yourself and Baldwin to divide between yourselves. I then asked what he was paying you now. He said $2.50 per day. I would not ask him whether he paid Baldwin the same. I thought he might tell me that was none of my business."

This letter suggests a father who is not at all reticent about using his influence to promote his son's career. It also suggests a certain distance between father and son: The elder Leisenring seems to have learned more details about his son's work (such as his son's pay) from Douglas than he learned from young John himself.

It was about this time that the LC&N decided to construct the Lehigh & Susquehanna Railroad between White Haven and Wilkes-Barre, since a canal over that mountainous stretch wasn't feasible. Notwithstanding the concerns about his carelessness, at the age of eighteen John Leisenring was promoted to assistant engineer, and he and his colleague Baldwin were placed in charge of the new railroad's eastern division, as Douglas had hinted to John's father. After young John's section of the work was completed in 1840, he was loaned to the Morris Canal & Banking Co., which was enlarging its canal from Easton across New Jersey to Jersey City. Here again he was assistant engineer in charge of a division lying east of Dover, New Jersey. He also assisted Douglas in the survey of a railroad along the Delaware River for the Belvidere-Delaware Railroad Co. in 1838.

In 1843, the LC&N began rebuilding and extending its mine railroads west of Mauch Chunk, converting the original switchback system into a combination that used both gravity and steam power. The cars were hoisted up steep planes by stationary steam engines, then coasted downhill to the foot of the next plane, losing a small percentage of the elevation previously gained to cover a long horizontal distance.

Young John Leisenring moved to Ashton (now Lansford), Pennsylvania, in 1843 to be close to this construction project, which he worked on with his brother-in-law Isaac Salkeld and Robert H. Sayre, the son of another LC&N official. John and Sayre were both credited with building the new Mauch Chunk switchback railroad (although Douglas, Josiah White, and Erskine Hazard made all the important technical decisions), and somewhere around this time residents of the Lehigh Valley began routinely referring to young John Leisenring as "the boy wonder of the anthracite region."

Leisenring's growing influence was reinforced in 1844 by his marriage to nineteen-year-old Caroline Bertsch. Her Pennsylvania German father, Daniel Bertsch Sr., had arrived in Mauch Chunk in 1827 to run the blacksmith shop adjoining the Mansion House. Like many small-town entrepreneurs, Bertsch made a living by wearing several hats simultaneously. Within a year after his arrival, he was working as a construction contractor on the Lehigh Canal as well as the Lehigh & Susquehanna Railroad, both projects of Josiah White and Erskine Hazard that brought Bertsch into frequent contact with the senior John Leisenring. In 1833 he built a Mauch Chunk hotel, the Broadway House, which he leased to a series of proprietors. The hotel put Bertsch briefly in competition with the senior John Leisenring, but Caroline Bertsch's marriage essentially put the two hotels under a common family ownership, and the Bertsches and Leisenrings began exploring other joint business ventures. A year after the marriage Bertsch started operating some of the LC&N's mines in the Summit Hill–Lansford area as an independent contractor, usually in partnership with the Leisenrings and a few Mauch Chunk businessmen.

Sayre, young John's colleague on the switchback railroad project, subsequently went on to a distinguished career as a railroad builder and executive. He seems to have been promoted by the LC&N to resident engineer ahead of young John Leisenring, but when Sayre accepted the post of chief engineer with the Lehigh Valley Railroad in 1852, John became Edwin Douglas's second in command at the age of thirty-three.

At this point young John appears to have realized that by staying with the LC&N he would remain an employee, and that no matter how far he would advance, the company's ownership was fixed in other hands. Fortunes in coal and railroads were being made all around him. Asa Packer, having arrived in the anthracite region as a seventeen-year-old in 1822, launched the Lehigh Valley Railroad, built a mansion in Mauch Chunk every bit as large and elaborate as Josiah White's, and had enough left over to endow Lehigh University in 1866. Moncure Robinson, chief engineer for the Allegheny Portage Railroad, turned to English financiers in the 1830s to create the Reading Railroad that linked Philadelphia with Reading; he acquired coal lands along the route, made a fortune by exploiting immigrant coal miners, and in 1846 retired at age forty-five to a Philadelphia life in which, for the next forty-five years, he and his family considered themselves, in the words of his grandson, "something like royalty, and looked down on all Philadelphians as rather middle class and stuffy."

Leisenring's own father, having made the transition from LC&N company man to entrepreneur, had retired as Mauch Chunk's postmaster in 1847 and then retired from his hardware business four years later before dying in July of 1853. To an ambitious young man like John, whose father had just died at age sixty, thirty-three probably didn't seem all that young, and now he drew upon his family connections and his insider ties at the LC&N to seize new opportunities. The LC&N at this point was more interested in hauling coal than mining it, and so it was opening new mines and leasing them to contract operators. Many of the lessees were old business associates of the Leisenring family, among them John's father-in-law, Daniel Bertsch. Soon after his father died, even as he remained in the LC&N's employ, young John went into business with his father-in-law as the contract operator of an LC&N mine at Summit Hill. This firm, called Bertsch, Leisenring & Co., was probably his first experience as an entrepreneur.

In 1854, four of these contract operators formed a new mining partnership called Sharpe, Leisenring & Co., to work land they leased from the estate of Tench Coxe of Philadelphia, who had acquired 80,000 acres of coal lands as a speculative venture in the eighteenth century. Leisenring moved his family there to direct the work as superintendent, and here he and his partners created the company town of Eckley, named for their landlord, Eckley Coxe. By this time John had grown a close-cropped beard that later endowed him with a resemblance to President Abraham Lincoln (Lincoln hadn't yet grown a beard in the 1850s). The Eckley operation proved a great financial success, and Leisenring and his associates subsequently formed a series of anthracite mining companies scattered through Pennsylvania's Eastern and Western Middle Fields, from Eckley and Upper Lehigh on the east to Wilburton on the west.

Each of these mines was technically owned by a separate firm, and each firm was launched with the profits of its predecessor wherever an advantageous site could be bought or leased. The firms were kept independent so that the failure of one wouldn't drag down the others, but in practice most of these companies were controlled by the same owners, usually related to each other within the Leisenring family by blood or marriage. John's brother-in-law Daniel Bertsch Jr. ran John's Upper Lehigh Coal Co., and he was later succeeded by John's son Edward B. Leisenring.

This pattern was common among medium-sized operators throughout the coal industry. Then as now, most businesses in America as well as the rest of the world were family companies. Entrepreneurs, especially immigrants, tended to go into business with their relatives because they didn't trust anyone else. They were especially cautious in the nineteenth century because most businesses were partnerships, and a partnership was much like a marriage. Whereas shareholders in a corporation stood to lose no more than the amount they had invested, in a partnership each of the partners could be held personally liable for the partnership's debts. Like a drop of ink in a glass of water, a single rogue partner could ruin not only the business but his individual partners as well. (The notion of a "limited liability partnership" wasn't developed until late in the twentieth century.) This unlimited liability explained why most partnerships consisted either of blood relatives or of very close friends. In a family business, often the safest expedient was to admit only partners who were already relatives or spouses, or to marry one's daughters only to one's partners.

Sometimes these relationships bordered on the incestuous. John Leisenring Sr.'s nephew Walter Leisenring, for example, married Mary Ann Kemmerer, the widow of Charles Kemmerer, whose son Mahlon Kemmerer married John Leisenring Jr.'s daughter Annie. That is, Mahlon Kemmerer married a first cousin of his half-brothers and half-sisters. Meanwhile Mary Ann Kemmerer's brother married Harriet Bertsch, a daughter of Daniel Bertsch Sr., who was also the younger John Leisenring's father-in-law.

Such inbreeding might have been a recipe for a weakening of the Leisenrings' gene pool, but John Leisenring Jr. seems to have been fortunate in his siblings' and children's choice of spouses. One of John's sons-in-law, Dr. John Shriver Wentz, was born in 1838, graduated from the University of Pennsylvania's medical school and served as an army surgeon during the Civil War. After the war, probably about 1867, he moved to Eckley, where his older brother George was already the company doctor for the LC&N. George had married young John Leisenring's sister Anna Maria in 1858, and he remained a doctor in coal mining communities until his death in 1903 at the age of seventy-four.

John Wentz similarly married into the Leisenring family, but unlike his brother he seized the first opportunity to give up medicine and go into the coal business. Four years after marrying young John Leisenring's daughter Mary Douglas (named, of course, for John Leisenring's mentor Edwin Douglas) in November of 1871, John Wentz became a partner in John Leisenring & Co. and superintendent of its mine at Eckley. Eventually his father-in-law set him up in his own coal business as J. S. Wentz & Co., in partnership with John Leisenring's son Edward B. Leisenring, to operate coal mines at Hazle Brook and Black Ridge, near Hazleton.

John Leisenring Jr.'s oldest daughter, Annie, was married in 1868 to one of her father's protégés, Mahlon Kemmerer, who since 1862 had worked under John as an assistant engineer in the repair of the LC&N's canal and railroad. In 1876 John set him up in another business, M. S. Kemmerer & Co., which opened mines at Sandy Run, just south of Upper Lehigh. Although Mahlon Kemmerer subsequently drifted apart from the Leisenrings and moved to New York after Annie died in 1888, he continued to invest in the Leisenrings' ventures while building up his own parallel set of coal companies, which his son and grandson expanded until their successful sale in 1981.

This family/business support group was still in its early stages in the late 1850s, when John Leisenring was still living at Eckley. He prospered sufficiently to send his oldest son, Edward, born in 1845, to a private school in Eckley and then to take a course at Philadelphia Polytechnical College. But in 1859 Edwin Douglas died, and the following February the LC&N offered to hire John back to fill Douglas's old post: chief engineer and superintendent, at Douglas's annual salary of $3,000 plus housing and fuel.

The honor and challenge were too great to resist, even for an entrepreneurial spirit like John Leisenring. The job put John in charge of all the company's field operations. Because of the time demands of this job, he resigned from the Eckley partnership (while maintaining his stake), hired his teenage son Edward—now known familiarly as "Ned"—into the LC&N's engineering corps, and moved his residence to "Whitehall," Josiah White's old mansion in Mauch Chunk. This spacious frame residence, set on a hillside surrounded by beautiful, well-tended grounds and gardens, was little changed from the time when White and his wife had lived there thirty years earlier. In effect, within the space of a year the younger John Leisenring had occupied the position and then the home of Mauch Chunk's two most accomplished and inventive citizens: Edwin Douglas and Josiah White. The "boy wonder of the anthracite" was about to reinforce his nickname in a field where both his technical and his entrepreneurial ambitions were sorely needed.

In the two decades leading up to the Civil War, America's coal production increased tenfold, from two million tons in 1840 to twenty million in 1860. Nearly three-quarters of that coal was mined in Pennsylvania's anthracite region. By the Civil War, some 200 coal mines operated in northeastern Pennsylvania, employing about 25,000 of the nation's 36,500 miners. But the advent of the war in 1861 brought into the market a new coal customer whose appetite was larger than any other: the federal government. Washington needed bigger, faster, and farther-reaching railroads to carry troops and supplies; it needed iron mills to produce guns and cannon balls. It needed steam to power the engines of U.S. naval ships. When Robert E. Lee led the Army of Northern Virginia into Pennsylvania in the summer of 1863, his purpose was very specific: to strike at the Pennsylvania Railroad and, in the process, cut off the North's access to its coal supply. Had Lee not been stopped at Gettysburg, some thirty miles south of the Pennsylvania Railroad's line connecting Philadelphia and Pittsburgh, the North's effort to quash the Southern rebellion might have ended then and there.

Young John Leisenring's first major challenge at the LC&N occurred three years after his appointment, in 1862, when a flood completely destroyed the Lehigh Canal from White Haven to Mauch Chunk and partially washed away many locks and banks from Mauch Chunk to Easton. John was charged with rebuilding the canal as quickly as possible. After surveying the loss of life and property caused by the flood, he concluded that there was no sense in rebuilding the canal and risking a repetition. Instead, on his suggestion, his two-thousand-man crew built a railroad alongside the canal's path. This new line from Mauch Chunk to White Haven provided Mauch Chunk for the first time with a rail link to the coal fields around Wilkes-Barre, thirty miles to the north. In the process John modernized the original portion he had built twenty years earlier. And now John extended his ambitious vision even further.

The canal from Mauch Chunk to Easton, on the Delaware River, normally shut down in the winter when the canal waters froze. Why not, John argued, seize this opportunity to extend the rail line all the way to Easton—fifty more miles—so the LC&N could have direct year-round rail-and-water access from its coal mines to New York and Philadelphia?

The rail line he subsequently constructed there between 1865 and 1868 was one of America's first railroads built entirely with steel rails rather than iron, and it remained in service for decades.

At the same time, between 1864 and 1867, John expanded his private operations by developing the Upper Lehigh Coal Co. just north of Eckley. This was the first mining venture in which he held the major interest, and it was even more successful than his earlier Eckley operation, especially since in this case he and his associates also owned the land.

Another Leisenring concern, the Nescopec Coal Co., which John created in 1864, brought him into a partnership with David Thomas, the great iron master who had first smelted iron with anthracite coal. John Leisenring's relationship with Thomas probably stretched back to 1840, when the LC&N financed the organization of the Lehigh Crane Iron Co. for Thomas. Thomas had subsequently set up his own Thomas Iron Company in the early 1850s. But the Nescopec partnership appears to have been the first financial connection between John Leisenring and Thomas. To John's already extensive network, Thomas brought perhaps the most sophisticated understanding of iron smelting in America.

One of John Leisenring's most monumental projects during this period was the construction of three inclined planes at Solomon's Gap, four miles above the Ashley Planes in Wyoming County. The original Ashley Planes, a world-famous project begun in 1837, were intended to move coal cars up an elevation of sixteen hundred feet from the anthracite basin on the floor of the Wyoming Valley to the rail line at Solomon's Gap. As John shrewdly extended them a generation later, the project consisted of four separate inclined plane railroads. Passenger and freight cars alike were raised and lowered along five- to fifteen-degree inclines by cables powered by steam engines. The planes were said to be capable of raising two thousand cars daily, carrying a total of between ten thousand and twelve thousand tons. These planes saved the railroad about four-fifths of the cost of hauling the coal in railroad cars for thirteen miles over the mountains, and they remained a critical part of the passage from Pennsylvania's third anthracite basin to Solomon Gap, and from there to all points south, until 1948.

When his major LC&N construction projects were finished in 1868, John resigned his post with the LC&N to devote full time to his own coal mining operations (although the LC&N, reluctant to lose his services altogether, at this point elected him to a seat on the LC&N's board of managers). In 1868, John bought three thousand acres of coal land in Newport Township, southwest of Wilkes-Barre, and organized the Lehigh and Luzerne Coal Co. A year later he sold it for a profit to the LC&N.

John presciently left the LC&N at the very time that Josiah White's old company was being driven close to bankruptcy. The LC&N was going deeply into debt to expand its coal operations as protection against a new class of corporate predators, who were themselves gobbling up coal lands with money they didn't have. The enemy was no longer impassable mountains and unnavigable rivers. Now it was railroad men—in particular the men who ran the Reading Railroad.

CHAPTER **5**

Souls in Darkness

From its birth in the 1830s through the Civil War, the Reading Railroad had been merely one link in a chain of many small rail and barge lines delivering anthracite coal from Pennsylvania's Schuylkill Valley to Philadelphia. But a shrewd and tenacious operator named Franklin B. Gowen thought he had a better idea.

Gowen was the son of middle-class Irish Episcopalian immigrants to Philadelphia. He attended private schools there and grew into a handsome youth, wiry and strong, with what his contemporaries described as an almost hypnotic charm. His father had made money speculating in anthracite coal, and in 1856, when he was twenty, Gowen spent a year managing his father's mine in Shamokin, Pennsylvania. Then he moved to nearby Schuylkill County, where he and a partner bought a small mine. That partnership went bankrupt after two years, but Gowen eventually paid off the company's debts, gave up coal, and turned instead to the practice of law. In November of 1861, at age twenty-five, he was elected district attorney of Schuylkill County. While still in office he began representing the Reading Railroad on the side—a conflict of interest that would be frowned on today but was then accepted as a good way to supplement a public official's meager salary. When his two-year term was over, Gowen became the Reading Railroad's general counsel and then, in 1869, its president.

Confronted with competition from other railroads and with the Reading's dependence on business from unreliable coal mines along its routes, Gowen sought to convert the Reading Railroad into a road that, in his words, "owns its own traffic, is not dependent upon the public and is absolutely free from the danger of the competition of other lines." In practice this meant buying up coal properties all along the Reading's lines in the Schuylkill Valley, mostly with borrowed funds. Similar business monopolies of all sorts were spreading across the country in the decades after the Civil War with the tacit consent of government, which perceived large organizations as more stable and efficient than small ones. But coal was one commodity that differed in quality and usefulness from one mine to the next, so it was impossible to standardize and unlikely to benefit from economies of scale the way that, say, manufacturers of ships or rail cars or iron bathtubs did. But that perception is evident to us only in hindsight.

By the mid-1870s the Reading Railroad had grown into a large regional network that controlled not only the delivery of anthracite coal but much of the anthracite industry itself. It owned 150,000 acres of coal lands—more than 60 percent of the total in the anthracite region. For that matter, by 1873 six corporations owned most of Pennsylvania's anthracite mines and their connecting railroads.

In the scramble for the necessary critical mass to survive these consolidations, management of many of these companies passed from hands-on engineers and adventurers to absentee owners, hard-nosed accountants, and lawyers like Gowen. Utopian paternal mining towns like Mauch Chunk were replaced by sullen company towns peopled by miners in thrall to company stores and company landlords. And it now became clear that, whatever blessings anthracite coal may have bestowed upon civilization, the mining of this coal was one of humankind's least civilized tasks.

Then and now, the "room and pillar" system of mining was an engineering exercise designed to leave enough coal still standing, like a bearing wall in a house, in order to keep the mine from collapsing. Each section of a mine consisted of a series of long parallel tunnels beneath deep subterranean shafts, as well as shorter tunnels that cut across them. Timbers were extended from the bottom to the roof—not to support the tunnel, but as an early-warning system of potential collapse: "When the timbers start talking," went the saying, "miners start walking." Anthracite mining involved the delicate business of "undercutting"—digging out an opening perhaps two feet high and six feet long at the bottom of a vein of coal (just enough space for a miner to squeeze in), then blasting the coal above the cut to break it, hammering it loose, and loading it onto a cart headed for the surface, usually pulled by mules. Undercutting required a miner to work on his stomach in a claustrophobic tunnel-like space, hammering out a narrow slit in the coal by hand.

Standing or walking in these narrow spaces was impossible; miners moved around by crawling. Blasting required precision and care, to minimize waste and avoid collapsing the roof or walls. Hammering and shoveling required time and strength. And a miner worked in constant danger of being crushed by a roof collapse or by a runaway loaded cart. A blast explosion could shatter walls or ignite unseen pockets of natural gas. If the giant fans installed at the surface broke down, miners could be asphyxiated. The cages that carried the miners up and down some mines could fray and snap, sending men plummeting hundreds of feet to the bottom. The pumps that kept the mines dry could fail, drowning miners in groundwater floods. Their working conditions underground were tolerable only by comparison to the mules who worked alongside them, pulling the heavy carts loaded with coal: Once placed in the mines, the mules never saw the light of day again, for fear the shock would drive them mad.

Cutting or blasting the coal loose from its foundations was merely the beginning of the challenge: Only about three-tenths of mine workers actually worked underground. Loaders were needed to scoop the coal into the heavy mule carts pulled underground by mules. Once these carts reached the surface, "breakers" were needed to break the irregular-sized coal chunks into useful, marketable sizes—at first by hand, and later by large "breaker" machines in which coal chunks were broken down while passing over the machine's revolving pinions. And once the coal was broken down into a half-dozen or so marketable sizes—from "pea" coal to "egg" and "stove" coal—more hands were needed to clean the coal, picking out pieces of slate and shale and other waste products as the coal flowed into troughs from the rollers of the breakers.

Someone had to perform this monotonous and dispiriting work. The rising demand for coal after 1840 transformed the industry into a continuous year-round enterprise. At first the need was filled by importing experienced miners from England, Wales, Scotland, and Ireland, who were tired of the depressed conditions of their lives and cherished a hope that America might offer them something better. Some two thousand miners arrived from Britain during the 1840s and 37,000 more during the 1850s.

Unlike their part-time predecessors, these Britons were skilled miners who thought of themselves as professionals; unlike the Eastern European immigrants who followed them, they spoke English. They frequently operated as independent contractors who entered into individual contracts with an individual mine owner. They were piece workers, paid for the coal they mined by the bushel, ton, or carload, and as such they determined their own working hours. Because their piecework rates were low, usually this meant that they worked underground for ten to fourteen hours a day in temperatures that averaged about 45 degrees. They entered the mouth of the mine before sunrise, carrying their tools, their lunches, and their lamps, and often didn't emerge until the sun had set.

Before the Civil War, these skilled miners usually worked in pairs within a room of a mine. These mining "buddies" in turn hired a day laborer to load the coal onto wooden wagons and transport it to the surface. The miners themselves were responsible for checking for air quality and explosive gases that might have accumulated while they were gone. They were also responsible for maintaining the mine's roof—a task they voluntarily performed without pay because the collapse of a roof was a leading cause of mine workers' deaths during that period. In those days of unfettered free enterprise it did not occur to either operators or workers that a more efficient system might be feasible. The top priority of each was independence. Or as one British miner put it, "No damned foreman can look down my shirt collar."

Miners often began their careers as children, pressed into service out of necessity following the loss or dismemberment of a father or brother. They typically started as "breaker boys," earning extra pennies by bending over the coal trough and sifting slate from the chunks. By adolescence a young miner would be underground, loading coal onto mule-drawn carts. Eventually he would graduate to a miner's assistant, learning the trade firsthand in exchange for a small share of the miner's wages. By thirty he might be a miner in his own right, if he survived. And once a miner, he would most likely find it impossible to leave: Although mining was demanding work, it involved few skills that could be applied to any other work.

This sullen subterranean life was a far cry from Josiah White's first surface mines. In the 1830s a visitor had described Mauch Chunk as "a paradise," but by the 1840s this paradise had already become the scene of labor disputes and strikes, most of them quickly suppressed and consequently bitter. In the anthracite region the first collective action occurred in July of 1842, when workers from Minersville, Pennsylvania, marched on the Schuylkill County seat of Pottsville to protest their low wages. The first known miners' union, the Bates Union of 1848, enrolled five thousand members but collapsed when the president, John Bates, absconded with the union's funds. In 1853 the Delaware & Hudson Canal Company's miners struck successfully for an increase of 2 and a half cents per ton in their piece rate, but they failed to establish a lasting union presence at the company.

The expansion of the coal industry after 1860—together with the labor shortage caused by the Civil War—required massive new importations of workers. By 1870, forty-four thousand Irish Catholic immigrants were working the mines in northeastern Pennsylvania alone. Soon coal operators were casting their nets even wider. Industrial companies dispatched agents to Europe to fill their growing demand for cheap and unskilled workers. Among Slavs and Hungarians, Italians, Poles, and Slovaks, they promoted America as a Utopia and offered to pay the cost of the trip from southern and eastern Europe to the mine or factory.

What these immigrants found when they arrived was, of course, no Utopia. As unskilled workers who spoke no English, they were easily ex-

ploited. The cost of the voyage to America which their employer had paid
was deducted from the workers' pay. To house the new arrivals, coal operators constructed "company towns" where all the real estate—homes, stores,
schools, even the churches—belonged to the company. These towns arose
for eminently practical reasons: Mining sites were inaccessible, so operators needed to entice workers there by providing housing and amenities.
The company town also provided a mechanism for reducing the company's labor turnover and exerting control over its labor force. Private
merchants and service industries, for their part, were rarely willing to invest in homes or businesses in a mining town for fear the town would be
abandoned once the mine closed or the coal seam was exhausted. So coal
companies operated towns of their own as part of their routine cost of
doing business.

This arrangement essentially reduced the relationship between operators and miners to that of feudal lords and serfs. The rights and protections granted to other Americans under the U.S. Constitution were largely
nonexistent in company towns. As one former Pennsylvania miner later
recalled, the mine superintendent "was mayor, council, big boss, sole
trustee of the school, truant officer, president of the bank, in fact he was
everything." A tenant's lease could usually be terminated with five days'
notice, and it permitted the company to evict tenants for almost any reason. The company constable could enter a miner's home at any time, usually without a search warrant. Miners who went on strike risked losing
their homes.

Although mining is an especially grimy occupation, at first these towns
provided virtually no bath facilities. Before state statutes required mine
companies to provide bathhouses, miners had to bathe every evening in a
large tub in the kitchen. (As late as 1922, less than 3 percent of all miners'
dwellings nationwide had bathtubs or showers.)

The company's housing monopoly was replicated in the local "company
store." This widely despised institution often exploited its local monopoly
by charging higher prices than comparable independent stores. Even if
miners had access to less expensive stores elsewhere (which they usually
did not), some company towns required miners to trade at the company
store. Miners short of cash (as most were) were given scrip—a certificate
issued by an employer in lieu of cash wages, usually redeemable only at the
company store—to purchase supplies; the cost was then deducted, with interest, from their pay.

To the coal operators, these arrangements seemed eminently reasonable: Operating a store was an investment that no independent retailer
would make, and the company deserved a return on the capital it risked.
But to the miners, the company store could be a formula for sucking them
into an inescapable spiral of debt.

Most of the immigrant miners—newly arrived in a strange land, speaking no English, some illiterate even in their native tongues, indebted to the company for their passage—were incapable of articulating the misery of their condition. But a miner's son, writing many years later, suffered no such incapability—and had a sharp memory. "I lived under the regime of the Henry Clay Frick clan," wrote Ben Shedlock in a letter to the *Pittsburgh Post-Gazette,*

> in a little town called Trotter, Pa., near Connellsville and my Dad and brothers worked in the coal mines and coke ovens of the H.C. Frick Coal Co. . . . That little town was under the control of the Coal and Iron Police, hired by the company. They were like the Gestapo. As a youngster I had to be in the house by 9 p.m., no more than five boys were allowed to gather in one place.
>
> One had to buy from the company store or lose his job. Nothing was bought with cash. Purchases were put on a slip and taken out of your paycheck. Prices in the company store were always very high. . . . They robbed the people blind in those stores.
>
> The men in the coal mines were slaves. They worked twelve hours a day. I remember my Dad coming home in the winter, his clothes frozen to his body from working in the water in the mines. Of course, if you were killed in the mines, the company paid your family nothing.

Miners occasionally walked out of a mine en masse to protest low wages or poor working conditions. But collective bargaining was an alien concept to miners, who prided themselves on their autonomy and their stoicism. It was a matter of pride among miners (who were compensated on a tonnage basis) that they set their own hours, resisted supervision, and refused to complain about their lot.

Coal operators, for their part, took it as a given that mining was dangerous and unhealthy work with unpredictable pay; but in their view, the real risk of mining was assumed by the operators, who sank fortunes into mines that often failed to pan out. "The miner is free and can protect himself," one operator told the Ohio Mining Commission, "for he can engage in mining or not." By the same reasoning, operators perceived mine safety as the responsibility of the miners who voluntarily chose that line of work. Labor organizers were perceived, in the words of coal historian Carmen DiCiccio, as "demagogues who were too lazy to work themselves and unwilling to permit self-respecting men to work." Equally alien was any notion that government should interfere with negotiations between operators and miners.

Both parties took a fatalistic attitude toward their mutual co-dependence, and some acknowledged as much. "Coal operators and men are very much like a man and his wife," the anthracite coal operator Eckley Coxe testified to the U.S. House of Representatives in 1888. "They quarrel and fuss but they have got to live together." Later Coxe admitted that the first time his own workers struck, "it almost made me sick. [But] it is like a man when his first child has a tooth, he thinks it is dreadful, but when the third child has the fifteenth tooth, he does not think so much of it."

Where the first mass labor movements arose, coal operators were quick to discourage them—by legal means if possible, by force if necessary. When miners mounted a protest march in Pittsburgh in 1877, the mayor summoned the National Guard, which opened fire on the marchers, killing twenty-six men. During a march in Scranton that same year, the mayor—goaded by newspaper editorials demanding that he prevent a "revolution"—mobilized the police, handed out arms to private citizens and gave orders to shoot to kill; a single volley killed three of the march leaders and wounded many more. Ten years later, when Pennsylvania anthracite miners marched from Hazleton to Larimer to dramatize their demands for more money, they were met by the sheriff and his deputies and ordered to disperse. When they refused, the lawmen opened fire, killing eighteen marchers and wounding forty others.

This violence rarely restored order, of course. On the contrary, it fueled the miners' anger and convinced them that violent retaliation was necessary to protect themselves. The Molly Maguires, a militant secret organization, arose in the early 1860s in Pennsylvania's anthracite region. Its members were Irish-Catholic immigrants, and their namesake was a legendary Irish heroine who led poverty-stricken farmers in revolt against the brutal bailiffs who collected house and cattle rents for absentee English landlords. The Mollies advocated terror as a legitimate weapon, but they lacked specific goals or any interest in promoting labor unions: They simply exacted revenge against any employer who fired a Molly, or vented their rage at a system that reserved the most lucrative skilled jobs for Protestant miners of Welsh, Scotch, and English descent. Many of their victims were mine superintendents, foremen, or colliery supervisors who had angered a Molly or a friend of a Molly. Often, before the crime was committed, the victim received a crude, anonymously served warning, a "coffin notice," signed by a "Son of Molly Maguire." But because the Mollies' membership was secret, a miner or operator could rarely be certain whom he had offended or why.

The result was an almost total breakdown in law and order for more than a decade across three Pennsylvania counties and parts of two others. During this period the Mollies were accused of 42 murders, 162 felonious assaults, and myriad destructive acts against property, but the numbers may have been much higher. Schuylkill County alone experienced 142 unsolved homicides and 212 felonious assaults between 1862 and 1875.

The man who ultimately brought down the Molly Maguires was none other than the Reading Railroad's tenacious president, Franklin Gowen. After his brief and unsuccessful career as a young coal operator in the late

1850s, Gowen had graciously attributed his failure to his own lack of judgment. But in private he blamed rising unionism, which he said was forcing independent mine operators out of business. Only close cooperation of all mine owners, he argued, could defeat the threat posed by unions. As a district attorney hampered by a limited budget and small staff, he had been forced to stand by helplessly in 1862 while eleven mine supervisors were murdered in his county after discharging or quarreling with miners. Gowen's own client, the Reading Railroad, had been a major victim of the Mollies' sabotage.

But by the mid-1870s, as head of the largest coal operator and transporter in the anthracite region, Gowen was a force to be reckoned with. He also found himself subject to tremendous pressure: Although he was no longer a district attorney answerable to voters once every two years, now he was a manager answerable to nervous and skeptical investors every day. The extent of the Reading's rail and coal network had come to affect not only the Reading's own stock price but millions of dollars' worth of stocks and bonds of dependent trunk line railroad systems and coal companies— the engines of the U.S. economy. Gowen's ambitious expansion plans depended heavily on his ability to raise more funds from investors in Philadelphia, New York, and especially London, America's greatest source of capital at that time. Many of these wealthy and conservative men were terrified by the endless tales they heard of the Mollies' seemingly uncontrollable violence.

Unlike with the Mollies, emotion played no part in Gowen's calculations. He had no interest in revenge; he bore no personal animosity toward the Mollies and even expressed admiration for their courage and tactics. But he perceived them as an obstacle and came to see the removal of that obstacle as his mission in life. Gowen also shrewdly perceived in the Mollies a unique opportunity: By tying the Mollies to legitimate unions in the public mind, he could discredit the entire union movement.

Gowen and the Reading had already induced the state legislature, in 1865, to pass an act permitting a company to form its own private police force. Now, in October 1873, he sought a provable connection between the Mollies and the unions. He hired the Pinkerton detective James McParlan, himself an Irish immigrant, to infiltrate the Maguires' inner circle under the assumed name of "James McKenna." McParlan never did establish a link between the Mollies and the unions—and, to his credit, he refused to manufacture one—but after two years he had enough evidence to identify and try the Mollies' ringleaders and to testify about the group's inner workings.

Although it was widely understood that the law in coal regions was beholden to mining companies, most operators maintained at least the pretense of subservience to government. But Gowen, in his zeal to nail the Mollies, was unwilling to delegate even that small modicum of authority. At this point he took a temporary leave from the Reading's presidency and offered his services without pay as an assistant to the Schuylkill County district attorney. When the first murder trial of four Mollies began in Pottsville in 1876, the Mollies' greatest private nemesis and victim was sitting in the prosecutor's chair.

Gowen began the trial by inducing the judge to exclude Catholics from the jury so that the jury, in the words of a contemporary historian, "might not be influenced by sympathy." He ended the trial with a spellbinding seven-hour address in which Gowen insisted that the terms "Molly Maguires" and "labor agitator" were synonymous and interchangeable. Gowen recounted his past fruitless efforts, as district attorney, to bring the Mollies to justice. "I made up my mind," he told the jurors, "that if human ingenuity, if long suffering and patient care and toil could succeed in exposing this secret organization and bring well-earned justice to the perpetrators of the awful crimes, I should undertake the task."

The verdict was a foregone conclusion. By 1880, through a series of trials in Pottsville, Mauch Chunk, and other coal region county seats, nineteen men had been hanged and others imprisoned as Molly Maguires. The Molly Maguires were finished, but organized labor had escaped being tarred with their brush.

Were the Mollies alone responsible for the mayhem they spread? Reasonable minds even then raised the question. "If a soul is left in darkness," the great French social commentator Victor Hugo had already noted, "sins will be committed. The guilty one is not he who commits the sin, but he who causes the darkness." Isolated from the rest of the world, lacking transferable skills, unable to articulate their grievances (or, increasingly after the Civil War, even to speak English), coal miners represented the dark underside of the bright new day that their product had ushered in. The power and rhetoric of a Franklin Gowen could not allay the growing suspicion as to who was causing the miners' darkness.

"If a miner with a family of eight or ten . . . once gets into debt with the 'pluck me' [company store]," the *Philadelphia North American* commented in 1881, "he remains in debt until his sons have grown up, and then their earnings, perhaps, will help to decrease their burden. . . . There are families who for ten years have never received a dollar in cash from their labor."

The Leisenring group's labor policies, by comparison, made at least some efforts to light candles in darkness, even in the heart of the Molly Maguires' domain—less as an act of kindness than as an enlightened deterrent to unionism. Unlike many absentee operators who ran their mines from Rittenhouse Square, Fifth Avenue, Back Bay, or even Leicester Square in London, the Leisenrings and a few others (such as Asa Packer) were still working and living (albeit in fine mansions) in Mauch Chunk in the heart of the anthracite country. The Lehigh Coal & Navigation Co. conducted periodic studies of its workers' welfare. One review found that its miners had averaged the relatively high compensation of $2.72 a day between January 1886 and the end of June 1887; it also found that workers owned 72 percent of the 1,613 houses in the LC&N's four company towns. When strikes did shut down the mines, the company routinely put miners to work in its machine shops and at other tasks—a rare case of battling unions by killing with kindness.

In 1884 the LC&N introduced a "beneficial fund" to care for injured miners and other workers. The company contributed 1 cent to the fund for each ton mined; the miners (who joined voluntarily) contributed 1 percent of their wages, and other workers one-half percent. Each injured worker was entitled to receive half his weekly wages; if a worker died, the fund paid $30 for his funeral expenses, and one-half of his wages to his family for a year. The benefits may seem minuscule today, but they were very progressive in their day, and almost all the workmen signed up. By the time John Leisenring's son Ned became president of the LC&N at the beginning of 1894, the fund had accumulated a balance of nearly $47,000. These devices allowed the Leisenrings to perceive of themselves as beneficent employers. But that conclusion hadn't yet been truly tested.

CHAPTER 6

A Road Not Taken

Ultimately the consolidation of coal companies after the Civil War failed to protect the new coal combines from the cyclical nature of the U.S. economy, not to mention coal prices. Amid the deflationary U.S. economy of the 1870s and 1880s, coal prices dropped sharply along with everything else, and in 1880 the Reading Railroad—which by then had a total bonded debt of more than $95 million from all its real estate purchases—was forced into the first of the three bankruptcies that the company would undergo over the next thirteen years.

Amid the chaos of coal bankruptcies and consolidations after the Civil War, John Leisenring's unique combination of technical vision, coal expertise, and entrepreneurial shrewdness was much in demand. In 1871 the LC&N leased its railroad system to the Central Railroad of New Jersey, which completed construction of the LC&N's line from Easton to New York. The Central had no direct prior experience in the coal industry, so it relied on local experts like John. He was made a director of the Central, and in that capacity in 1872 and 1873 he helped the Central to assemble a mining subsidiary, the Lehigh & Wilkes-Barre Coal Company. The L&WB capped the process by leasing the entire coal property of the LC&N in 1873. These large expenditures, financed by bonds, sent the Central itself into receivership in 1877. When that happened, the LC&N lease was cancelled, and in the reorganization John lost his seat on the Central New Jersey board.

Meanwhile, John had also been elected in 1871 to a five-year term as an associate judge of the Pennsylvania District Court. This single term in a public office accorded him the honorific title of "Judge" Leisenring, which his descendants used forever after to distinguish him from his father. And his service on the bench did not interfere with his coal empire building. When the Carbon Iron Company in nearby Parryville failed in 1876, John Leisenring's extended family group reorganized the property as the Carbon Iron & Pipe Company.

The Mauch Chunk *Democrat* of February 22, 1879, regaled its readers with a description of the "magnificent edifice just completed by our worthy townsman, Judge Leisenring." The Judge, nearing sixty, already had his own magnificent home, which had originally been built by Josiah White. But in this gilded age a mansion was a reflection of one's success, and this new stucco two-and-a-half-story home and its five-acre park, built for Judge Leisenring's daughter and son-in-law Annie and Mahlon Kemmerer, surely quashed any notion that the Leisenrings had fallen into Asa Packer's shadow. The nineteen-room house, the newspaper proclaimed, had converted "what was once a bleak and unsightly piece of land into a yard, park and garden, which will be a crowning ornament to our town." A "fine piazza" ran along the entire front of the house; halls were lighted with sunlight reflected through stained-glass windows; the seven bedrooms on the second floor were "flanked by two nicely furnished bathrooms"; and "electric call bells, speaking tubes and all modern conveniences abound."

The wealth of coal men and other plutocrats of the post–Civil War era was defined not only by their mansions but by another phenomenon of the Gilded Age: the growing network of private boarding schools and colleges that helped to define membership in a rapidly expanding upper class. In a more complex and mobile age, the sons of the new and old rich, from Boston, New York, and Chicago to boom towns like Mauch Chunk, mingled amid the secluded halls and homogeneous values of a few dozen boarding schools, most of them in New England. Although the ancient Phillips Academies at Andover and Exeter, Massachusetts, had been founded in the eighteenth century, and St. Paul's had opened before the Civil War, boarding schools opened and grew most rapidly in the decades after 1880. Exeter's enrollment, for example, increased from some two hundred boys in 1880 to more than four hundred by 1905. St. Paul's graduated forty-five boys per year in the 1870s and more than one hundred per year by 1900. Groton was founded partly in response to this demand in 1884. Some of the most famous New England prep schools—Taft, Hotchkiss, St. George's, Choate, Middlesex, Deerfield, Kent—were all founded in the two decades after 1890.

The Leisenrings were typical of the newly rich families that fed this growth: It took perhaps one generation to acquire the wealth necessary to afford a private school education, and one more generation to perceive the social and intellectual advantages such an education could provide. The first two John Leisenrings had no formal education to speak of. Judge John Leisenring's oldest son, Edward, born in 1845, attended a private school in Eckley and then studied engineering at Philadelphia Polytechnic College; but his younger brother, John, born in 1853, was sent to Schwartz's Academy in Bethlehem and then to Princeton. Edward's son and grandson attended Hotchkiss and then Yale, and Edward's brother-in-law Dr. John Wentz sent his son Daniel to Andover and Harvard.

Judge John Leisenring's new mansion and his sons' schools may have suggested the extent of his wealth, but social status was not his primary concern that spring of 1879. Through his relationships with iron manufacturers like David Thomas, the Boy Wonder of the Anthracite reached the conclusion that anthracite's time was passing. Anthracite had been unmatched as a home heating fuel, and its superiority to charcoal as a fuel for smelting iron ore had played a role in reviving the iron industry. But now a stronger, more durable metal was at hand. In the mid-1850s the Englishman Charles Henry Bessemer and the American kettle maker William Kelly of Pittsburgh, working independently and more or less simultaneously, had both invented a pneumatic process for converting pig iron into steel in thirty minutes or less, a task that had previously required three months. Suddenly steel was a feasible product, barely more costly to produce than iron. When the Cambria Iron Works produced America's first two and a half tons of steel rails in 1867, the *Pittsburgh Courier* declared that the steel would last twenty years, compared to the three-year life of the iron rails it would replace. "The day of iron is past!" declared Andrew Carnegie before launching his steel works at Braddock, twelve miles from Pittsburgh, in 1873. "Steel is king!"

And just as iron was about to be overtaken by steel, so anthracite coal was about to be superseded by a new miracle fuel for powering the blast furnaces that produced steel: a solid residue of coal called *coke*. Like anthracite coal, coke was strong enough to reduce a load of iron ore and limestone in a blast furnace; but coke alone was sufficiently porous to respond quickly to the sudden blast of air that activated the smelting process.

Although coke was itself made from coal, anthracite's hard quality made it especially unsuitable as a source for coke. Coke was produced by heating coal to a high temperature until practically all the volatile matter had been burned off, leaving only carbon and other mineral matter. But the hard anthracite coal resisted this process. The softer bituminous coal of other regions, on the other hand, had too much phosphorous and sulfur content for smelting iron—but it made an ideal coal for coking. And where anthracite's reserves were limited and restricted to a narrow region, bituminous coal abounded in western Pennsylvania, Virginia, and many other parts of the United States.

These developments at first glance seemed part of a natural and logical progression. Coal was inexorably replacing wood as the nation's primary energy source: In the 1850s wood provided nearly 91 percent cent of the nation's energy consumption, but by 1870 that figure had declined to barely 73 percent, and in 1883 coal would take the lead over wood for the first time. In transportation, similarly, canals had replaced roads, railroads had replaced canals, and steam-driven trains had replaced rail cars pulled by horses or pushed by gravity.

But the coal industry itself now found itself, curiously, turning full circle to the past. Anthracite had replaced bituminous coal as a heating and industrial fuel, but now coke was replacing anthracite—and coke was made from bituminous coal. The way to wealth no longer lay in the ability to mine and transport coal per se, but to mine bituminous coal and cook it into coke. And coke could be produced with relative simplicity wherever coal was mined. The first brick "beehive" ovens for making coke from coal—so called because the interior was shaped like a beehive—appeared in the western Pennsylvania town of Connellsville in the 1830s, and by the 1860s dozens and even hundreds of these dome-shaped ovens could be seen clustering wherever bituminous coal was mined in the Connellsville district.

Connellsville sat atop the "Pittsburgh seam," a rich, thick vein of bituminous coal said to be the best coking coal in the world. Connellsville coal didn't need to be washed to remove impurities. "There is no other seam that can compete with it in cheapness of production," remarked an observer in the late 1870s. "There is no other coal so regular in form; so uniform in quality; of so convenient a thickness; or so easily mined." And this Connellsville coke could be shipped easily and quickly by rail or river to iron-and-steel-producing blast furnaces throughout Pennsylvania, Illinois, Michigan, and Ohio. By 1880 the seven thousand beehive ovens in and around Connellsville were producing two-thirds of the nation's coke. What Mauch Chunk had been in the 1830s—the energy capital of the nation—Connellsville had now become.

Judge John Leisenring had thrived by seizing opportunities where he spied them. The fact that he was now past sixty—the age at which his father had died—made no difference. After careful examination, in 1880 John purchased eighty-five hundred acres around Connellsville from two coal and iron operators, the brothers Abraham and Christian Tinstman, and organized the Connellsville Coke & Iron Co.

At a club dinner in Pittsburgh in December 1879, the hot topic of the moment was not steel—which hadn't yet penetrated most people's consciousnesses—but the current boom in iron products and how to satisfy the demand. The speaker that night, newly arrived in Pittsburgh from Virginia, was eager to address their concerns. John D. Imboden, then fifty-six, was a former Confederate general who had fought at Bull Run and Gettysburg. After the war he had become a lawyer and politician in Staunton, Virginia. But the Civil War had left him impoverished and bitter over his loss of wealth and status, not to mention the deaths of his first three wives.

"I don't care a damn about the truth or falsehood of history so far as that war was concerned," he remarked in a letter around the time he appeared in Pittsburgh. "I know it ruined me financially—and nobody thanks me for my efforts in a common cause then, and never will unless I get rich." Since the war, Imboden had tried to interest northern and foreign investors in his mineral and railroad projects in western Virginia, traveling as far as London without finding many takers. Now he had hooked up with another Civil War veteran, a young Virginia lawyer and promoter named Rufus A. Ayers, who was convinced that he had found the next Connellsville.

That night in Pittsburgh Imboden told his audience about the remarkable abundance and purity of iron ore deposits in the woods and hills of Wise and Lee Counties, in the extreme southwestern corner of Virginia, near the borders of Kentucky and Tennessee. The land, he said, was available for a song. A new railroad was being built from Bristol, Tennessee, to develop these properties, he added. Imboden lacked the necessary capital to act by himself; he was looking for partners with the money and vision to seize this opportunity.

It happened that Imboden's audience included the same two Tinstman brothers who were then dickering to sell some of their Connellsville coal lands to John Leisenring. The Tinstmans were sufficiently impressed to give Imboden $500 plus expenses to travel to Virginia that month and bring back samples of iron ore. During his visit Imboden also learned of a rich vein of bituminous coal, ideal for coking in order to make steel.

Unlike oil prospecting, which was a matter of drilling test wells into the ground, there was no scientific way to search for coal. But thick coal seams invariably left outcroppings on the surface, so the search was largely a matter of talking to farmers, walking in woods, and poking into the sides of hills and cliffs. Imboden could hardly believe his good fortune: this rich seam he had discovered lay beneath a 48,000-acre tract of supposedly worthless land that had been purchased at a delinquent tax sale in 1836 for just four dollars and ninety-two cents.

When Imboden returned to Pittsburgh in February of 1880 he found the Tinstman brothers more interested in the coal lands than the iron ore properties—presumably because they perceived, like Carnegie and John Leisenring, that iron was about to be overtaken by steel as the world's superior metal. Imboden returned to Virginia in March 1880 and arranged to acquire more than 27,000 acres at a price of thirty-five cents per acre. In the process he became the Tinstmans' Virginia agent and partner and settled at Three Forks, which later became the town of Big Stone Gap.

Over the next few months the newly formed Tinsalia Coal & Iron Co.—named for the Tinstmans—acquired mineral rights to more than 40,000 additional acres as well as a controlling interest in the narrow-gauge railroad being built into western Virginia from Bristol, Tennessee. Imboden attempted to extend this railroad to Big Stone Gap and westward to the breaks of Cumberland Mountain. The frontier hardships of the region—as well as its potential payoff—were never far from his thoughts.

"You can form some idea of the isolation of this place," Imboden wrote the Tinstmans' Pennsylvania agent on June 17, 1880, "when I tell you that your interesting letter of the 25th of May reached here this day a week ago, and today is the first chance to reply to it, as our horse mail will pass here this afternoon." On the other hand, he noted, "We have a coking coal here equal to Connellsville, and 80 million tons within five miles of this gap. Have bought one pit of it, and it is splendid. . . . This will be a second Connellsville in five years in coke, and a Johnstown in iron and steel."

To his sick fourth wife, Annie, more than two hundred fifty miles to the east in more civilized Staunton, Virginia, Imboden again blamed his long absences on his poverty. "Had I the money to meet all imperative demands," he explained, "it would be easy of solution, for I would quit all else and give all my time to you. But if I do this we shall be dependent on charity for mere existence."

A week later he again beseeched her understanding: "I do want so much to put you in more comfortable circumstances. . . . Oh! What a relief it will be when eternal money strain is ended, as it will be after my Company is fairly at work."

Whether wealth would have brought Imboden happiness is a question modern psychologists may validly raise, but of course only with the benefit of hindsight. Imboden himself, convinced that his problems would end when he found his pot of gold at the end of the rainbow, waited impatiently for the proceeds of the Tinstmans' bond offering with which to pay his expenses in Wise County. "Till that is done," he wrote the Tinstmans' agent, "it is a 'hand-to-mouth' matter with me. I only receive what funds I am obliged to have to live on and keep my family economically."

But by the spring of 1881 the Tinstmans, hundreds of miles north in Connellsville, had become alarmed by the money Imboden was spending on the railroad—probably $25,000 at that point, just for masonry and cross-ties. At this point the Tinstmans sold their entire interest in Tinsalia to another one of their partners, Edward K. Hyndman of Connellsville. But Hyndman had no interest in constructing mines or railroads. He was simply a middleman who had already helped the Leisenrings acquire their Connellsville land, and he sensed that the Leisenrings might be interested in the Virginia ground as well.

After a meeting in Philadelphia and an examination of deeds in Virginia county courthouses, in the summer of 1881 the key members of the Leisenring group—Judge John Leisenring, his brother-in-law Daniel Bertsch Jr., John's son Ned Leisenring, and John's sons-in-law Mahlon Kemmerer and Dr. John S. Wentz, as well as their wives, along with Hyndman and Imboden—made the long train trip from Mauch Chunk to Philadelphia to Bristol, Tennessee, and from there traveled by hack and horseback forty miles to Big Stone Gap. Here they made a cursory examination of the properties for sale, and what they found was indeed impressive. "There is enough coal above the water level on this land alone," an article remarked at the time, "to supply the market with one million tons a year for a thousand years." Shortly afterward, E. K. Hyndman deeded to the Leisenrings the entire former property of Tinsalia Co. plus additional lands—some seventy thousand acres in all, as well as control of the unfinished railroad between Bristol and Big Stone Gap. This meant that General Imboden, once the Tinstmans' minority partner, now became a minority partner of the Leisenrings in the newly christened Virginia Coal & Iron Company.

That visit was the only time Judge John Leisenring would see the property: By 1884 he was suffering from Bright's disease, a disorder of the kidneys. And for the moment the Leisenrings had too many other things on their plate. Their anthracite mines in the Lehigh Valley would continue to heat homes and factories in the burgeoning cities of the Northeast well into the twentieth century. But in western Pennsylvania they were committed to the entirely different challenge of mining bituminous coal and cooking it into coke. To Imboden's dismay, as Judge John Leisenring's health failed, he temporarily shelved the Virginia development. In April of 1884, to defray the expenses Imboden had run up, shareholders of Virginia Coal & Iron were assessed $750 for each $5,000 interest. The lion's share of these assessments fell on the Leisenrings themselves, but for Imboden the extra $750 charge was enough to break him. "I think it was these assessments," Imboden's son later noted, "that ruined my father."

When Imboden asked the dying John Leisenring to buy out his interest for $5,000, he heard instead from Leisenring's son Ned. "My father is no better, and still a very sick man," Ned explained, "and the physicians do not allow us to talk business matters of any kind to him, therefore cannot present your letter to him." But Ned did agree to buy Imboden's stake. "Unless there is a decided improvement in general business within the next year or two," Ned wrote, "it is doubtful whether any income could be derived from the property for a long time."

Less than a month later, Judge John Leisenring was dead. "He grew rapidly worse," Ned wrote to Imboden, "becoming weaker day by day, passing away without suffering pain." For Imboden, more serious news came in another letter from Ned in early January of 1885: Because of "the great and general depression," Ned said, work on the Virginia properties would be suspended. "While I personally am anxious to see the [rail]Road built and mines opened," Ned wrote, "[I] do not feel like bearing more than my share of the burden." Then came the kicker: "I regret that at this time I cannot purchase the balance of your interest in V.C. & I. Co. Our family has now as much as they care to carry. I hope you will be able to retain it, as when we do go on with the development, we want you to be interested."

Imboden, strapped for cash, sold his interest to another investor and died an embittered man in 1895, never having made the fortune that seemed so nearly within his grasp. His name remained on the rich coal seam he had discovered, but to Imboden that sort of immortality was a small consolation. In retrospect the contrast between Imboden and the Leisenrings is striking: Imboden lacked the Leisenrings' technical expertise, their family support network, and their capital resources (John Leisenring's wealth was estimated at $1 million when he died), but he lacked something else that was even more important: a passion for coal and coal mining. To Imboden coal was a messy and undignified business, to be tolerated only as a necessary means to wealth; to the Leisenrings it was a way of life. If their Virginia investment could not pay off in the present, Ned's note presumed, perhaps it would pay off in the future.

Yet whatever Imboden's failings, he had kept his word; Ned had not. Ned had tabled his Virginia commitment because his attention and capital were already committed to Connellsville, which in many respects was like entering a whole new industry. Few other anthracite operators before or since would attempt this transition, and the Leisenrings would soon discover why. Unlike the Lehigh Valley, where they had arrived first on the scene and had been revered as technical innovators, in Connellsville they faced the prospect of battling established operators. Like Imboden, many of these operators had entered the business solely for the money; unlike Imboden, some of them were hardheaded captains of industry who possessed deep pockets as well as an advanced corporate vision that the Leisenrings lacked. One of these operators—a relative of the Tinstman brothers, as it happens—had already concluded that the surest way to wealth was the systematic elimination of his competitors. His name was Henry Clay Frick.

PART **II**
Connellsville

CHAPTER 7

The Ambitions of Henry Clay Frick

Henry Clay Frick's contemporaries always felt there was something not quite human about him. "No man on earth could get close to him or fathom him," the extroverted steel magnate Charles M. Schwab told an interviewer in the 1930s, long after Frick died. "He seemed more like a machine, without emotion or impulses. Absolutely cold-blooded. He had good foresight and was an excellent bargainer. . . . His assets were that he was a thinking machine, methodical as a comptometer, accurate, cutting straight to the point . . . the most methodical thinking machine I have ever known."

In some respects Frick was the nineteenth-century forerunner of an executive type that became widespread in the late twentieth century: the "pure entrepreneur"—like, say, Saul Steinberg, Charles Bluhdorn, or Jack Welch—whose pursuit of growth and profits is unfettered by sentimental attachment to any particular industry, company, or product. But Frick was so far ahead of his own time that he baffled and intimidated almost everyone who crossed his path—including his parents and, as we shall see, the Leisenrings. Even today it is difficult to pinpoint the precise forces of heredity and environment that created Frick's exceptional business personality.

He was born in 1849, the year of the California Gold Rush and a watershed moment in the psychology of the Western world. Gold, unlike coal, possessed little intrinsic value at that time. But if only by virtue of the glittering spell it cast upon people, gold had served as a universally favored backing for currency since ancient times. Its supply had long been presumed to be finite, but in the five years after the Gold Rush began, more than half a billion dollars would be taken out of California; over the next twenty-five years, more gold would be mined in the world than in the previous 350 years. A cautious world of finite growth suddenly seemed to many people like an exuberant world of limitless opportunities.

Frick's parents, of German-Swiss descent, had settled near West Overton, in southwestern Pennsylvania. His father was a not-very-successful farmer, but his maternal grandfather, Abraham Overholt, had in 1810 established a nearby distillery, and its "Old Overholt" whiskey had made him the wealthiest man in the area by the time he died in 1870. Not surprisingly, grandfather Overholt became Frick's chosen role model.

Soon after Frick began working on his parents' farm it became clear that he was neither suited for nor inclined toward physical work. His frame was small and slight, his face was pale, his features somewhat delicate (aside from an unusually prominent jaw), and his constitution generally sickly. He engaged in no active sports as a boy and appears to have had no friends. Even as a child he was a silent, lonely, methodical fellow who rarely revealed his emotions, even under great stress. To the extent that he demonstrated passion, it was a passion for efficiency and neatness—in his dress and appearance as well as in the precise manner in which numbers could be aligned in an accounting ledger.

One of young Clay's childhood pleasures was to ride in his grandfather Overholt's fine carriage, which contrasted starkly with his parents' humble lifestyle. His dream, it was said, was to make a fortune like his grandfather. By the time he was sixteen he had left home to live with his maternal uncle Christian Overholt, a leading merchant in Mount Pleasant and president of the First National Bank there. Frick attended college briefly, but the only subject that interested him was mathematics. While working as a retail clerk in Pittsburgh, he demonstrated a talent for keeping the books, which prompted his grandfather to hire him, in 1869, as chief bookkeeper at the Old Overholt distillery.

His starting salary was a thousand dollars a year, a comfortable income for a nineteen-year-old at a time when most Americans earned less than $500. But right from the beginning Frick had higher goals. One night at the distillery, when the conversation turned to future ambitions, Frick remarked laconically, "I see no reason why I should not become a millionaire during my lifetime." At a similar age, Josiah White had challenged himself to put aside $40,000 so he could devote the rest of his life to good work. But to Frick the getting of money seemed an end in itself.

That such a meticulous man would seek his fortune in a messy and unpredictable business like coal seemed ironic to everyone save a small group of businessmen who shared Frick's instincts. Frick was the precursor of a new generation of industrialists about to assume industrial power—cold, colorless men who succeeded precisely because of their talent for imposing order on chaotic enterprises. These men were indeed creative—not as engineers or inventors, but in their astute understanding of the value of market domination and of the uses of credit to achieve it. In an age when few men borrowed money, Frick suffered no such compunction. Like his fellow bookkeepers Henry Phipps, Collis Huntington, Jay Gould, and especially John D. Rockefeller (of whom it was famously said, "He had the soul of a bookkeeper"), Frick enjoyed the single-minded ability to look beyond the sweat and chaos of a given industry and steer it inflexibly toward his clearly envisaged (if narrow) financial goals.

The Overholt distillery stood in the Connellsville region, about forty miles southeast of Pittsburgh, almost directly over the great seam of bituminous coal which, by 1870, had already attracted the notice of the Leisenrings and other coal men. Ugly little coke ovens—the critical tool for producing the coke necessary to smelt iron ore into steel by the Bessemer process—were beginning to dot the landscape. Frick's relatives too were buying up coal acreage in the area. Frick—as the twenty-one-year-old overseer of his uncle Martin Overholt's scattered investments—saw the opportunity to seize control of a key resource in the new industrial age. He resolved to jump into the game himself.

In March of 1871, with help from his relatives (and by anticipating a $10,000 inheritance), Frick and three partners made their first purchase: 123 acres of coal land for more than $50,000. Over the next few months he bought hundreds more coal acres, staking his father's credit and borrowing at banks against every security he could possibly pledge. These properties formed the basis of H. C. Frick & Company, his newly formed partnership, in which he held a one-fifth interest.

On this land Frick constructed a series of coke ovens, and their fiery blasts soon lit the sky at all hours of day or night. His extraordinary management skills—organizing production on a large scale, assiduously weeding out waste, unifying his mining operations with his shipping and selling operations—whipped a disparate operation into an efficient machine. He was one of the first coal operators to open up company stores in his mining towns—and where other operators at first opened stores for the miners' convenience, Frick saw the stores from the outset as sources of profit. "Efficiency was his idol," remarked one historian, "and all that was weakly human was to be stripped and flung aside."

When a faulty excavation at one of his mines caused an explosion that buried thirty miners alive, Frick's frightened uncle Overholt quit the business. But Frick was undeterred. In mining as in war, he understood, death and dismemberment were necessary evils. Just as great generals did not shrink from ordering soldiers into battle, great captains of industry did not shrink from sending men deep into the mines or exposing them to the heat of the coke furnaces. (Frick differed from generals in the sense that some military commanders, at least, loved their troops.) Drawing on his first profits as well as new promissory notes, Frick bought out his uncle's interest.

Late in 1871 Frick approached the Pittsburgh banker Thomas Mellon for a $10,000 loan with which to build fifty more coke ovens. Mellon was a retired judge who had left the bench and opened a bank only the previous year, at age fifty-six, in order to have a business to leave to his sons. Mellon's father had been a friend of Abraham Overholt's, and Mellon himself had known Frick's mother when she was a girl. Mellon was thirty-five years older than Frick, but the two men were soulmates in their fastidiousness. Frick explained to Mellon in painstaking detail how the coking process was essential to the fabrication of steel. Mellon was impressed but refused to take Frick at his word. Instead he hired a business associate named J. B. Corey to investigate. Corey subsequently reported his confidential assessment: "Lands good, ovens well built, manager on job all day, keeps books evenings . . . knows his business down to the ground."

Still Mellon hesitated to make the loan. He sought what businesspeople over the next half-century would seek without success: some insight into Frick's seemingly inscrutable character. The obliging Corey surreptitiously broke into Frick's bachelor quarters in one of his company houses. On the basis of that inspection, Corey subsequently reported to Mellon that Frick was a person of meticulous habits and, in his opinion, a worthy risk. Frick got his loan, repaid it out of earnings soon after, and subsequently borrowed increasing amounts from Mellon to expand his holdings and buy out his partners.

Frick's determination surfaced again to his advantage during the Panic of 1873. This major depression paralyzed businesses across the country and forced most steel mills and mines to shut down. The Pittsburgh coke dealers who had previously disposed of Frick's coke went out of business as well. But Frick—who by this time owned four hundred acres of coal lands and two hundred coke ovens—refused to close down. In lieu of wages he offered miners his own scrip (called "Frick dollars") for goods they purchased at Frick's company stores. He set up his own sales office in Pittsburgh and continued to sell Connellsville coke at any price he could get. Each day he arose at 6 A.M., spent an hour inspecting his mines and ovens, three hours on the train to Pittsburgh, five hours calling on customers in Pittsburgh, and three hours on the train back home. From 6 P.M. to bedtime he worked on his books. It should have been a killing regimen, even for a twenty-four-year-old. Frick never complained; he had a business plan, and he intended to follow it.

Meanwhile, Frick borrowed still more money to buy up more coal lands at bargain prices. Thomas Mellon and his two sons extended $100,000 in credit to Frick during the depression, even though the Mellons' own bank was briefly forced to close. Now Frick was snapping up not only the lands of desperate farmers—who learned nothing from Frick about the future value of coke—but of some of his largest Connellsville competitors. When A. S. Morgan & Co. proposed a merger, Frick remained distant and non-committal; before long, Morgan failed, and Frick acquired Morgan's two thousand acres at foreclosure. By 1879 the depression was over, the steel mills had resumed full-blast, there were forty-two hundred coke ovens in the Connellsville district, and Frick controlled four-fifths of their output. More than a thousand men were digging coal for him, then baking it for forty-eight hours into gray lumps of coke, then shipping more than ninety carloads of coke a day to the steel mills along the Allegheny and the Monongahela in and around Pittsburgh.

Frick was now in a position to set the price of coke to his liking. During the depression, coke had dropped as low as 90 cents a ton. Now Frick fixed the price at $3.60 and eventually raised it to $5. His customers, the steel-masters, had no choice but to comply: Not only was Connellsville coke far superior to any other substance for fabricating high-quality steel; it was also the only coking coal then available to them by rail. "We found that we could not get on without a supply of the fuel essential to the smelting of pig iron," Frick's largest customer, the steel baron Andrew Carnegie, wrote in his memoirs.

On December 19, 1879, when Frick turned thirty, he marked the milestone by totaling up his accounts and learned that he was worth a little more than $1 million. He had achieved the lifetime goal of his youth in less than fifteen years. But of course he was not yet satisfied.

What Frick lacked at this point was a wife. But in his customary businesslike fashion, and with minimal wasted time, he soon filled that void. In the late spring of 1881 Frick met and fell in love with Adelaide Howard, daughter of Asa P. Childs, a wealthy Pittsburgh footwear manufacturer and importer. Within three months they were engaged; by December they were married. Whether or not Frick planned it that way, Adelaide's appreciation for the demands of business made her an ideally supportive partner. Indeed, it was on their honeymoon trip to the East Coast that Frick first met Andrew Carnegie and the two businessmen began to think of themselves as partners rather than rivals.

In many respects Frick was a younger, tougher incarnation of Carnegie himself. Carnegie had been born in Scotland in 1835; his family brought him to the United States when he was twelve and settled in Pittsburgh. By the time he was eighteen, Carnegie was private secretary to Thomas A. Scott, then superintendent of the Pennsylvania Railroad's Western Division. At that point the Pennsylvania was just six years old and Carnegie's salary was only $35 a month, but everyone in that division office seemed headed for bigger things: Eventually, in the 1870s, Scott would become the Pennsylvania Railroad's greatest president: the man who built the Pennsy from a struggling experiment into the world's largest corporation twice over. But Carnegie's name would be remembered long after Scott's and the Pennsylvania's were forgotten.

When Scott became vice president of the Pennsylvania in 1859, Carnegie was promoted to Scott's old job as superintendent of the railroad's Western Division. Shortly afterward, when Scott was called to Washington at the outbreak of the Civil War to help run the Union Army's railroad service, Carnegie followed him there. By that time Carnegie was in his mid-twenties and already investing his savings in companies that were growing with the railroad boom. Most of these were iron companies, but one was a company that built steel bridges—Carnegie's introduction to the manufacture of steel. In 1865, when he was thirty, Carnegie resigned from the Pennsylvania Railroad and combined two firms to form the Union Iron Mills, which was rolling the iron and steel beams for the bridges over which the Union Pacific Railroad would soon lay its tracks westward across the continent.

Like Frick, Carnegie exploited the Panic of 1873. He positioned what became his Carnegie Steel Company to be the dominant player in the steel industry, just as Frick was doing in the coke industry. Like Frick, Carnegie was a hard-driving boss, constantly pushing his executives and workers to produce more steel at lower cost. He didn't hesitate to cut wages when the market was soft. And nothing fueled his anger more than his lack of control over the supply of coke that was so vital to his mills. In a letter in 1872, Carnegie accused Westmoreland Coal Company of selling him the dregs of its product: "We are all weary," Carnegie wrote, "of being apparently considered as only poor pensioners upon your bounty, dependent upon your own sweet will for any crumbs you may in your own good nature see fit to shower upon us."

Yet Carnegie still followed the old patriarchal approach to business rather than the new faceless corporate model. Unlike Frick, in some corner of his mind Carnegie fancied himself a humanist and cherished his image as a friend of the working man. His professed ambition had been set down just before New Year's Day of 1868, when he was thirty-two:

> By this time two years I can so arrange all my business as to secure at least 50,000 per annum. Beyond this never earn—make no effort to increase fortune, but spend the surplus each year for benevolent purposes. Cast aside business forever except for others. . . .

Man must have an idol—The amassing of wealth is one of the worst species of idolatry. No idol more debasing than the worship of money. Whatever I engage in I must push inordinately therefore should I be careful to choose that life which will be the most elevating in its character.

The misgivings about wealth that Carnegie harbored (in contrast to Frick, who had none), ironically made him amenable to partnerships with men who suffered no such qualms and were happy to perform his dirty work while Carnegie himself looked the other way. One such partner was the ruthless Henry Phipps—another accountant by training—whose company had developed the Bessemer process for fabricating steel. Carnegie bought him out in 1867 for stock that made Phipps a major partner in Carnegie Steel. By the time Carnegie met Frick in 1881, Carnegie was forty-six and unwilling to place himself at the mercy of a supplier like Frick—or of anyone else, for that matter.

Frick, like Phipps, was stubbornly impervious to threats, flattery, or cajolery. When Frick refused to lower the price Carnegie paid for coke, Carnegie resolved to buy him out as well. Carnegie's investigation convinced him that, as he explained in his memoirs, "The Frick Coke Company had not only the best coal and coke property, but that it had in Mr. Frick himself a man with a positive genius for management."

But Frick refused to sell. Instead he permitted Carnegie to buy a minority of H. C. Frick & Co. stock. Carnegie believed this investment would bring him leverage with Frick, but he quickly discovered that his stake brought him no influence with Frick whatsoever. On the contrary, Frick soon announced a $1 million increase in the coke company's capitalization, effectively diluting Carnegie's investment. When Carnegie wrote Frick an angry letter demanding that he cancel the capitalization, Frick replied, "I do not like the tone of your letter," and proceeded with the capitalization as he had planned. Frick also acquired three thousand more coke ovens, about one-third of them from the Thaw family of Pittsburgh. As a minority shareholder in Frick & Co., Carnegie benefited from Frick's coke monopoly. But the damage he suffered as Frick's largest customer was far greater.

In 1883 Carnegie made Frick the inevitable offer Frick couldn't refuse: He proposed that Frick become general manager of all Carnegie properties. In the process, Frick would sell his company to Carnegie and eventually become a stockholder in Carnegie Steel (as Phipps had done previously). Frick thought it over for a while and casually accepted.

In his new job he methodically set out to do for Carnegie Steel what he had done for Frick & Company: eliminate competition by acquiring rival plants and molding these disparate mills into a single compact, harmonious machine. Like John D. Rockefeller in the oil business, Frick and his equally tightfisted partners Carnegie and Phipps were determined to build a juggernaut capable of dominating and controlling the coal, coke, and steel industries. Frick was well on his way to achieving that goal when the Leisenrings arrived in the Connellsville district to launch their Connellsville Coke & Iron Company.

CHAPTER 8

At War in the Coke Fields

Judge John Leisenring and his relatives all believed they had struck the equivalent of gold in Connellsville. The Connellsville coking coal basin was about thirty miles long by an average of two and a half miles wide, and the Leisenring group's property occupied about six miles in length at the very heart of this basin. "The coal is very unlike that in the adjacent basins," their Connellsville Coke & Iron Company explained in its first annual report to stockholders, in February of 1881. Whereas coal produced elsewhere required cleaning and crushing to remove the sulfur content before shipping, "the coal contained in your property, owing to its moderate percentage of sulfur, is taken directly from the mine and dumped into the ovens, without any desulfurizing process whatever. The cost of producing Connellsville coke is therefore at least fifty cents per ton less than that of the neighboring regions."

With a capitalization of $1 million, work on the Connellsville Coke & Iron Co.'s plants and mines began on March 27, 1880, using the advanced techniques the Leisenrings had developed in the anthracite region. The first plant, called Leisenring #1, boasted the deepest shaft in the Connellsville seam: It descended 371 feet beneath the surface. "The deeper the coal is buried," the company's annual report explained in February of 1881, "the purer and better it is found."

At first the Leisenring group's contact in the Connellsville region was Edward K. Hyndman, who had been born in Mauch Chunk and had served with Ned Leisenring in the engineer corps when they built the Lehigh & Susquehanna Railroad. Hyndman had subsequently moved west to the Pittsburgh & Connellsville Railroad as its chief engineer, and it was there that he had snapped up eight thousand acres of coal land for the Leisenring group.

The company built 501 coke ovens and produced its first coke in April of 1881. The prospects for this coke seemed unlimited—or, as the company informed its stockholders, "limited only by the means of transportation." That obstacle was removed the following month when the Pennsylvania Railroad built a branch from New Haven, Pennsylvania, near Connellsville, up to the company's ovens at the company town of Leisenring. Even before those tracks were set down, the company advised its stockholders that the property's value had roughly doubled. "Doubtless it is among the best tracts of coking coal land in the State, and probably in the world," the annual report declared. "Your property has the elements for one of the best future paying enterprises in the country."

Despite this rapturous prognosis, Hyndman left the Leisenrings' employ as superintendent in June of 1881 to become general manager of the Pittsburgh & Western Railroad, a division of the Baltimore & Ohio system. The reason appears to be that Hyndman, like many another ordinary man exposed to potential extraordinary wealth, had evolved from an engineer into a deal maker (he was already involved as the middleman who sold the Leisenrings their undeveloped coal lands in Virginia).

From the East the Leisenrings dispatched Hyndman's replacement, the man who actually designed and operated the first Connellsville plant. John K. Taggart had been born about 1851 in Northumberland in central Pennsylvania, the descendant of Irish immigrants who had emigrated to Philadelphia in the 1740s. He was probably attracted to anthracite coal mining at some point (a James Taggart, who may have been a cousin, operated a coal mine nearby at Tamaqua in the early 1850s). In contrast to the opportunistic Hyndman, Taggart was a highly regarded engineer who seemed to possess just the right combination of iron willpower and quiet, unpretentious manner needed for such a job. He was, according to one contemporary account, "cautious without being slow, and was very reserved in his business matters." In many respects Taggart was the ideal surrogate for the Leisenrings, who were three hundred miles to the east and thus heavily dependent on his judgment.

A second shaft was begun at West Leisenring in 1881 and opened in 1883. This shaft was about four hundred feet deep and was accompanied by five hundred more coke ovens. By then the Connellsville Coke & Iron Co. had four hundred men producing about a thousand tons of coke daily, with a potential capacity of five thousand tons. In the fall of 1882 the company proudly issued a circular promoting the unique virtues of its coke: "We respectfully call attention to the great purity of the coke; its tenacity in retaining its shape in the Cupola or furnace under the most intense heat;

its non-clinkering properties; its large percentage of Carbon and small amount of Ash; its freedom from Sulfur; its cellular structure; and its adaptability for sustaining heavy burdens."

What difference would this make to iron and steel smelters? The CC&I circular had a ready answer: "Running on anthracite coal," it noted, "a furnace of Bethlehem Iron Co., Bethlehem, Pa., made 428 tons of pig iron in one week. Running on Connellsville coke, Isabella Furnace #1, Aetna, Allegheny County, Pa., made 702 tons pig iron in one week."

Yet even Connellsville coke might vary in quality from one seam to another, the circular warned. "Owning, as we do, the coal lying furthest below water level, we claim superiority of coal in that regard," it claimed. "As we mine from one vein only, uniform quality and preparation can always be depended upon."

A single flaw threatened to undermine this rosy scenario. The feverish coke production by the Leisenrings as well as other operators eager to cash in on the coke boom had created a buyers' market. Thanks to this overproduction, by 1883 the price of coke had dropped back down to 90 cents a ton—its level during the Panic of 1873. The rise of new competitors like the Leisenrings had impaired Henry Clay Frick's ability to dictate coke prices, and he was determined to recover that power. At this point Frick was still the region's largest coke producer, operating 2,784 ovens. But he was no longer dominant: Now there were more than ten thousand coke ovens in the Connellsville district. To push the price up, Frick would need help from his competitors. (Frick had another problem as well: His coke company was now fifty percent owned by the steelman Andrew Carnegie, who preferred to keep the price of coke low.)

Frick's solution was to induce his three largest competitors to form a syndicate—with Frick as its president—to reduce production and thus stabilize the price of coke. This was the kind of price-fixing restraint of trade that was outlawed by the Sherman Anti-Trust Act of 1890, but it was perfectly legal in 1884 when Frick created what was known simply as the Coke Syndicate. Its four participants—including the Connellsville Coke & Iron Co., the smallest of the four, with 764 coke ovens—together accounted for 5,474 ovens. Eighteen other operators in the Connellsville district, with nineteen plants and 1,421 ovens, agreed to let the Coke Syndicate market their coke. So in one way or another the pool accounted for 6,895 ovens, or more than two-thirds the capacity of the Connellsville district.

From the coke operators' perspective, the pool was simply the best way to assure continued production amid unpredictable market forces. The collaborators agreed never to charge less than $1 a ton for coke. They would operate at full capacity but would curtail their operations—shutting down perhaps one day a week—if the syndicate's management ordered them to.

Thus was born what was probably the most extensive price-fixing experiment the U.S. economy had ever witnessed up to that point. It also represented the ultimate test of Henry Clay Frick's manipulative management skills. On the one hand, he had to hold in line the syndicate's collaborating independent producers, whose very survival would be jeopardized if they were forced to cut production. On the other hand, he had to hold off the steel and iron manufacturers (including, of course, Frick's own partner, Andrew Carnegie), who hoped to drive the price of coke downward.

The agreement worked effectively at first: Despite a mild recession that year, the price of coke rose immediately to $1 a ton, and then to $1.25. But in late January 1884, the independents broke ranks, and the pool broke up. A month later, a new accord was reached that went into effect on April 1. This time the Big Four syndicate members actually restricted their production. At first they shut their ovens and mines down for one day a week. But as the recession deepened, they reduced their production to less than two-fifths of capacity in order to keep prices at $1.20 or higher. As he had done during the Panic of 1873, Frick also seized the opportunity to buy out many of the region's smaller producers, in the process expanding his control to one-third of the region's total productive capacity. But for a man as obsessed with control as Frick, these solutions were merely temporary stopgaps. Eventually he would need to find some more permanent way to deal both with his large competitors in the Connellsville district and with his partner Andrew Carnegie.

In the midst of these market manipulations, on February 20, 1884, a gas explosion in the Leisenring group's #2 shaft claimed the lives of nineteen miners. The subsequent investigation blamed the disaster on the negligence of the mine boss, Thomas Jenkins, who had failed to inspect the workings every day, and on the state mine inspector, who had never set foot in the mine. Connellsville Coke & Iron promptly installed one of the largest fans in the state and made other improvements. This was the sort of carelessness that Judge John Leisenring in his prime, with his careful attention to detail, would not have tolerated. But now the Judge was dying of Bright's disease; more to the point, his family's operations were growing so far-flung that no individual could closely supervise them.

When John Leisenring died six months later at age 65, his wealth was estimated at more than $1 million. His will created partnerships for both his sons-in-law but was careful to leave his money to his daughters and grandchildren, just in case their husbands' businesses should fail. The task of dealing with Frick, with the steel masters, with coke prices and mine safety now passed to the judge's two surviving sons, Edward and John, then thirty-nine and thirty-one, and his sons-in-law, Dr. John Wentz and Mahlon Kemmerer, then forty-six and forty-one. But it was John Leisenring's eldest son—Edward or "Ned," as he was familiarly known—who succeeded to most of the judge's offices as head of the Leisenring group.

If any man was qualified to compete with Frick in the Connellsville coal fields, it was Ned, Judge John Leisenring's oldest son and the leader of the third generation of coal-mining Leisenrings. Unlike Frick, Ned had been marinated in the coal business. He was born in Mauch Chunk in 1845, was educated in a private school nearby at Eckley, and took a course at the Philadelphia Polytechnic College before joining the engineering corps of the Lehigh Coal & Navigation Co., where his father was superintendent and so many of his relatives were employed. In the fall of 1868 he was placed in charge of the LC&N's mines at Newport, Pennsylvania; a year later he was formally promoted to superintendent there, at a salary of $2,500 a year. In 1869, his first full year on the job, Ned supervised the digging of three different shafts, each to a depth of four hundred feet; the construction of a coal breaker capable of breaking four hundred tons of coal daily into marketable chunks; the installation of a large steam pump to keep the mines free from flooding; and fifty blocks of what Ned called "comfortable miners' houses" to augment the eleven blocks that existed when the LC&N acquired the property. He was all of twenty-four years old at the time, and many of his future business associates came from this engineer corps (unlike Frick's future partners, who were, above all, numbers men). In 1877 Ned branched out independently and for the next seven years ran coal mining operations at Audenried under contract to the Lehigh & Wilkes-Barre Coal Co. It was a profitable venture that produced nearly half a million tons of anthracite coal annually.

In 1873 he had married Mary Middleton, daughter of a Philadelphia iron-and-steel merchant, but both their daughters died shortly after birth, and Mary herself died in 1876, at the age of twenty-two, from complications of the second childbirth. Rather than sulk about an empty house, Ned consumed most of his waking hours with his business. "Whatever business he undertook," one of his obituaries remarked, "he went at it with a will, and rarely stopped until he had become entirely familiar with all its details."

Eventually Ned became president of all of the coal companies organized by his father and his uncles, as well as the First National Bank of Mauch Chunk, the Mauch Chunk Heat, Power & Electric Light Company and, ultimately, the Lehigh Coal & Navigation Company itself. He supervised all of these companies from his office in Mauch Chunk, communicating by mail (using a different letterhead for each company) with his superintendents in the field, and he saw no reason why the Connellsville Coke & Iron Co. should be handled any differently. But there were important differences: The CC&I was a new and as yet unprofitable operation. It was the only one of the Leisenrings' concerns that had to contend with Henry Clay Frick. And it was making its way into the world at the very time that miners were discovering the power of collective bargaining.

Frick was ahead of his time in his perception that coal operators needed to organize, and at least some miners recognized the need for workers to organize as well. It slowly dawned on miners and operators alike that coal and coke were labor-intensive industries, which meant that miners could enjoy a powerful negotiating tool—if they could organize themselves. The first fitful local attempts at unionization by miners in the 1840s and '50s had failed to achieve any of their goals for long. In 1861 the first national miners union, the American Miners' Association, was launched in St. Louis, where delegates from four states heard a passionate plea from the union's first president, Thomas Lloyd: "Men can do jointly what they cannot do singly. . . . How long, then, will miners remain isolated? Our unity is essential to the attainment of our rights and the amelioration of our present condition; and our voices must be heard in the legislative halls of our land."

That union too collapsed from internal dissension in 1867. But the Knights of Labor, formed in 1869 by nine Philadelphia tailors, hit on a practical formula for applying Lloyd's ringing principles. Its founding president, the scholarly Uriah Smith Stephens, operated the Knights as a secret organization—like the Masons, to which he belonged—in order to protect its members from harassment by factory owners. "An injury of one is the concern of all" was their slogan.

This union welcomed all gainfully employed workers of any trade (except for "parasitical" occupations like bankers, lawyers, investors, gamblers, and stockbrokers), and by the late 1870s the Knights were strong enough to go national and public. By the early 1880s they were organizing in the bituminous coal fields of western Pennsylvania. That is, the Knights were organizing the miners and coke workers around Pittsburgh and Connellsville at the very moment that Henry Clay Frick was organizing their employers in a syndicate to prop up the price of coke.

At first there was no conflict between these activities. After all, the operators' ability to charge higher coke prices reduced the danger that miners' wages might be cut. But as the Coke Syndicate succeeded and the price of coke rose, miners began to wonder why they shouldn't share in the increases. Two labor organizations—the Knights of Labor and the Mine Laborers' Amalgamated Association—moved into the neighboring Monongahela River District in the summer of 1885. The following January, workers in the coke regions struck for a 10-cent-per-ton raise, and also to correct two other grievances: the alleged use of false weights (which caused miners to be paid less than what they had produced) and the inflated prices charged at company stores.

Frick responded by ordering strikers evicted from their homes. His miners responded with violence and vandalism. On February 9, 1886, the syndicate's Big Four companies ordered a total shutdown of operations, and by the latter part of February, three-quarters of the coke ovens around Connellsville were idle. This action sharply drove up the price of coke, with an unexpected side consequence. Many independent operators, eager to make a killing at the new higher prices, agreed to the workers' terms and resumed operations. Eventually one of the Big Four operators, Colonel James M. Schoonmaker, himself broke ranks with the Big Four and offered to settle with his workers. But perhaps the greatest pressure on the operators came from Frick's partner Andrew Carnegie, whose first priority was maintaining a steady flow of coke to keep his steel mills running. "Of course," Carnegie wrote to Frick, "you won't let us stop again at Bessemer [steel works] *if possible to prevent it.* We do want to go along there regularly now."

The result was a virtually total victory for the union, perhaps the first of such magnitude in American history. By the end of February the strike was over; the operators had conceded a 10 percent increase together with a pledge that the workers' complaints about company stores and short weights would be addressed later. A local grand jury refused to indict strikers who had been jailed by Frick's security police for violence; instead, Frick was required to pay the court costs. A permanent conciliation board was set up to handle other grievances.

The settlement raised the cost of coke by 7 cents per ton, but the iron and steel industries were able to absorb the increase and pass it on to their customers. In any case, Frick's Coke Syndicate maintained prices through 1886, encouraging the Leisenring group to acquire five hundred more acres in April; by December the group was beginning to dig its third shaft in the Connellsville district.

But now the balance of power had changed through a new perception on the workers' part: The coal and coke operators were so reluctant to disappoint their customers that they would grant small concessions to workers rather than suffer a strike. Consequently, the remainder of the year 1886 degenerated into a succession of walkouts and slowdowns whenever a grievance arose.

Unlike Colonel Schoonmaker, Ned Leisenring was based in Mauch Chunk, more than three hundred miles to the East, when the Knights of Labor struck. The distance afforded him a degree of insulation from the pressures and emotions of the coal fields, but it also, of course, removed him from a real sense of what was going on. His daily letters to Taggart, his superintendent at Connellsville, reflect a man who believed in the justness of the operators' cause and was astonished by the miners' ingratitude, but who was equally reluctant to deal harshly with them. They also provide what is perhaps the best insight we will ever have inside the mind of a coal operator of the 1880s as he came to grips with a revolt among his workers.

"I regret to hear of so much nastiness at the mines which has curtailed shipments," Ned wrote to Taggart on May 3, 1886. "Now think we ought to ship every ton possible. The only plan is to quietly get rid of those men who neglect their work[,] and get in a set which you can depend on." A month later, prior to a meeting between the operators and the striking miners, Ned wrote, "I hope the meeting with the men today will be a satisfactory one. And the Syndicate will convince them their new demands are unfair and unjust and cannot be granted."

After receiving a telegram from Taggart in August, Ned replied with a letter that reflected both his dependence on Taggart and his trust in him: "I see you have resumed work again, and I am very glad you did so without making concessions, although I have not yet received your letter and do not know the cause of the strike; but you know my opinion on matters of this kind." Ned's letter also referred to a union demand to station a checker at the tipple—the huge wooden structure at the mouth of the mine, where the coal cars from the mines were unloaded—to measure the coal tonnage credited to each man. Ned's response provided a succinct statement of his philosophy:

> I will not have anybody that will interfere with our work, whom we do not employ, anywhere about it; but [I] have no objection, if the men insist on putting a man there with the understanding that he has nothing whatever to say to our men; and if he does interfere in any way with your directions as to how the cars are to be loaded, docked or dumped, or anything else, he must be ejected from the place at once. I prefer not to have a man there at all; but if the men insist I would not offer objections upon the above conditions.

In another note to Taggart, dated October 15, Ned dealt with the miners' demand for a closed union shop:

> If any [miners'] Committee asks you to discharge any man because he will not join the miners' union I think the best answer you can make is to tell them to put the demand in writing and sign it, that you will refer it to the company. This they are not likely to do, as they can be prosecuted for intimidation. I find Schoonmaker has done this and never heard anything further about it.

Meanwhile, Ned was grappling with another labor problem: retaining competent managers. "The former employer of your #2 mine boss Anderson wants to take him away from you," Ned wrote Taggart on October 15,

> but Anderson says he will not leave us without our consent. His employer called here yesterday with Anderson and will call again today. I told Hyndman

[the Connellsville C&I's agent at Pittsburgh] to say to him that you had entire charge of the mines and he would have to see you, and that you had been to considerable trouble to get him and were satisfied with his work. He offers more salary than we could pay, but I think you can prevail on Anderson to stay, if you want to keep him.

By October 21 Ned was no longer so confident: "I hope you will succeed in retaining your #2 mine boss as he seems to be a competent man," he wrote to Taggart, "but Mr. Stabler [a competitor] told me he was bound to have him and would outbid us on salary to get him—you better have your eye on someone to replace him at any rate."

A walkout of unspecified cause at the Leisenring #2 shaft in late October of 1886 prompted Ned to write Taggart:

I am very sorry the men are so foolish at #2 and should strike for so trivial a matter, and particularly now when furnaces are so hard up for coke—but I should never have them dictate to me who I employed and [who] is changed, and would recognize no committee of employment. It seems to me we all should have to make a fight this winter, and settle all these questions, and the more demands they make, the sooner the fight must come.

The fight came soon enough. By 1887 the other major coke operators, goaded by Frick as Ned Leisenring had been, were eager to make a stand. In the spring of 1887, the region's miners and coke workers demanded a 12.5 percent wage increase. The operators—convinced of the justness of their cause, or of their influence in the legal community—agreed to submit the issue to an arbitrator, John B. Jackson of Pittsburgh.

"Mr. Atcheson [John F. Atcheson, secretary of the Coke Syndicate] writes me that the arbitration papers are all in the hands of Mr. Jackson, the umpire," Ned wrote to Taggart on April 12, "and the Syndicate members seem to think that the decision will be entirely in our favor, and that no advance will be granted. As the comparative statement of wages paid by Connellsville and other regions was not accepted as evidence, I am afraid that we shall have to pay an advance of from five to ten per cent; but I hope it will be otherwise."

Ned added that he expected to be in Pittsburgh the following week. "I would like very much to come to the mines," he remarked, "but if the strike is still on I think perhaps I better not do so. What do you think about it?" Taggart replied that a visit by Ned to Connellsville might send the wrong message—creating the impression that he was eager to compromise.

On May 1 Taggart cabled Ned with the good news that the arbitrator had rejected the unions' demands for any raise. But this was a case of "Be careful what you wish for; you may get it." Although the unions' national officials accepted the arbitrator's decision, the local workers refused to abide by it. It was the first example of a pattern—a national union's inability to impose its will on local miners—that the Leisenrings and other coal operators would still be grappling with ninety years later.

"[T]his a.m. I saw in the papers that the men will demand an advance and strike in six days if it's not given," Ned wrote that day. "Papers said they want 12½%. I thought Jackson would give an advance of 5 or 10% and did not believe men would accept if they got nothing." Ned suggested that a 5 percent or 10 percent increase would have been a reasonable price for long-term labor peace; but now, he implied, the union's far greater demand had stiffened his will: "I earnestly hope now the Syndicate will be stiff, and not make any compromises, and give nothing. Let the men strike if they want to, and get enough of it. . . . If any meeting of [the] Syndicate is held and you attend on the wages question: I want you [to] talk and vote against an advance [that is, an increase] or compromise. If the men will not accept [the] decision of umpire there is no use dealing with them."

Just as the papers had predicted, on May 7 the unions, without even the pretense of waiting for a response from the operators, struck throughout the region. Under Frick's direction, the coal operators brought in strikebreakers, along with a force of 150 security guards from Allan Pinkerton's private force to protect them. Ned Leisenring remained in Mauch Chunk, utilizing the benefits of operating at a distance. "I saw the whole account of the riot at Jimtown in the papers," he wrote to Taggart on May 23. "I will not be out [to Connellsville] unless I am telegraphed for to come, as I do not want to come near the works while we are on a strike, it might create an impression, as you say, that our Company are very anxious to have the men go to work and want to compromise."

Bolstered by Frick's determination, the operators continued to present a united front against any compromise whatever. But each day brought fresh doubts among them as to how long the others could hold out.

"I have a notice that the Syndicate have agreed to have a conference with the leaders of the men today," Ned Leisenring wrote to Taggart on June 2. "I was decidedly opposed to this, and I regret that they agreed to see the men at all; and earnestly hope that no promises were made to them, except that they would have to go to work unconditionally. If there is any yielding now on the part of the Syndicate I do not think I will have anything more to do with it."

Ned seems to have been reading Frick's mind. At this juncture Frick was under pressure to compromise from another source: his partner Andrew Carnegie, who cared more about fulfilling his steel contracts than about holding the line on pay for coke workers and coal miners. The strike had forced Carnegie to shut down seven steel mills, at an estimated loss of $250,000 per hour. From Scotland, Carnegie cabled Frick to grant the pay increase in order to keep the mines and coke ovens operating. Frick replied that he could not do so—that he had given his word to his fellow operators and his honor was involved. Carnegie replied that if Frick felt that way, he could resign. Frick did just that. "I do not feel like standing in the way of you managing the property as your judgment and interests dictate," he wrote Carnegie, but "I object to so manifest a prostitution of the Coke Company's interests in order to promote your steel interests."

With Frick gone, the Frick Coal Co. settled the strike on the workers' terms, granting a 12.5 percent raise. The other three coke operators were furious at what they saw as Frick's betrayal.

But the workers felt equally demoralized: As the three operators whom Frick had abandoned continued to hold out against the strikers, divisions opened up between the Knights of Labor and the Amalgamated union. The strike had exposed the vulnerability of a "primary producer" industry like coke when it was controlled by a finishing operation like steel. But the devastation to the miners was far more severe. By late June 1887, the operators had lost $42,000 in potential profits; the miners had lost $689,000 in wages.

The struggle continued into the summer, punctuated by negotiations, rallies, some evictions of miners from their homes, and periodic violence, most notably at the Leisenring shaft. By June 27 Ned Leisenring was in Pittsburgh and contemplating evictions. "You can find out . . . the expense . . . to eject each tenant," he wrote to Taggart. "I should not think it would be much. However, we don't want to do it unless compelled to."

In another letter to Taggart written the same day, Ned addressed the issue of property destruction:

> I wish you would read over the insurance policies on the tipples and bins at each place and see whether there is a clause in them by which the insurance companies are not liable for any loss by fire in case of riots etc. I think it's usual to have such a clause in them, which would prevent any payments in case of fire. I only mention this so you will understand the necessity of using extra precautions about fire when the attempt to resume [work] is made. I hope the men themselves in the whole region will conclude to give up the fight next week.

Some of the men did indeed return to work the following week. "All quiet," Taggart cabled Ned on July 7. "I have working today 34—men working above ground four, working underground thirty." Ned wrote back: "I hope to hear tomorrow you have 134 at work and 100 of them underground." He also instructed Taggart in the importance of secrecy in the use of cables: "Make your telegram here and to Pittsburg short and to the point and careful in using the cipher so it will not confuse us."

Another letter to Taggart that day found Ned taking a more cautious approach to eviction of miners:

> I see by this morning's papers that there is a report that we propose to evict the men living in our houses immediately. Of course I understand that this is only a rumor, and without authority. I do not want to undertake anything of the kind unless absolutely necessary for two reasons. First, it will create a very bitter feeling against us hereafter. Second, it will cost us a great deal of money to make the evictions. You will therefore have to exhaust every effort to get our own men at work before we do anything of the kind; and I trust to hear tomorrow that you have succeeded in getting a good many men in quietly.
>
> I also see in the papers that Jimtown has about fifty men at work. I hope somebody else will make a break this week. The Pinkerton men were all carefully selected, and in case of trouble you can depend on them to stick.

Taggart resolved to evict six families at the Connellsville C&I's Number 2 works, and Ned—after consulting the company's lawyers—gave his cautious approval to proceed. "I want you to be very careful in making these evictions that they are strictly made under the letter of the law, and that there is no question about the signatures and the witnesses to the leases signed by the parties who are to be turned out." The six families were evicted a few days later without apparent incident, but Ned's colleagues in the Coke Syndicate pressed him to evict more. "The Syndicate seem dissatisfied with the evictions going on so slowly," Ned wrote Taggart on July 21, "but in this matter I desire you to carry out my instructions, and do not want evictions made unless they are necessary."

Still the strike continued. "I thought they would decide at Saturday's meeting to resume [work]," Ned wrote Taggart that same month. "I understand the K. of L. [Knights of Labor] have a meeting today to decide whether to resume or not."

In his search for alternative labor sources, Ned traveled to New York to personally investigate the European immigrants streaming through Castle Garden on the Battery, where the federal immigration center operated between 1855 and 1892. After Taggart cabled him to "Send 50 men at once," Ned replied: "I find there are no men to be had at Castle Garden today. The steamers coming in for the past month are only bringing in 100 to 200, while they usually bring 1000 to 1500 if they can. I will see what can be done and think I can arrange to send them latter end of week but if you find they are not needed wire me promptly."

Two days later Ned was more encouraging. "You may expect 50 to 75 Swedes and Germans by Friday or Saturday," he wrote Taggart.

> I take it for granted that Mr. Donnelly [the company's agent in New York] has made all arrangements to have them sent up to you. Another batch will be sent into the region to arrive Sunday or Monday. As I wrote you from New York

men are extremely scarce; but will be plenty in two or three weeks, so the Superintendent of Castle Garden tells me; and I would very much prefer to get the men from there than through a Labor Bureau. . . . Of course if the men should determine to go to work anytime you will at once notify Mr. Donnelly so as to have the shipment of men from New York stopped, as it is expensive to have them sent so far.

The notion that the superintendent of Castle Garden—a federal government official—would have the time or inclination to help Ned Leisenring with his labor problems is remarkable in retrospect, and indeed seems to have struck Ned the same way: The following year Ned told Taggart that he had proposed "to pay him [the superintendent] for his trouble in sending the men out to you; but as Mr. D. [Donnelly] told me he was a man we could not give money to, I told him to supply him with the coal for his own use." What later generations would perceive as a bribe, Ned perceived simply as a kindness.

The trainload of ninety-two immigrants arranged by Ned left New York headed for Greensburg Junction in western Pennsylvania on the night of Thursday, July 14. "As it was very late when I received notice of their departure," Ned wrote Taggart on July 15, "I wired Donnelly to arrange to have policemen meet them, and he wired towards evening to have them sent via Pittsburgh; but it was too late to change them, and that Pitcairn [one of the company's agents] sent them up in a special train. You will have to see that they are taken care of, and keep them together."

On July 21 Ned reported to Taggart that he had mustered more immigrant workers:

> I am today making arrangements with a Labor Bureau to send out 50 or 75 Germans and Swedes, and expect to consummate the matter by this evening. I have written to Mr. Donnelly asking for instructions how to ship them, either via Greensburg or via Pittsburgh; and he will notify you when and by what route you may expect them. I have never seen so few emigrants in New York as this season. The Superintendent of Castle Garden says that within two or three weeks they will begin to arrive very rapidly, and we can get through him selected men, and a very much better class than through the Labor Bureau. It is pretty expensive work sending these men so far out there [to Connellsville] from New York; and I need not therefore urge upon you the importance of getting all the men about there you can. I am afraid you have been oversanguine in your expectations to get our men to work, and I trust that there will be no failure about it. . . . When you are advised these men will arrive you must make preparations to feed them until they get to work.

With the benefit of more than a century's hindsight, Ned's approach to his workers—natives and immigrants alike—appears callous and shortsighted. Preoccupied with his own problems, he seems oblivious to theirs. (When a miner was killed in an accident, Ned declined to make a company donation to the miner's family, lest that precedent threaten the viability of a company that wasn't yet profitable. Instead, he instructed Taggart, "ascertain whether the family are very poor or not; and if they are absolutely in want of anything you can send them $100 for me personally.") He perceives his company as threatened, just as the miners see themselves. He sees clearly the disruptions to his company, his industry, and himself personally, but he seems blind to the proportionally much greater disruptions in the lives of the strikers and the immigrant strikebreakers. "I suppose accurate accounts are being kept of all expenses incurred by feeding the police men, new men, &c, so that when settlement day arrives we can have a full detailed account," he wrote Taggart on July 21. "I have myself paid several thousand dollars sending men out, policemen, &c." His unspoken assumption was that the miners had a vested interest in the financial health of his mining operations, and the miners' failure to perceive this led him to regard them as foolish. Ned was indeed concerned about his workers' anger but saw himself as powerless to improve their living conditions, short of assuring them steady work if only they would accept his pay scale.

Although he presided over a small empire of mining companies, Ned was clearly frustrated by his need to rely on the field lieutenants (such as Taggart) to whom he had delegated authority. "I asked you in several letters whether you were doing anything at all at No. 3, and you have not replied," Ned chastised Taggart in one letter. "I want you to keep me posted so that I know exactly how things are going on at all the places." In another letter from Mauch Chunk, Ned wrote, "You promised to telegraph me at Philadelphia yesterday but failed to do it, and I wired you from here last evening, and at this hour, 9:30, I have no reply. . . . You also promised to get me the actual number of men at work in the mines and out before I left; but I suppose forgot to do it."

Often Ned's best sources of information were public newspapers, and his perceptions were heavily influenced by wishful thinking. "I see by the papers this morning that there was a man killed by the policemen at Jimtown," Ned wrote to Taggart from Mauch Chunk on July 15. "I hope nothing of this kind will occur at our works. I also see that Youngstown had fifty men at work yesterday, and I do not believe the [union] leaders will be able to hold them together but for a few days longer."

While reading his morning newspapers on July 20, Ned learned that the coke workers had voted to end their strike and would resume work the next day. This was the first he knew of this momentous news. When he telegraphed the Coke Syndicate's office in Pittsburgh to ask if it were so, the syndicate's secretary replied that the Knights of Labor had voted to go back to work and that the Mine Laborers' Amalgamated Association was expected to do the same later that day. The next day Ned instructed Taggart: "You are not authorized to accept any terms from the men, except that they shall go to work at the old wages unconditionally." He also directed Taggart to make sure that all strikebreakers retain their jobs: "As I feel now, any new men who have gone to work [during the strike] will not be discharged under any considerations, as the strikers had every opportunity to take their places on the same basis that the new men went in on."

The strike against CC&I and the two other holdout operators was settled on July 27 on the operators' terms—with no pay increase at all—but the operators' victory was largely empty: Since Frick, on Carnegie's orders, had already agreed to a pay raise to get his men back to work, all workers in the Connellsville district had the option of moving to the Frick works for a 12.5 percent pay increase. Equally important, Frick's resignation meant the end of the Coke Syndicate and consequently the end of high coke prices.

With the strike over, Ned focused his attention on regaining the miners' goodwill. "I think now that the men have all got to work again, that you better try and get up a good feeling among them and show that we have no animosity against them, and propose to treat them fairly if they do the same to us," he wrote Taggart on August 2. In another letter, he asked Taggart, "Have you taken many of the evicted men back, and if so have they houses, and have the others removed their goods from the side of the road?"

Another letter from Ned advised Taggart, "If a large majority of the No. 3 men desire to be paid once a month, and will sign a petition to that effect, you can make the pay monthly instead of semi-monthly, notwithstanding there may be one or two kickers." Similarly, when a despised checkweighman named Hamill was arrested for intimidating the miners, Ned proposed to discharge him, presumably as a peace offering to the men. "I have no doubt that Hamill is a very bad man," he wrote Taggart; "and I think you better not allow him to act as checkweighman at all[;] in fact, if you can possibly do so, [do] not have any checkweighman about the places. I do not know what has been done by the other operators on the subject." Ned remained apprehensive about holding any discussion of pay scales with the miners. "My impression is that we ought not to have any meeting unless the men in the region are fully represented, both associations," he wrote Taggart. "And it would be a great deal better if you would not recognize them at all. The [Coke] Syndicate, without my consent, promised a [wage] scale, and that is the only reason that I am willing to agree to it, otherwise I would prefer to make our own wage arrangements with our own men."

The miners had returned to work, but the issue of their pay scales remained unresolved. Without Frick's forceful leadership, the operators were hard-pressed to present the miners with a united front. One of the operators "proposed to offer his own men a scale today, which would be 5 to 6¼ per cent higher than the one the committee offered," Ned wrote Taggart on August 15, "and if he succeeds in having his own men accept it . . . some others, rather than have any trouble, may follow suit. I hope nothing of the kind will be done, and that we all will have one scale, otherwise it will cause endless trouble." Ned pledged to abide by the wage scale offered by the operators' committee "and that I would agree to no other scale which would concede an advance in wages to the men. . . . Unless we fix up something satisfactory on the scale question this week I think we will join with the outside producers and offer a scale to the men on our own account, or else agree to have no scale at all. I am tired and disgusted with this meeting committees and arriving at no conclusions."

Unspoken in this correspondence is the central issue that would plague coal operators, the most independent and competitive of men, over the next century: how to present a united front to anyone on any issue.

What the faithful superintendent Taggart thought about all this is not known; his letters to Ned Leisenring have not survived, and even if they had, his advice was rarely solicited. His role was to exercise his judgment on the scene, keep Ned abreast of everything, and follow Ned's orders when directed. A rare personal exchange between the two men occurred in October 1887 when, with the strike crisis over, Taggart revived an old request to have the company build houses for him and two foremen. Ned showed the letter to his in-laws and business partners, Daniel Bertsch Jr. and Dr. John Wentz.

"Neither of them would say positively we should go on and build it," Ned advised Taggart, "but as you have waited patiently and I appreciate your position in the matter I am willing to expend now $3500 on a house for you above the foundation, but this will be the extreme limit, and must include out-houses and kitchen, if you desire them." He declined to spend anything that year on houses for the two foremen. The reason, Ned explained, was the uncertainty of the coal business: "If I was sure that we should have continuous work and make some money on our investment out there [in Connellsville] between now and next Spring, I should be willing to do it, but you understand, as I told you, the situation of affairs, and stock-holders will growl and grumble if they get no return whatever for their investment." Ned said he had no objection to Taggart's moving his family into the town of Connellsville until Taggart's house could be built the following spring. "I want you to understand that it is my desire that you and your family, and in fact everybody connected with the concern, shall be as comfortable as possible, and if our Company was a dividend paying one"—that is, if it were turning a profit—"we should be inclined to be very much more liberal to all of you."

This concern was reflected in Ned's detailed specifications for the foremen's houses. "I desire," he wrote, "that the location shall be made at such a point that the other two houses can be placed about in a line, and within a convenient distance of yours, say 200 to 300 feet apart, and that it should be located so that you will not have to build an expensive wagon road with steep hills to get up to them. . . . The style of architecture of the three houses should be a little different so that they will not be exactly alike." Ned closed the letter by implying that he had larger problems on his mind: Although the Connellsville strike had ended, "Our strike in the Lehigh region still goes on, and I do not believe that a particle of work will be done before the first of January next, or perhaps until next spring. The operators have all placed themselves in a state of siege, cut down their expenses to a minimum, and even suspended some of their clerks and foremen and sent their mules off to pasture."

When Carnegie returned to America from Scotland that fall of 1887, he persuaded Frick to resume his old position at the head of H. C. Frick & Co. But Frick's company had withdrawn from the Coke Syndicate when Frick himself resigned, and now the remaining syndicate members needed him back if they were to continue propping up the price of coke.

"I am afraid," Ned Leisenring lamented to Taggart in January 1888, "that the Connellsville Exchange"—that is, the Coke Syndicate—"will never be as successful as it was the past three or four years, unless the Frick Coke Company come in. I find that they [Frick & Co.] are cutting prices all over in the East. At least it is so claimed by our Agents."

But Frick refused to rejoin the syndicate unless it repaid about $90,000 that he maintained was owed to him—money that the syndicate had withheld on the ground that the Frick Company had violated its agreement by paying a wage increase instead of acting in tandem with the other operators. After lengthy negotiations, Frick got his money, but now he (and, to be sure, the other three major coke operators) raised a new condition: The new price-fixing pool would have to include all the operators in the Connellsville district; the Big Four operators were tired of seeing their coke prices undercut by smaller independents. "It seems to me very strange," Ned Leisenring remarked in a letter to Taggart, "that the coke operators should be fighting among themselves continually, and not be able to regulate wages and prices so that there is a decent profit in the business."

But some of the larger independents refused to join the pool. In the meantime, all the operators battled each other in unfettered and (to Ned's way of thinking) unhealthy price competition. At a meeting in Philadelphia in mid-February, Frick talked to Ned about reducing the price of coke to $1.50 per ton. "I told him I was decidedly *not* in favor of so doing, unless wages were reduced ten per cent all around," Ned wrote Taggart.

He told me that there would probably be a meeting to discuss the matter last week and he would wire me, but he did not do so. If a meeting is called for to-morrow, and you attend it, you can say that we are satisfied to reduce coke to $1.50 next month, provided a reduction of 10 per cent in wages is made at the same time; but if this is not done, and the price is reduced to $1.50, we will not agree to it, and will sell the coke at the best figure we can, to get steady work, and reduce the wages ourselves when we get ready to do so.

After sending his chief sales agent on a general trip through the East and Midwest, Ned was dismayed to learn that his competitors in Connellsville were deriding him to their customers as "the man who made two-dollar coke." This ridicule "disgusts me very much," he wrote Taggart, "and shows what little men some people can be."

Unlike Frick, who clearly grasped the value of market domination and economies of scale, Ned does not seem to have considered that reducing the price of coke per ton might increase his company's sales volume or its market share. Instead he presumed that any decline in price would bring a concomitant decline in revenue and must therefore be offset by a proportionate reduction in wages. Although his workers doubtless perceived him as a powerful coal baron, Ned reveals in his letters a man who consistently saw himself as the helpless victim of forces beyond his control, which in some respects he was: As an operator raised in the engineering and production side of the industry, he lacked the grasp of the larger picture enjoyed by financial minds like Frick's.

By March 1888 the Connellsville price-fixing pool was abandoned altogether. Without the pool, the price of coke fell, the hard-fought wage gains that the miners had won in 1887 were wiped out by reductions, and Frick—convinced anew of his need to increase his control of his companies' destiny—resumed his efforts to buy out his competitors. Slashing the price of coke was a key tactic in his strategy. In May Ned complained to Taggart that Frick was cutting prices to 90 cents and even 80 cents. Ned refused to peg his price below $1, but his sales agents felt heavy pressure to do just that.

"Whitney [a sales representative] told me that Frick's Agent offered one of our customers in New Jersey coke at 85 cents," Ned wrote Taggart, "and he had to accept that figure to prevent him from leaving us." The only alternative, Ned felt, was to cut wages. "If any of your men complain about the reduction in wages," he advised Taggart,

you can use the argument that we propose to run full if we can, and give them steady work, while other people are only running part time; and without this reduction of wages we would probably order a shutdown until the price of coke advanced; and you can also say that even at these wages we cannot make coke for what we get for it, and we are simply keeping our works in operation in order to keep up the organization and give them employment.

Even as Frick cut the price of coke, late that summer he further tightened the screws on his Connellsville competitors by granting a 5 percent wage rate increase to his miners and workers—a move that triggered strikes demanding similar increases at rival firms, including the Leisenring group's Connellsville Coke & Iron. But Frick's machinations were beyond the ken of Ned Leisenring, who seems to have believed that everyone he dealt with was as honest and straightforward as he.

"You say that Mr. Frick is kicking against the cost of coke with his superintendent," Ned innocently wrote Taggart on September 4. "I therefore do not understand why he should advance his men employed at the Southwest Coke Company's Works, five per cent, as I noticed in yesterday's paper he had done voluntarily. If you know what his object was please advise me."

A week later Ned acknowledged that "the advance given by Frick has caused considerable uneasiness among the men in the region"—that is, the workers at Frick's rival companies—"and the leaders are quietly making efforts to re-organize the labor organizations." To Taggart, Ned expounded his own benign (and, given the cutthroat nature of his competition, naïve) philosophy: "We want to work right straight along and do as big a business as possible, not because there is any money in it at all, but because it is cheaper for us to work very full than to be idle; and secondly because if there is an active fall trade, and prices go up, and we are idle, we would lose our customers, and have to undersell the market in order to get the trade." The bottom line, Ned said, was that the company would settle small demands from the unions but would "continue our old policy and fight them . . . if there are any demands made out of all reason." Ned professed himself "very glad indeed that the men have got the idea that our Company always maintain [its] rights, even at the cost of a strike; but at the same time I want them to have the impression that we are always honorable and just to our employees."

But honorable impressions couldn't suffice to stem the heavy turnover among miners at CC&I. When Ned placed a help-wanted ad in a Pittsburgh Hungarian-language newspaper, the editor helpfully suggested why Ned's company was unable to hold on to workers. "I am informed," Ned wrote Taggart,

> that one of the troubles at our No. 2 [mine] with the men is that they do not earn any money, and that this is the reason why they will not stay there. The editor of the Hungarian paper at Pittsburgh says that a great many of our No. 2 men have called at his office, and they claim that it is a very bad place to work, and that the charges for goods in the store are too high. Now about this I do not know anything; but if Howell [the storekeeper] does charge more for his goods there than he does at No. 1 *he must immediately* reduce the prices. In fact the prices ought to be, if anything, a little lower than at No. 1, in order to keep the

men there; and I want you to see him *at once* about this, and I suggest that he reduce the price of all classes of goods and particularly that class used by the Hungarians, five per cent less than they are at No. 1. The store there is only a secondary consideration.

Ned similarly suggested raising pay rates "so that the men have not the excuse that their earnings are not as much as at other places. . . . There is no use whatever in trying to get men in there by scouring the country for them, and then have them leave after working a short time because they cannot earn as much there as in other places in the region where similar wages are paid." As an afterthought, Ned added another point that would have been second nature to Henry Clay Frick but that had apparently only lately occurred to Ned:

I think you should be careful to file all letters you receive, and put them away somewhere. I do not like to write you on matters that might be of importance to us, and have other people see the letters. I wrote you some time ago, and asked you whether you opened all letters yourself that came from me, and if you did not whenever I had anything that I wanted to write you about confidentially I would mark the letter 'Personal.' Please let me know about this.

The same letter reported that Ned had met Frick in Philadelphia and discussed a strike at one of Frick's plants. "Mr. Frick is very friendly to us," Ned added, "and he and I understand one another perfectly."

But Ned had badly underestimated his rival. Somewhere around this time Ned decided to deal with Frick as Carnegie had done six years earlier—that is, he would eliminate Frick as a competitor by selling Frick a small interest in the CC&I. "My object," Ned explained to Taggart, "is to get Mr. Frick personally interested with us with a view of our working together and helping one another in the trade wherever an opportunity offers and to so interest him. . . . Mr. F. has long desired to associate himself with me and if I can get him to do so would gladly have him in our Co."

No lamb ever entered so eagerly into negotiation with a lion. Ned's hopeful strategy suffered from a critical defect: Frick, frustrated by his subservience to Carnegie, had resolved never to place himself in a minority or subordinate position again. If he was to invest in a firm, he would do so on an all-or-nothing basis. To that end he was busy positioning himself and the price of coke so that his Connellsville competitors would be forced to sell to him on his terms. If he seemed receptive to Ned Leisenring's proposal, he did so only to acquire confidential information about CC&I that he could then use to maneuver against Ned.

When Frick asked to inspect the CC&I property to determine its value, Ned readily acquiesced. "I saw him yesterday a few minutes," Ned wrote Taggart after the inspection, "and . . . from a few remarks he made [he] is not unfavorable except as to the condition of the ovens . . . that our ovens were badly constructed and in bad condition and very bad brick used. . . . However, I hope all this will be remedied." When Frick casually surmised that CC&I could ship about seven hundred thousand tons of coke in 1889 if the plant ran full steam and had all the rail cars it needed, Ned eagerly replied, as he told Taggart, "Yes at least that much if not more."

Less than two weeks later the newspapers reported that Frick had bought the controlling interest in CC&I, and Ned moved frantically to quash the rumor. "There is not one iota of truth in that report," he wrote Taggart, "and I have no idea where it sprang from. . . . [A]s yet Mr. Frick has not one cent's worth of interest in our concern, and I do not propose to give him a controlling interest. Our idea was that I would be able to buy up, for him, some of the stock of outside small holders, and get him interested in that way."

Nothing in Ned's correspondence from those months suggests a man who is thinking of pulling out of the Connellsville district. He declined a chance to acquire coal lands in the South, telling Taggart, "You are aware we have a large body of coal lands in southwest Virginia, and my impression now is, if we could make a good profit out of our investment there by selling, we would do so rather than develop [it]." When Taggart contemplated leaving the CC&I to rejoin his family's mining operations in the anthracite region, Ned—who had previously refused to engage in bidding wars for talent—made an exception in Taggart's case: "I shall be very sorry to have you sever your connections with the Company," he wrote Taggart on March 26, 1889, "and hope that you will not commit yourself positively, until I can see you personally, and have a talk with you on the subject. . . . I will try to come out sometime next week." Nearly two months later Ned was still entreating Taggart to stay:

> Our relations have been so pleasant, and being reluctant to make changes in the management, I write to ask you if you have not positively committed yourself to your father, whether an increase in salary would be any inducement for you to remain with us. As I told you I should long ago have voluntarily given you an increase had the business been a paying one to the stockholders, but you know exactly and as well as I do what has been the results [sic]. Of course I do not desire to influence you in any way, and if you cannot remain with us, you will have my best wishes for your success and prosperity, wherever you go.

But the coming of 1889 had put Frick in an even more powerful position. On January 14 he was appointed head of all operations for Carnegie Brothers, in effect becoming a major power not only in coke but in steel as well. Exploiting his leverage in the marketplace, Frick now dedicated himself to streamlining the Carnegie properties into a new industrial phenomenon: a "vertical" operation that would coordinate every step in the production of steel, from raw materials to finished product.

By late March Frick had driven the price of coke back down to about $1.10 per ton and was threatening to push it lower. Wrote Ned to Taggart: "I think Mr. Frick is making the low price to worry the Col."—that is, Colonel Schoonmaker, another major Connellsville coke operator—"as there seems to be a very bitter feeling between them. I have not seen Mr. Frick for sometime, nor do I know what he proposes to do about his wages." The notion that Frick might also be maneuvering against Ned, or that Frick was not one to let feelings influence his business decisions, doesn't seem to have occurred to Ned.

The month of May 1889 found Ned abjectly apologizing to Frick: "I regret very much to hear that our Company have trespassed on your property, and removed any coal. I was not aware that it had been done, and your communication was the first I had heard of it. . . . We certainly are willing to reimburse your Company for the coal removed."

When Ned failed to persuade Taggart to stay on as superintendent, he sought and obtained Frick's consent to hire one of Frick's employees as Taggart's replacement. Why Frick would give such a consent so readily, again, does not seem to have occurred to Ned. On June 18 Ned sent Taggart instructions about the transition, asking Taggart to arrange for the new superintendent to "spend as much time with you as possible, so as to familiarize himself with our Works, and post himself on all matters that will require future attention, particularly as I shall be away for a couple of months; and I would also like him to enter our service permanently on the first of July if he can do so, and if you can arrange to remain until the 15th of July it would be very satisfactory to me."

This does not sound like a man who is negotiating to sell his entire Connellsville Coke & Iron Co. works to Henry Clay Frick. But that is exactly what happened a few weeks later. On July 10 Ned notified Taggart that "we had sold to Mr. H.C. Frick all of our Real Estate, Personal property, &c., in and about our mines in Fayette County, and that I had agreed to deliver the property to him on the first day of August." The new superintendent, formerly in Frick's employ, would remain in Frick's employ after all.

Ned provided Taggart with no explanation for the abrupt sale of his Connellsville works, which by that time had expanded to nine thousand acres of coal and 1,500 coke ovens. And indeed no written explanation from his hand or in the corporate records has ever been found. The story passed orally over the next several generations of Leisenrings held that Ned, having been squeezed into a corner by Frick, went to Frick and asked if he could work something out. Frick in turn asked how much money the Leisenring group had invested in its Connellsville properties; when Ned gave him a figure, Frick replied, "That's what we will pay you." The price, according to some published accounts, was more than $3 million.

This account is supported to some degree by a history of the region's 1880s coke boom that appeared in the *Connellsville Daily Courier* in 1914. "The price paid by the H.C. Frick Coke Company was said to have barely covered the investment of the owners with interest to the time of sale," the *Courier* wrote. "Mr. Frick got a bargain." Ned Leisenring and his partners, the article added, "were disgusted with a business that had been so harassed by strike and labor disturbances, and whose product had been compelled to meet such competition that there was up to that time no prospect of any dividends."

Shortly afterward, Frick took over the holdings of two other major Connellsville rivals: Colonel Schoonmaker, with five thousand acres of coal and 1,500 coke ovens, and J. W. Moore, with two thousand coal acres and 509 ovens. Although Frick in his entire career had built only two small coke plants himself, he now dominated a coke industry that had grown in gross revenues from less than $4 million in 1880 to nearly $8 million in 1889. His competitors had developed the region for him and then removed themselves as competition. Now Frick's only obstacle was the unions, and he would deal with them even less politely.

In April 1892, in an attempt to break the union at Carnegie Steel Company's Homestead Works south of Pittsburgh, Frick provoked a strike by imposing a wage cut. In the subsequent riots, state troops were summoned, and fourteen men were killed and 163 seriously wounded. Soon after, the Homestead Works reopened with a completely non-union work force.

Three months later, as Frick sat in his office on Fifth Avenue in New York, a labor activist name Alexander Berkman broke in and shot him twice, in the neck and the left ear, before being subdued and arrested. Frick, though bloodied, remained impervious, continuing to work at his desk in bandages until the end of the day. While doctors removed the bullet to his neck, Frick refused any anesthetic, explaining, "I can help you probe better without it." That night, from his home, Frick issued a statement. "I do not think I shall die," it read. "But whether I do or not, the Carnegie Company will pursue the same policy and it will win."

Frick did not die, but something else did. The assassination attempt, together with the death the previous year of Frick's beloved six-year-old daughter, Martha, pierced the façade behind which Frick had always concealed his emotions from others. In 1895 Frick suffered a nervous breakdown. In the process he seems to have lost forever his effectiveness as an impenetrable human business machine. He remained in charge of Carnegie's company long enough to steer it into its merger with the newly created United States Steel Corporation in 1901. Over the remaining eighteen years of his life, as his wife, Adelaide, sank into severe depression, Frick accumulated a remarkable collection of masterpieces by Vermeer,

Manet, Rembrandt, Goya, and other great painters. Toward the end Frick would routinely descend at night from his wood-paneled second-floor bedroom to the ground floor of his Fifth Avenue mansion, there to sit for a while in the darkness, cigar in hand, staring at one of the paintings—lit only by a small spotlight—before moving on to another painting and repeating the process.

After Adelaide died in 1935, their surviving daughter, Helen Clay Frick, opened their mansion to the public. In 1998, after ten years' of research into his collection as well as modern grief theory, a Frick great-grand-daughter contended in a book that Frick's collecting was driven neither by connoisseurship nor by a hunger for status symbols, but by grief over the loss of his favored first daughter. Through the paintings, she said, he "displaced his feelings for Martha." The paintings, then, became the outlet for the emotions that Frick had never seemed capable of sharing with other people.

But by the time of Frick's transformation in the mid-1890s, the Leisenring family would be long departed from western Pennsylvania. The Leisenrings had recouped their original Connellsville investment, plus interest, but had nothing else to show for their nine years there. (Twelve years after their departure, the three deep shafts and the coking operation they had developed around Connellsville would become the backbone of the newly created United States Steel Corporation.) Still, they had learned an important (albeit expensive) lesson. They had been drawn to Connellsville by the high quality of its bituminous coal and by their naïve faith in their ability to develop it. They had failed to consider that Connellsville's rich coal deposits would attract other, rougher competitors as well.

But now the Leisenrings had another chance. They had $3 million in payoff money from Frick. They had fresh wisdom acquired from hard experience. And they had those seventy thousand acres of coal lands in the hills of southwest Virginia—eight times the size of their acreage at Connellsville—that John Leisenring had acquired seven years earlier and that his relatives, in their preoccupation with Connellsville, had neglected ever since. Surely, here was an place so inaccessible to the rest of the world that they could develop it undisturbed.

PART **III**
Big Stone Gap

CHAPTER 9

Starting Over

Long before the southwestern corner of Virginia was discovered by coal men, this land of gently rolling hills, ridges, and winding streams was valued as a safe place for another vital human activity: hiding. As General Imboden learned in the late 1870s, its mountains and forests and narrow valley floors had cut this region off from the rest of the world. Early homesteaders found the lack of level ground so uninviting that land could not even be given away: After the Revolutionary War, many war veterans received government land grants in these mountains for their service in the Continental Army, but few of them actually settled there. Politicians had isolated the region politically as well: The state capital in Richmond was a greater distance from Wise County than the capitals of eight other states.

Many of the first white settlers in Wise, Lee, and Scott Counties liked it this way. Their eighteenth-century ancestors in many cases were impoverished English, Scotch, Irish, and Welsh natives who had been locked up in British debtors prisons for failure to pay their bills and then, as an alternative to jail, shipped to eastern Virginia to open up the new land as indentured servants. Their working conditions on Virginia's tobacco farms were barely better than those of slaves, and eventually many of them slipped off under cover of night and headed westward until they stumbled upon places where they were unlikely to be found. In these pursuits they were joined by another group of outcasts: the "Melungeons," a dark-skinned people of mixed race who were variously said to be descended from American Indians, Greeks, Portuguese, Negro slaves, or even Virginia's original lost colony at Roanoke. Eventually all of these fugitives settled into hollows and coves and scrabbled for a living by fishing, hunting game, picking fruit and berries, and raising corn and livestock in whatever bottomlands they could find.

Whatever their origins, long after the rest of the nation had developed an interconnected society, these Appalachian mountaineers continued to lead the kind of simple, self-sufficient agrarian life that inspired one observer, as late as 1899, to dub them "our pioneer ancestors"—that is, the last surviving remnant of America's original rugged individualists.

Among these early settlers was a family named Givens whose origins have been lost to posterity, either because they were unmemorable or because family members chose to conceal them. The name may have originated as "Gibbons," but the earliest known family member was a sharecropper and drifter named Bob Givens, who was born in Lee County sometime after the Civil War. A similarly impenetrable background attaches to an itinerant farmer named Warden who appears to have arrived in Scott County, just south of Wise County, sometime before the Civil War. Here he worked on a farm and raised a family, including a son named James, who was born about 1862. When timber men began harvesting the rich forests of Wise County after the Civil War, the Wardens gravitated there to work for M. C. McCorkle, who bought two thousand acres of land and set up a sawmill, commissary, and about a dozen dwellings for his employees just about the time the Leisenrings were opening their first Wise County coal mine.

None of these settlers—the Givenses, the Wardens, the Melungeons, or the descendants of indentured servants and debtors prison inmates—had gravitated to western Virginia to mine coal. But they constituted a rare class of people who were likely to perceive coal mining as an upward step on the economic ladder. The Givens and Warden families were typical of the hill people who would ultimately provide generations of miners for the Leisenrings in Virginia. Long before they thought of themselves as miners—long before the mines were opened—they called themselves "hillbillies," and for good reason: The hills and caves provided their refuge from those social enforcers who would track them down and return them to the legal obligations of an unforgiving social order.

In such places, outsiders inevitably arouse suspicion. So when two men on horseback—their saddles loaded with maps, books, papers, and rain slickers—began trotting daily through the countryside in the fall of 1890, the local farmers and squatters were quick to notice. The horsemen were not simply passing through from one point to another. On the contrary, they spent days in the woods, sometimes camping there overnight, and they stopped frequently to jot their findings, sketch pictures, and redraw their maps.

One of the horsemen was familiar to some of the locals: He was Sam Wax, a former deputy marshal recently hired as the Leisenrings' land agent. The other was a newcomer to Virginia but a familiar figure to the Leisenrings: John K. Taggart, their indispensable superintendent from Connellsville, who had returned to the family fold.

Though Frick's purchase of the Leisenrings' Connellsville properties had brought them $3 million with which to develop their Virginia coal lands, it had also brought them a credibility problem within their industry. For the past eight years the Virginia Coal & Iron Company had been nothing more than a legal entity that owned real estate. To transform this rough, barely accessible terrain of hills and bogs into a viable mining operation, Ned Leisenring needed an experienced superintendent. But what qualified manager was likely to uproot himself to work in such an unknown part of the world for a family that had just departed so abruptly from Connellsville?

Ned's first choice for the job was Erskine Ramsey, the highly regarded superintendent of the Tennessee Coal & Iron Company. Ramsey was making $3,000 a year at TC&I, and Ned tried to entice him by offering $4,000. But Ramsey explained that money wasn't his primary concern: From his perspective, the Leisenrings had pulled out of Connellsville when the going got rough, and the same thing might happen in Virginia, leaving him stranded there.

There was really only one man ideally suited to build the Virginia works from scratch—the man who had already proven himself in Connellsville, the man in whom Ned had confided during the darkest days of the labor strikes and the coal price wars—and Ned resolved to pursue him. Taggart was not merely an experienced coal man; he was a man of diverse interests and abilities. He knew engineering; he was proficient in title work; and his outgoing personality seemed ideally suited for dealing with the suspicious people living in the Virginia mountains. A man who made friends easily would prove an invaluable asset to Eastern investors invading this part of the world.

Taggart, from his new post in central Pennsylvania, clearly understood his bargaining position and his value to the Leisenrings. He was willing to return to them, he said, but he asked for an annual salary of five thousand dollars. That was what superintendents were making at large going anthracite concerns like Lehigh & Wilkes-Barre Coal Company in Mauch Chunk and the Reading Railroad's Reading Coal and Iron Company, both of which, Ned wrote, were mining "millions of tons" a year. Ned felt he couldn't justify a similar salary at a startup operation. Nevertheless, he sensed that the mutual trust that existed between the two men might be a useful bargaining tool.

"Since you have been frank at giving me your ideas of salary, I will do the same," Ned wrote Taggart. "To be candid I think $5000 a year too much to pay now, for the following reason, our company has carried the property for ten years and [is] only just beginning to develop it. And by paying you a large salary at the start, it would only cause a feeling among the stockholders that I was favoring you. . . . The usual salary for smaller corporations . . . is $2500. I think therefore you better make the figure $4000."

Ned explained that "after we begin making some money, so we can give the stockholders a dividend which should not be over a year after we start," he would be "perfectly willing" to consider a raise, "and if I can do you a good favor in any other way would be pleased to do so. . . . You now know us well enough to be able to judge whether you will be treated right or not. . . . I have written this frankly as if we were discussing the subject personally." In the margin, Ned scribbled: "If we can pay good regular dividends, the salary of superintendent will not be thought of[,] as stockholders are always satisfied with the management when they can get good regular dividends."

Taggart accepted the deal and arrived in Big Stone Gap in June of 1890 to look over the properties, conduct surveys, and choose locations for mines and ovens. Ned, penny-wise as always, wrote to advise his company secretary in Philadelphia: "I have written to Mr. J.K. Taggart to employ the necessary men and purchase materials required for opening the Virginia property. . . . I wish you would at once have a sight draft book printed, with say 300 to 400 drafts, and send to him by mail to Big Stone Gap. . . . Do not get an expensive book now to start with. . . . Send Taggart some Virginia Coal & Iron letter paper, except make address Big Stone Gap, Virginia."

Taggart's first order of business was to investigate all the company's deeds and titles and prepare maps for the company's abstractors and surveyors. These were no minor matters. The Leisenrings' revived interest in developing their Virginia land had focused new attention on Wise County. The prominent Louisville editor Charles E. Sears had arrived to launch Big Stone Gap's first newspaper, the *Post*, proclaiming in his first issue: "Seat of Empire! Such Big Stone Gap will surely be unless an earthquake swallows it." Such euphoric rhetoric drove up land values, with one inevitable consequence: Now local people found it worthwhile to challenge the Leisenrings' titles to the supposedly worthless land they had acquired in 1882.

Under the complicated system of Appalachian land tenure, squatters who had settled along the creek bottoms could claim title to their plots after ten years' continuous possession. Hyndman, the agent who had negotiated for the land ten years earlier, had compromised with many squatters by buying up only the mineral rights to the land. Under these deeds, the Virginia Coal & Iron Company owned all the coal beneath the ground but no level surface site on which to build collieries or coke plants. In theory this meant that the company had the right to dig beneath the surface as long as the landholder was fairly compensated for the subsequent cost of reclaiming the land. But in practice many squatters didn't comprehend the terms, and the company made little effort to enlighten them.

By 1890 several of the squatters, angry about intrusions on what they believed was their land and hoping to extort a greater payoff from the Leisenrings, had joined forces to lay claim to twenty thousand acres, more than one-quarter of the company's holdings. At least some of these claims were bolstered by illegally backdated deeds. Taggart's delicate task was to buy up the surface rights to the land wherever he could—preferably before the matter reached court—while simultaneously quelling animosity among local residents who might one day be working for him.

(To be sure, public relations concerns of this sort became secondary when matters did reach court. On one occasion a local squatter sued the company, and the company—reluctant to argue the case before an unsympathetic Wise County jury—responded by incorporating a "dummy" company in Pennsylvania, which filed its countersuit in a federal court, since the case now involved an out-of-state firm.)

Taggart spent much of this time surrounded by maps and deeds in the company's office in Big Stone Gap. But often he and Wax would disappear on horseback into the woods for days at a time. Here he would analyze firsthand the information he found on all those land patents and deeds, drawing new maps of his own or correcting old ones. He liked to sketch drawings on these maps, either to illustrate a problem he encountered or to emphasize some puckish point he hoped to make. One Taggart map showed a fisherman sitting on a bank and pulling in a fish; on the fish, Taggart wrote, "The sucker who buys lands without investigation of titles."

Beyond the land titles lay the stickier question of the value of the coal beneath it. Although a rudimentary test coke oven had been constructed in Wise County even before Taggart arrived, over the summer of 1890 Taggart constructed on Mud Lick Creek the district's first beehive coke oven, made of stone and lined with firebrick. On August 25 the first batch of local bituminous coal was cooked into coke.

Taggart was astonished by the result: The coke burned in his first oven, he reported, was the best he had ever seen—remarkably strong, with a bright metallic luster and unusual cell space, and clearly superior to the Connellsville coke, which had previously set the standard for the industry.

Within a month an analysis performed by more objective experts in Pittsburgh confirmed Taggart's opinion: The Virginia C&I coke was the most nearly perfect coke yet made. Newspapers quickly circulated this analysis to amazed coal and iron men across the nation. To the coke opera-tors back in Connellsville, the implications were startling: The manipulative Henry Clay Frick, it now seemed, had been too clever by half: He had out-smarted himself by acquiring a monopoly on a second-rate coal in Con-nellsville while driving the Leisenrings into first-rate coal lands in Virginia.

The news was so astounding that Ned Leisenring himself and five of his excited directors immediately made the long trip to Big Stone Gap to in-spect the mine openings for themselves and give Taggart their approval to proceed with mine shafts and ovens. Speed was necessary because the Leisenrings weren't the only coal people suddenly flocking to Wise County. "As soon as the news reached the coal districts of Pennsylvania," the *Big Stone Gap Post* reported on the night before the Leisenring group's arrival, "a number of capitalists interested in that section determined to send a prospecting party to examine and, if possible, purchase other coal lands near the Gap."

> The two Broadheads and Mr. Tompkins were here Monday night and went to examine the coal deposits down the valley. They would not talk freely of their purpose, but the fact was developed that it was the coking coal they wanted. . . . Other parties are expected to arrive within a few days. Newspapers and manu-facturing journals are discussing the matter, though it seems difficult for them to realize that here, midway between the teeming and industrial populations of the Atlantic coast and the Ohio and Mississippi Valleys and so near the con-suming centers, are vast and inexhaustible deposits of the finest coking coal in the world. . . . There is every indication that a boom will be started this fall, the proportions of which cannot now be estimated.

By this time three different railroad companies had extended their tracks to the edges of Wise County and were rapidly converging upon Big Stone Gap from three different directions—the Louisville & Nashville from the west, the Norfolk & Western from the north, and the South At-lantic and Ohio from the south. The Virginia Coal & Iron Company expe-dited their arrival by deeding many miles of right-of-way through its property, at no cost to the railroads. The South Atlantic won the race, send-ing its first train to Big Stone Gap in February 1891, but by April the Louisville & Nashville and the Norfolk & Western were both operating in Wise County as well.

As yet there was no rail connection to the mine's opening on Mud Lick Creek. The closest rail line, the South Atlantic and Ohio, was four miles away. This meant that the mine's only connection to the outside world was a bumpy wagon road up Callahan Creek. Yet over this miserable road Tag-gart had a sawmill moved, and here he began sawing timber with which to build a tipple and sixty dwellings. He named the community and the mine "Pioneer" to reflect its status as the site where the VC&I would mine its first coal and produce its first coke. The railroads would come soon enough, Taggart said, and when they did, he would be ready.

By September 12 Taggart's workers had driven two openings two hundred feet deep into this Imboden seam, as it was called in honor of its original promoter. Ground was being graded for coke ovens, and brick was being quarried. Above Possum Trot Creek, the company's agent Sam Wax reported, "Some five or six houses are under construction. Double entries are being driven into the hill for mining coal, and work on a tipple"—that is, a huge wooden shed for loading and unloading coal cars—"is being done."

Better news was still to come. The Louisville & Nashville Railroad agreed to build a branch line up Callahan Creek from its main line to the mouth of the mine at Pioneer; the VC&I, as its part of the bargain, agreed to condemn some of its own land in order to deed the right-of-way to the railroad. Then, while prospecting on Preacher Creek one day in the early fall, Taggart found the outcropping—that is, the surface remnants—of a previously undiscovered seam of coal. This seam, later named for Taggart himself, lay four hundred feet north of the Imboden seam and contained a similar high grade of coking coal.

What had once happened to Mauch Chunk and Connellsville, Pennsylvania, was now happening to Wise County, Virginia: Thanks to coal, a previously forlorn and inaccessible spot had become, almost overnight, a commercial and tourist destination. By early October the curious visitors to Big Stone Gap included the Duke and Duchess of Marlborough, accompanied by the writer John Fox Jr., later author of the coal-country novel *The Trail of the Lonesome Pine.*

Back in the company's headquarters in Mauch Chunk, the excitement was tempered by the Leisenrings' memories of Connellsville. Now, borrowing a page from Henry Clay Frick, the Leisenrings moved to co-opt their potential competitors by buying up as much land as they could before the railroads inevitably arrived. In Mauch Chunk, Ned Leisenring wrote to the company's secretary in Philadelphia:

> Please let me know how much money you have in the Treasury. We expect to buy another piece of property shortly; and will also now need considerable money to open the mines, build ovens, houses &c., and will have to make preparation for it. . . . Let me know whether you have the agreement, or subscription paper, signed by all the stockholders by which they agreed to take the $100,000 of additional stock at par for improvements. This will be our first resource. Have you the Virginia Coal & Iron Company's books in shape now, so that we can go on and keep our accounts regularly?

Within a generation, mines and coal towns blanketed almost the entire southwestern corner of the state. But the Virginia Coal & Iron Company remained the area's first and by far its largest operator, its mines and towns often distinguishable by their eastern-oriented names, such as Andover and Exeter, for the prep schools favored by the Leisenrings, Wentzes, and Kemmerers.

The VC&I's towns stood out from other company towns as well, even if they gobbled up much of what little level farmland existed in the region. "No hunter's cabins if you please," declared the local Republican Party chairman, W. S. Rose, "but large and commodious residences of brick and tile and hundreds of them, all equipped with the latest comforts and conveniences."

The relatively few local men were insufficient to work all these mines, and before long the local operators were importing miners from throughout the South and Midwest. But to keep wages—and therefore prices—competitive with Connellsville coke, the Leisenrings needed to do in Wise County what Frick had done in Connellsville: pre-empt the competition.

In the fall of 1893 Taggart, sounding very much like Frick, reported to Ned Leisenring that the Virginia C&I now owned 89,514 acres and was negotiating to buy another 1,464 acres from a neighbor named J. J. Kelly Sr. Kelly's tract, Taggart explained, was mostly virgin forest devoid of coal, but it was a necessary right-of-way for transporting timber to and from the Virginia C&I's lands. More important, if Virginia C&I owned Kelly's lands, Taggart wrote,

> it would do away with all competition on Coal and Coke on the S.A. & O. Railroad, and Bristol and East Tenn., Virginia and Georgia points and we would have but little competition on L&N Railroad from Norton County. . . . In addition to controlling water power and doing away with all competition, and thereby being able to maintain a fair price for our product, there is another point just as important, and that is the establishment of wages on a proper basis, so that we are at least on the same plane with Connellsville . . . and that can readily be done when there are no competitors to make concessions. If we purchase the J. J. Kelly, Sr. lands we are masters of the situation.

By this time, though, Ned Leisenring had other matters on his mind. Like many captains of industry of his generation, Ned enjoyed eating and drinking and took little exercise. As a consequence, by his mid-forties he was seriously overweight. But in an age that equated corpulence with prosperity, his girth merely enhanced his appeal as a prospective husband. In March of 1892, after sixteen years as a widower, he had married Annie Wickham, daughter of a prominent New York family. After a six-month wedding tour of Europe and the Holy Land, the newlyweds returned to Mauch Chunk. But in May of 1893, when Ned added the presidency of the Lehigh Coal & Navigation Co. to his duties, they moved their home (as well as the Leisenring group's headquarters) to Philadelphia.

This move represented more than a concession to his new wife's sensibilities; it represented a conscious recognition that the coal industry had changed in two important respects: Anthracite was becoming less important than bituminous coking coal, and the mining of coal was becoming less challenging than selling it in an intensely competitive marketplace. From this point on, coal executives needed to be closer to their customers than to their workers.

At first Ned and Annie lived in a townhouse near Philadelphia's fashionable Rittenhouse Square, where Annie gave birth to their first child, a daughter named Mary. But in the early summer of 1894, with Annie expecting their second child, Ned purchased a mansion in the relatively suburbanized section of West Philadelphia. He was having it renovated into a permanent home when he was overcome by exhaustion brought on by a failing heart.

He resolved to recover his strength by spending a short season at the famous healing springs near Hamburg, Germany. Leaving his pregnant wife at home, Ned sailed on August 28 from New York, accompanied by his twenty-two-year-old nephew, Daniel Bertsch Wentz, and a friend. But on arriving in Hamburg he was so exhausted that he could not proceed further. He was put to bed in a Hamburg hotel, and his family was notified by cable that he could not be moved. On September 20 a small party of Leisenrings—Ned's wife, her parents, and Ned's brother, John—set out from New York for his bedside. But they were too late: By the time they arrived, Ned had already died, at the age of forty-nine.

With Ned's death, most of his posts were inherited by his brother-in-law, the physician-turned-capitalist Dr. John S. Wentz. At the same time, Ned's sister—Wentz's wife, Mary Leisenring Wentz—inherited the Mauch Chunk mansion of her father, Judge John Leisenring. Like Ned, Wentz would remain in the East. But to him would fall the burden of developing the Virginia Coal & Iron Company.

Immediately Wentz found himself grappling with an unexpected problem. The Panic of 1893 and the subsequent depression had left the nation's railroads strapped for cash. Although the VC&I had already deeded its right-of-way to the Louisville & Nashville Railroad for a branch line to the Pioneer mine, the railroad had decided not to build the branch.

From his office in Big Stone Gap, Taggart refused to be deterred. The necessary spur line to Pioneer involved only about four miles of track, he reasoned. The VC&I was already building its own mines, plants, ovens, houses, and roads. Why, in that case, couldn't the VC&I build its own railroad track?

His proposal brought John Wentz to Big Stone Gap in May of 1895 to investigate and pick the brains of executives and engineers of the L&N's two rival railroads. Within two months, the L&N had agreed to transfer its right-of-way back to the company, and work on the branch line began.

"All Aboard for Pioneer," announced the headline in the *Big Stone Gap Post* of July 18, 1895. "The development long looked forward to by the people," its editorial crowed, "is now a certainty."

Now all of Taggart's plans seemed to be coming together. In the fall of 1895 a firm in Birmingham, Alabama, was contracted to build five hundred coke ovens at Pioneer. By February 1896, thirty-five coke ovens had been completed. In March, the first coal was mined there. By April, one hundred coke ovens had been completed. For reasons since lost to posterity, at some point during these busy days the name Pioneer was changed to Stonega—a contraction of the words "Stone Gap." The growing town at this point consisted of some one hundred miners' houses, offices, a company store, and a stable. But in Taggart's mind this was only the beginning.

On May 23, 1896, a Saturday, Taggart was busy at Stonega as usual, supervising the construction of more coke ovens in the company of J. A. L. Minor, the coke oven contractor, and the company's land agent, Sam Wax. At a stone quarry about fifty yards away from them, some of Taggart's men had set a powder charge in order to break loose stone for use in building the ovens. The explosion was intended to blow the loosened rocks straight upward out of the ground. But instead it broke loose the side of the quarry's embankment and, like grape shot from an immense cannon, poured a deadly shower of stones straight in Taggart's direction.

Two miners were standing just twenty-five feet from the blast when it occurred. One of them, who was standing erect, was killed instantly. The other was spared only because he happened to be stooping over. Wax and Minor, as they saw the stones flying toward them, ran sideways from the blast. But Taggart turned to run directly away from it, stooping slightly as he ran. When a stone glanced along the back of his head, he too was killed instantly. Such was the force, observers remarked, that had the stone struck him squarely it would have knocked his head off altogether.

The news spread quickly through the valley. Crowds gathered on the street corners of Big Stone Gap. Men with horses galloped to the disaster scene. Rufus Ayers, the lawyer and real estate speculator who had helped found both the town of Big Stone Gap and the Virginia Coal & Iron Company, commandeered a dummy railroad engine and coach to the point where the VC&I's road intersected with the Louisville & Nashville. The company's physician, Dr. C. D. Kunkel, had left the Stonega works just before the explosion; now messengers from both Big Stone Gap and Stonega were dispatched to intercept him and bring him to the scene. But "nothing the physician could do," reported the *Big Stone Gap Post*, "could change the glassy stare of those eyes now closed in the sleep of death." At Stonega the two bodies were placed on a handcar, enshrouded in a canopy of white. "It was a scene," the *Post* reported, "to make the stoutest heart quail and brought tears to many eyes that had never wept before."

The bodies of the dead men were brought to Big Stone Gap on the L&N train; Taggart's remains were carried to his home by a large crowd of friends. "The father and husband who had left his family that morning had now returned," the *Post* recounted. "What a blow to his devoted wife, his daughter Helen and two sons, Jack and Ralph."

Like Ned Leisenring, John Taggart at forty-five was in the prime of his life when he died. After the Connellsville debacle, the Leisenring group's decision to move forward with the Virginia Coal & Iron Company had rested largely on the belief that both men would remain at the helm for years to come. Now both were suddenly gone.

Yet the product that had drawn them here transcended any mortal's involvement. The following month John A. Esser, a veteran of twenty-three years with the Leisenring operations in Pennsylvania, arrived in Big Stone Gap to assume the superintendent's duties. Daniel B. Wentz, the Harvard-educated twenty-four-year-old son of the company's president—the same Daniel Wentz who had accompanied Ned Leisenring on his final voyage to Germany—also arrived in Big Stone Gap from Pennsylvania that month to take up permanent residence as the company's land agent. Esser, reported the ever-optimistic *Big Stone Gap Post,* "is even a more sanguine believer in the outcome of this coal field than was Mr. Taggart himself." In its first full year of operation, the Stonega plant mined 214,000 tons of coal and shipped 149,000 tons of coke (as well as 31,000 tons of pure bituminous coal) to steel and iron plants as far away as Wisconsin, Illinois, and Alabama.

"The coke thus far turned out has no equal in the South and is just as good as the Connellsville coke in every respect," the *Post* remarked on November 19, 1896. Once again the paper repeated its contention that the Imboden seam would produce "the superior of any coke made in America or elsewhere."

For the Wentzes, the timing could not have been more propitious. The vast industrial expansion taking place across America since the 1870s depended entirely on the continued growth of two resources: energy supplies and the labor force. Coal was solving one of those challenges. America's per capita consumption of coal had doubled from one ton in 1870 to two tons in 1896. Thanks to the availability of coking coal, in 1892 American mills for the first time produced more steel than iron. By 1900 the United States would become the world's largest steel-producing nation, with production more than doubling that of the runner-up, Great Britain. As the iron, steel, and railroad industries expanded, so did America's appetite for coal. And by the mid-1890s another miraculous new phenomenon was demanding coal as well.

The challenge of capturing electricity for human use had fascinated scientists and inventors even before Benjamin Franklin famously identified it with lightning by flying his kite in a storm in 1752. Samuel F. B. Morse first harnessed electricity for communication in 1844 with his invention of the telegraph, which transmitted impulses through wires to points miles distant, and Alexander Graham Bell performed a similar feat for the human voice with his invention of the telephone in 1876. But the critical breakthrough in the development of electricity occurred in 1880, when the inventor Thomas Edison patented the first effective, long-lasting incandescent electric lamp. Thanks to the power and durability of Edison's electric lights, the waking hours of cities and towns were no longer restricted by the rising and setting of the sun. Now electric lights lit up streets, homes, and businesses long after the sky had gone dark. Factories could extend their hours and expand their capacities.

At first, the users of electric lights installed individual power units in their basements. But Edison realized that a central power station could increase the usage of electricity exponentially. In 1882 his recently formed Edison Electric Light Co. (later Consolidated Edison Co.) opened the nation's first power station—a huge complex of boilers, steam engines, and dynamos located on Pearl Street in lower Manhattan. These dynamos functioned by heating water into steam inside a boiler. Under high temperatures, the steam would build up pressure and burst through a turbine (the equivalent of, say, an airplane propeller). The turbine blades would spin, inducing an electrical current in a generator attached to the turbine. Electricity then flowed from the generator to the power lines that delivered power to a home or business.

Any fuel—coal, oil, natural gas—could be used to heat the water. But bituminous coal soon proved the most efficient and least expensive. And although most bituminous coal was unsuitable for coking, the vast majority of bituminous coal was highly suitable for powering steam plants. The Virginia Coal & Iron Co.'s high-quality coking coal came from two relatively thin seams. But its largest, most plentiful, most accessible seam consisted of highly volatile coal that could be used as steam coal. Suddenly the company, and other operators like it, was no longer quite so much at the mercy of the industry's roller-coaster market cycles: When the market for coking coal was saturated, the company could fall back on sales of steam coal, and vice versa.

Edison's first electric customers were New York department stores. Four months after his Pearl Street plant opened in 1882, Edison was supplying electric current for five thousand lamps. By 1890 his company was sending current to twenty thousand, and power stations were popping up in cities and towns across the country, reinforced by the superior "alternating current" adopted by Edison's rival George Westinghouse. Before long electricity became the indispensable element in modern American life, and coal became indispensable to electric power, just as it had previously been indispensable to iron, railroads, and steel. By 1910 coal was providing three-quarters of America's energy. More than 710,000 miners in some thirty states provided sustenance not only to steel mills and railroads but also to auto factories, offices, schools, and hospitals.

The Leisenrings had never anticipated electricity, of course; it simply fell into their laps. Less than a generation after Frick had chased them out of Connellsville, the Leisenring relatives found themselves in a unique position. Electricity and steel had become the new foundation supports of a whole new age, an age of ready-made clothes, cameras, and typewriters. Trolley cars, elevators, and bicycles were transforming American cities. Steam power was doing the heavy work in factories, so workers labored shorter hours. Homemakers enjoyed the benefits of sewing machines, egg beaters, carpet sweepers with revolving brushes and dozens of other labor-saving devices.

Nearly all these wonders depended on steel and electricity, and steel and electricity both depended on bituminous coal. The Leisenring group had beaten the rest of the industry to the world's richest bituminous coking coal for steel, and now it owned abundant reserves of steam coal for electricity as well. In the generation after John Taggart's death, the Virginia Coal & Iron Co. and later its operating affiliate, the Stonega Coke & Coal Co., would build nine more company towns in and around Wise County. By the time John S. Wentz died in 1918, at the age of eighty, his wealth would be valued at nearly $5 million, or five times the fortune that his father-in-law, Judge John Leisenring, had left in 1884.

Yet the coal industry remained as fragmented and disorderly as ever. In 1910 more than five thousand bituminous coal operators competed with each other in a bewildering array of mining conditions and markets. Virtually every other service or industry, from railroads to steel to automobiles to timber to cattle to fruit, was susceptible to corporate consolidation because the product could be shaped to consistent and predictable standards across the country. But coal mining was different: "It simply cannot be standardized," a coal leader observed many years later, "because nature has refused to standardize rocks, slate, coal or men."

For all the complex geographic reach of operators like the Leisenring group, coal remained the closest possible thing to an agricultural industry, and its operators retained the ornery independence of farmers. Working in caves had somehow endowed them with the sullen competitiveness of cavemen. As a result, even as coal came to dominate American life, the coal industry itself was incapable of speaking with a single voice to its customers, its workers, its investors, or to governments. The big difference between the new age and the past was that the Leisenrings and their fractious competitors could no longer cry poverty when their growing armies of workers demanded a quality of life equal to the quality of their coal.

CHAPTER 10

The Rise of John L. Lewis

The man who ultimately stabilized the coal industry—to the extent that anyone stabilized it—was neither an operator nor a government official but a working man, and only peripherally a miner. Just as Henry Clay Frick and General John Imboden were not so much coal men as opportunists, John Llewellyn Lewis was by nature a striver who just happened to see his opportunity in the organization of coal unions. Like Moses, like Joan of Arc, like Napoleon, Lewis was essentially an outsider who assumed the leadership of his suffering people without actually having suffered all that much alongside them. His strength lay not in his mining experience but in the passion, perspective, and eloquence he developed outside the mines. If he was not strictly speaking a miner, he was something better: a charismatic figure marinated from birth in the miners' world but barely infected by their pessimism and fatalism.

Lewis was born in 1880 in Cleveland, Iowa, one of several harsh southwest Iowa coal-mining towns where he and his five younger siblings grew up. His father, Thomas Lewis, worked irregularly as a miner, a farmhand, and a policeman—forced to keep on the move, John later claimed (although without any factual support), by his involvement in coal strikes. Like most coal wives, John's sweet-tempered mother, Ann Louisa, lost the bloom of her young womanhood quickly in a monotonous life of cooking, sewing, mothering, and cleaning—a losing battle in the sooty shadow of the mines. Her domain was a succession of cheap company houses, hammered together from raw lumber and covered with tarpaper, where two bedrooms, a kitchen, and an outdoor privy were expected to suffice for a family of eight.

Yet either through luck or determination, the Lewis family seems to have fared better than most miners. Despite the Lewises' poverty, insecurity, and nomadic lifestyle, their family remained intact. Despite the constant threat of disease, dismemberment, or death in the mines, Thomas Lewis remained physically healthy. Young John himself grew into a tall, robust, handsome youth with an outgoing personality and a flair for drama. It was John's special good fortune that, as he entered adolescence, his father joined the police force in Des Moines. As a result, this young man of multiple interests and talents spent four formative years in a comparatively worldly, bustling state capital city. Before the Lewis family returned to southwest Iowa, John had completed three and a half years of high school, a rare achievement for a working-class youth in the 1890s.

Once back in his native Lucas County, John went to work in the mines along with his father and his younger brothers. But he never resigned himself to remaining there. Because coal production rose and fell with the cycles of a boom-and-bust economy, work in the mines was sporadic—from 1890 to 1920 miners were idle an average of ninety-three working days per year, or nearly two days a week—and in his idle time Lewis performed in talent shows at the local opera house; he also doubled as its manager, booking traveling companies and exhibits. As he reached his majority at the turn of the twentieth century Lewis resembled not so much a typical miner as a typical small-town self–promoter out of the pages of Booth Tarkington or Sinclair Lewis. But how Lewis perceived himself—then as well as later in his long career—remains a matter of conjecture: Lewis was the least introspective of men. He kept his thoughts and feelings to himself. Even after his retirement he resisted all invitations to discuss his sixty-year legacy in memoirs or an autobiography. "My father can't have any friends," his daughter Kathryn remarked in 1949. "No great man can have friends."

It was during his Lucas years that Lewis first came in contact with an important new force in miners' lives. The United Mine Workers Union of America was the latest in a long string of attempts, stretching back to the 1860s, to patch together rival labor factions into a national miners' union. The dreams of earlier unions such as the American Miners' Association and the Knights of Labor—to speak to industry with the strength of a single unified national voice—had foundered for precisely the same reason that coal operators themselves were unable to get together: Since coal operators competed locally rather than nationally for customers and workers, and since wages accounted for about seven-tenths of the cost of coal, in effect miners in one mine competed against their counterparts in other mines. At the UMW's first national convention in Columbus, Ohio, in 1890, rival labor leaders, many of them veterans of earlier movements, once again swore allegiance to a new national union and endorsed a constitution that called for better pay, safer mines, regular mine inspections by state or federal agencies, an eight-hour day, laws against child labor, and one especially lofty plank: the hope that labor peace might be achieved through arbitration and conciliation, "that strikes may become unnecessary." "There is no truth more evident," the UMW's preamble announced, "than that without coal there could not have been such marvelous social and industrial progress as makes present-day civilization."

Although it began its life with just 20,000 members, the new union, unlike its predecessors, survived its infancy. When its president, John Mitchell, called for a nationwide shutdown in 1897, some 150,000 miners—half of the labor force in the bituminous coal industry—responded with a strike that paralyzed coal production in the Midwest (although it was less successful in Appalachia, where the Leisenrings operated). After three months the UMW won recognition as the miners' collective bargaining agent throughout much of the Midwest, and miners were granted an eight-hour day in union mines. These victories attracted many more new members, whose dues enabled the UMW to think about the tougher challenge of organizing the anthracite region of Pennsylvania, whose coal fields (unlike those of the bituminous industry) were controlled by a handful of seemingly invulnerable railroads and large mining combines. Two long and impressive UMW strikes there in 1900 and 1902 captured the nation's imagination and ended only when President Theodore Roosevelt intervened and brokered a settlement that called for a wage increase for workers. By 1903 the UMW had 173,000 members and a treasury of $1 million.

Although there appears to have been no history of labor activism in Lewis's family (notwithstanding his claims about his father), in 1901 Lewis became a charter member and secretary of the UMW's newly created local in Lucas County. But he didn't remain there long enough to make an impression. Later that same year Lewis headed west, where for the next four years he worked in several mines and on construction crews. These wanderings provided the basis for the unverifiable legends that subsequently spread (probably with Lewis's encouragement) along the miners' nationwide grapevine. In one tale, Lewis helped dig out the remains of 230 miners killed in a disaster in Hanna, Wyoming, in 1903. In another, Lewis killed a mad mule in a mine with his bare fist, and since a mule's life was then worth more than a miner's—it was the mule who pulled the coal cars underground—Lewis saved his job by covering the dead mule's wound with mud and telling the superintendent that it died of heart failure. Whether that story is true or not, it conveniently reflects the key characteristics that drove Lewis's subsequent career: dramatic action, resourcefulness, and, ultimately, ruthlessness and guile.

Returning home to Lucas, Iowa, in 1905, he worked briefly in the mines but continued struggling to avoid the trap of a miner's life. He took up amateur theatricals again, joined the Masons—a bastion of the middle class—entered into a brief feed and grain partnership, and in 1907 even ran unsuccessfully for mayor. That same year he married Myrta Edith Bell, a demure and soft-spoken schoolteacher whose education (a high school diploma, summer courses at Drake University) and family pedigree (a physician father, ancestral roots in colonial America) elevated his own uncertain standing. (Myrta's attraction to a coal miner's son may be explained by the fact that, at twenty-seven, she was on the brink of spinsterhood and had few other alternatives to choose from in a small town.) In later years Lewis promoted the legend of Myrta as the woman who introduced him to Shakespeare and the classics, much like the heroine of the Welsh coal-mining saga *The Corn Is Green;* but Lewis's biographers suggest that her true achievement lay in providing Lewis with a stable and nurturing home.

The following spring Lewis and his new wife left Iowa and moved two hundred miles eastward, settling in the new mining town of Panama, Illinois. Their reasons for moving are unknown, but it was an opportune decision for a man harboring political ambitions. The Sandy Shoal mine, where Lewis found work, was one of the largest and most mechanized coal operations in Illinois; and Illinois, with more than ninety thousand coal miners, was a far more important coal-producing state than Iowa. Panama was located only thirty-five miles south of Springfield, the state capital; and in a town like Panama itself, whose coal-mining inhabitants were mostly Italian, an English-speaking native-born American like Lewis could achieve the sort of recognition and power that had eluded him in Iowa and out West.

Later that summer, Lewis's father and brothers joined him in Panama and went to work at the Sandy Shoal as well. In short order this loyal and cohesive family unit came to dominate the United Mine Workers' Local 1475 there. By 1909, John was elected president of the local, while his father Thomas became a city police magistrate. By cultivating John Walker, the president of the UMW's Illinois district, John Lewis won an appointment in late 1909 as a UMW lobbyist at the state capital in Springfield—a comfortable job that exposed him to bigger possibilities. The following year Lewis ungratefully supported Walker's rival, John White, for the national UMW presidency—a calculated gamble that, had it failed, could have sent Lewis back to the mines. Instead, White's victory won Lewis a visible role at the union's national convention in 1911.

Here Lewis raised his visibility another notch by stoutly defending the most prominent name in the American labor movement: Samuel Gom-

pers, president of the American Federation of Labor, who had been accused of discriminating against black coal miners. Gompers was then
sixty-one and a heroic figure in labor circles: He had organized America's
first permanent federation of trade unions—the AF of L—in 1886 and had
served as its president ever since. By his eloquent public defense of Gompers, Lewis attached himself to a legend. When a position on the AF of L
organizing staff fell vacant that fall, Lewis was the consensus choice to fill
it, and he eagerly accepted. In barely two years he had risen from a small-
town UMW local to the national stage. His pay as an organizer was good—
$5 a day plus a generous expense account—and the work (unlike coal
mining) was steady. And of course the job kept him out of the mines. Now
he would have to prove his worth by persuading the men underground
that his union could do more for them than their employers could.

For many of the hardscrabble farmers and squatters of southwestern
Virginia, the mines represented greater opportunity and security than
they had ever known. James Warden, whose family had moved from a farm
in Scott County to cut trees for McCorkle Lumber Company in adjoining Wise County, went to work for the Leisenrings' operating company,
Stonega Coke & Coal, after the lumber companies moved on. Bob Givens,
an itinerant sharecropper and drifter in nearby Lee County after the Civil
War, saw all five of his sons go into mining; the choice, his grandchildren
later explained, was between mining and drifting.

Bob's third son, Charles Saylor Givens, followed his older brothers Albert and Lewis to work in Stonega's Arno mine in 1909, when he was just
fourteen. "Saylor," as he was known, started by driving a mule-drawn coal
cart, then became a full-fledged miner at Stonega's Imboden mine. Eventually he married James Warden's daughter Cora. After the wedding, the
company rented Saylor Givens a small two-room house, offering the hope
of something larger as he acquired seniority.

Ultimately Saylor's six sons, his four brothers, his three Warden brothers-in-law, his nephews, his cousins, and even his father, Bob Givens,
went to work for Stonega. One day about 1910 Saylor's eighteen-year-old
brother-in-law, Sam Warden, was in the mine, reaching under a loaded
coal cart to pull a switch on the track when the mules panicked and pulled
the cart over his arm, severing it three inches below the elbow. But even
this trauma failed to drive Sam from the mines. He continued to work
there for Stonega—his fellow miners said he used his stump as well as anyone with two normal arms—and the company compensated him by providing a house rent-free for the rest of his life. As the Givens men saw it, it
didn't take much to mine coal—"just a strong back and a weak mind," as
an old saying went. If you watched what you were doing, you wouldn't get

hurt. And mining was better money than farming, and more reliable than cutting trees.

While John L. Lewis worked his way through the ranks of organized labor and Saylor Givens learned the routines of Stonega's Imboden mine, two heirs to the Leisenring group's legacy were taking their first tentative steps into the ranks of management. Ralph Taggart had been nine years old on the day in 1896 when his father, superintendent John Taggart, died in an explosion outside the Stonega mine. After his father's death Ralph attended public schools in Big Stone Gap, but eventually he was sent to Phillips Academy in Andover, Massachusetts, the prestigious prep school where Stonega's president, Dr. John Wentz, had also sent his son Daniel. Exactly who paid Ralph's way to Andover is unclear, but it seems reasonable to surmise that, by virtue of his father's critical role in the Virginia operation as well as his father's supreme sacrifice for the Leisenring group, from the moment of John Taggart's death the Leisenrings and Wentzes treated this serious, long-faced young man as a member of their family. From the time he was sixteen, Ralph spent his summers working for the Stonega company, first on survey crews and later as an assistant in and around the mines. He joined the company permanently in 1906, when he was nineteen, and seems to have been placed on a management track from the day of his arrival. His marriage in 1910 to Virginia Bullitt, daughter of one of the founders of Big Stone Gap, reinforced his standing as a figure of rising prominence within the company. After serving apprenticeships in Stonega's inspection department and its engineering corps, Ralph rose through the ranks to coke yard boss, assistant superintendent, and colliery superintendent. By 1913 he was general superintendent of all ten of Stonega's collieries; by 1917, at the age of thirty, he was Stonega's general manager. That was already higher than his father had climbed within management.

The next Leisenring generation's route back to the family enterprise was even more circuitous. When Ned Leisenring had died in Hamburg in 1894, his pregnant widow had remained in Europe, giving birth the following January to a son named for his father and nicknamed "Ted." After a few years of Philadelphia widowhood, Annie Leisenring married Lyman H. Treadway, president of a bank in Cleveland, and moved there with Ted and his older sister, Mary. In that city on the shores of Lake Erie—far from the mines, from Philadelphia, from anything to do with the Leisenring group—Ted Leisenring received a typical upper-class upbringing of his day. His mother and stepfather sent him to the University School, Cleveland's leading private school for boys, through which he was introduced to Margaret Pierce, daughter of a local newspaper editor, who was attending the Hathaway Brown School, the leading Cleveland girls' school. As a youth he was fun-loving, modest, and popular, as well as a crack athlete. Eventually he was sent to Hotchkiss, the Connecticut boarding school, where he developed into an outstanding athlete: captain of the track and ice hockey teams and school golf champion. As a young man he was slender and tall—about six-foot-one—with brown hair and a handsome face that a decade later would remind people of the aviation hero Charles Lindbergh. He was quiet and modest, and people seemed to gravitate to him for that reason.

Ted Leisenring and Margaret Pierce were already sweethearts in the fall of 1913 when she went off to Smith College and he left for Yale's Sheffield Scientific School. There he continued his athletic exploits, running the hurdles for Yale's track team at Madison Square Garden. By this time his marriage to Margaret was all but preordained. The question of what he would do with his life was similarly settled even before he graduated when, in 1916, Ted turned twenty-one and inherited the proceeds from his father's estate, which included a major share of the Leisenring group's various mining companies.

Dr. John S. Wentz had by then installed his Harvard-educated son Daniel as superintendent in Virginia. Under the Wentzes, the company opened a second mine at Stonega in 1897 and a third in 1900. Eventually the Stonega colliery grew to include nearly four hundred houses, two stores, and an office—all constructed from lumber produced at the company's own sawmills—as well as 666 brick coke ovens. Like most mining towns, Stonega depended on the Wentzes for just about everything: Their Virginia Coal & Iron Co. and its operating successor after 1902, the Stonega Coke & Coal Co., built and operated the local bakery, the slaughterhouse, and the town's bottling plant. Between 1899 and 1901 the company installed electric lights in company buildings and many of the houses. It created two water companies to supply fresh water to its coal towns and the nearby town of Appalachia. At his own expense, Dr. Wentz put up a recreation building and library for employees in 1900. A hospital was completed in 1902.

If the Wentzes fostered a paternalistic environment in their ten company towns, that seemed preferable to the harsh living and working conditions suffered by miners in nearby Kentucky and West Virginia. And the Wentzes' company took steps to reduce the towns' isolation and, consequently, their dependence on the Wentzes themselves. To entice private merchants to their towns, they built a road over the mountain into Kentucky—and to further encourage merchants to settle in Stonega, in 1899 the company built a separate wholesale store to supply wholesale goods to local merchants. Ten years later, the Wentzes organized the First National Bank of Appalachia to provide credit sources for local merchants as well as the company's own miners.

The ten company towns ultimately operated by Stonega C&C were laid out by private contractors, ostensibly with "worker contentment" in mind. Newspapers described the miners' housing as "comfortable and convenient" and "much above the average of mining towns." Perhaps more important, the houses were superior to the one-room log cabins, with a loft and a shed addition, that local mountaineers had typically lived in before going to work for Stonega C&C. Although virtually all houses at Stonega lacked indoor plumbing, the company built an outhouse behind each structure (serving two families, like the houses themselves) as well as a coal shed (where deliveries were made for coal stoves) and a smokehouse (which was used by the miner's wife to cure meats). Since the company didn't build a common wash house for miners until the 1920s, many miners built backyard sheds where they changed their clothes before entering the house. A miner's wife had her hands full maintaining a clean household environment amid coal dust, smoke, and ash, but the presence of these wash houses suggests a conscious attempt by many miners' families to prevent the grime of coal from intruding into their homes.

As early as 1898, an excursion train ran to Stonega from Bristol, Tennessee, so visitors could admire what the promoter called "the marvels" of this industrial village carved out of the wilderness. Yet the very paternalistic nature of all company towns, where the company was the source of all blessings and all grievances, inherently sowed the seeds of antagonism between workers and employers. If anything went wrong in such a town, there was no scapegoating the government or one's neighbors or a rival racial or ethnic group; it could only be the company's fault. As one observer later remarked, the company town enabled rural mountain people to develop for the first time "that sense of group oppression necessary for class feeling and behavior."

The United Mine Workers made its first foray into Virginia in 1902, organizing locals and calling strikes against other companies nearby at Inman and Norton. But Stonega Coke & Coal, the largest operator in the area, was the union's prime target, and the Wentzes were determined not to let the union gain a foothold. Before the union had a chance to strike Stonega, the company sealed off its three mines—all located in dead-end hollows—and imported sixty freshly sworn "deputies" from Hopkins County, Kentucky. Seizing on a legal technicality—namely, that one of the targeted companies, in Inman, was in receivership—the local operators had the UMW's national organizers jailed for interfering with the decrees of the chancery courts. The strike was broken, and the UMW would not return to the region for nearly a generation.

Yet Stonega's response to union organizers was mild compared to most other operators. When Pennsylvania miners struck for union recognition in 1902, the coal and railroad magnate George F. Baer responded in an open letter to the press, "The rights and interests of the laboring man will be protected and cared for not by the labor agitators but by the Christian men to whom God, in his infinite wisdom, has given the control of the property interests of the country, and upon the successful management of which so much depends."

One such Christian man, the oil monopolist John D. Rockefeller Sr., acquired effective control of a coal mining operation called Colorado Fuel & Iron and installed his own management in 1907 with orders to do whatever was necessary to make the company profitable. Rockefeller, like George Baer, perceived of employment as a form of charity to workers. By this logic it followed that unionization, by distracting and restraining management, threatened the effectiveness of a business and thus threatened workers themselves (who would lose their jobs if their employers and investors lost money). Consequently, in the minds of owners almost any countermeasure against union organizers seemed justified.

Yet the employment CF&I offered at that point was barely removed from slave conditions. The company's miners were paid their low wages (about $1.68 a day) in scrip that was redeemable only at company stores that charged inflated prices. Most CF&I miners lived in cramped two-room shacks, provided by the company at high rents, from which they could be evicted on three days' notice. Their children attended company-run schools; their church services were conducted by company-hired ministers; their company libraries were devoid of books the company deemed subversive. Accidental deaths and injuries in the mines occurred almost daily because the company's political influence in Colorado enabled it to ignore even the minimal mine safety standards of the day with impunity.

In theory, such pockets of misery were ripe for the UMW's organizers, who provided the only outlet for the miners' resentment. But in practice, the CF&I spent more than $20,000 a year on a force of detectives, guards, and spies—many of them drifters and former frontier gunmen—to prevent union infiltration. The company's executives, determined to return a profit for Rockefeller, sincerely perceived union organizers as a threat to the company's viability. "When men such as these," said the company's chairman, LaMont Montgomery Bowers, "together with the cheap college professors and still cheaper writers in muck-raking magazines, supplemented by a lot of milk and water preachers . . . are permitted to assault the businessmen who have built up the great industries . . . it is time that vigorous measures were taken."

Back at Rockefeller's headquarters in New York, Rockefeller's lieutenant (and Bowers's nephew) Frederick T. Gates concurred: "The officers of the Colorado Fuel & Iron Co.," he declared, "are standing between the country and chaos, anarchy, proscription and confiscation, and in so doing are worthy of the support of every man who loves his country."

Armed with this conviction of their righteousness, in the summer of 1913 the CF&I's managers passed out Winchester rifles to their vigilante squads and prevailed upon local sheriffs to deputize these company enforcers. The UMW's organizers, equally determined to demonstrate the union's strength, stocked up on rifles and ammunition in small-town general stores throughout Colorado. On September 23, 1913, some nine thousand miners and their families—about seven-tenths of the CF&I's work force—walked out of the mining camps and settled into tent colonies that the UMW had set up nearby.

Four weeks later, company detectives in an armored car raced through the miners' tent colony near Forbes, Colorado, raking the area with fire from two mounted machine guns. The miners retaliated, and two weeks later Colorado's Governor Elias M. Ammons called out the National Guard to restore order. But as the strike wore on through the winter, the state—lacking the funds to support the Guard's payroll over a sustained period—turned to its business supporters (including coal operators) for help. In this manner the National Guard was effectively co-opted into a management army, escorting strikebreakers whom the CF&I had imported by rail from as far away as Pittsburgh and Toledo.

On the morning of April 20, 1914, a militia company that had repeatedly clashed with strikers took up positions on a rise overlooking the strikers' tent colony at Ludlow, Colorado. When a shot from an unknown source rang out, the nervous militiamen responded by opening fire at the strikers' tents, setting off a battle that would last all day. When some of the tents caught fire, the strikers and their families retreated to cellars they had dug under the floorboards. By nightfall, when the shooting finally ended, at least forty men were dead and countless others wounded. The following morning rescuers discovered the bodies of two women and eleven children who had suffocated in a cellar when the tent above them had burned.

News of this "Ludlow Massacre," as it came to be known, triggered retaliation by strikers throughout Colorado, who seized company towns and attacked company offices and collieries. The violence didn't end until President Woodrow Wilson ordered federal troops into the area.

When the Ludlow Massacre occurred, John L. Lewis was two years into his new job as an organizer for the American Federation of Labor. In effect the massacre accomplished what labor organizers had been unable to do: It dramatized the plight of coal workers for millions of Americans who hadn't previously understood that a problem existed. Perhaps more important, it created a visible villain: the heartless absentee operators like Rockefeller, impervious to human suffering and caring only for the bottom line. The job Lewis had taken was now a cause, but Lewis was as yet merely a green foot soldier. Over the six years beginning in 1911, as he crisscrossed the eastern three-fourths of the country on behalf of the AF of L, Lewis vastly expanded his network of contacts among local UMW officials. And he familiarized himself firsthand with miners' working and living conditions.

The conditions he found were truly appalling. Within the United States about a thousand men died each year in mining accidents—explosions, roof falls, haulage accidents, electrocutions—and the number of mine deaths was not decreasing but rising. An additional 12,500 miners were maimed each year. Miners who were not killed or maimed accidentally died more slowly, by inhaling dust from fine coal and sand within the mines—the silent "black lung disease" that clogged their lungs because operators refused to ventilate their mines adequately. Such was the magnitude of these health and safety problems that most owners and pit bosses simply closed their eyes and hardened their hearts, lest they be overwhelmed. On his first day in a mine near Pittsburgh in 1913, a young Philadelphian named Pemberton Shober Jr. was shocked to find that work proceeded after a fatality with barely a pause.

"Shouldn't we get the body out?" he asked his boss.

"Shober, you get your ass back to work," the boss replied, according to Pem Shober's recollection. "We'll take care of this at the end of the day. That's the way we handle these Slovaks."

Few people were better suited than Lewis to exploit these experiences, for Lewis possessed not merely the soul of a miner but the eye of an artist and the ear of a playwright. Lewis seems to have instinctively understood the dramatic impact of using words to paint visual images. The miners' shanties, he wrote a few years later, "lean over as if intoxicated by the smoke and fumes of nearby mills. . . . They cluster higgledy-piggledy in some gulch, along the banks of a polluted little stream, or squat on a pile of slate."

Lewis's travels did not take him to southwestern Virginia, where Stonega Coke & Coal Co. was rapidly expanding under the Wentz family's leadership, so we can only wonder how he would have compared conditions there to those he found elsewhere. Between 1910 and 1916 the Wentzes' company overhauled its existing mines and opened several new ones, although two of them failed to produce sufficient tonnage and were shut down. Because the local population couldn't supply enough workers, the company began recruiting immigrants, just as the Leisenrings had done in Connellsville in the 1880s. Eventually, Stonega C&C maintained agents at ports and other immigration centers who signed up their frequently uncomprehending recruits and shipped them off to Virginia by train. Even after immigration was curtailed by the war, the company's agents scouted industrial cities in the East and Midwest until this practice was outlawed by the federal government in 1918.

Between 1915 and 1920 Stonega Coke & Coal imported an average of 2,260 workers every year, although many of these imports didn't stay long. The company customarily paid a worker's travel expenses to the mines and then took out so much from the worker's paycheck each month until the travel costs were reimbursed. But many of these "transportation men" fled Stonega upon their arrival—usually at night, to avoid the company guards posted to watch them—and signed up with nearby mine operators to avoid paying back the transportation fee. During one ten-year period, from 1905 to 1915, almost 19 percent of all "transportation men" brought into the area left without working.

The company preferred to hire Hungarians, who had worked so steadfastly and uncomplainingly in western Pennsylvania. But in its hunger for workers, Stonega C&C now turned to another labor source: southern blacks, mostly unskilled laborers from the Cotton Belt. Usually these "transportation niggers," as they were called, were given the least skilled and least desirable jobs, such as the backbreaking work in the coke yards, where freshly baked coke had to be pulled from the ovens with long-handled ladles by hand.

In these days, miners still functioned largely as independent contractors. They were paid not by the hour, but by the amount of coal they loaded, and they provided their own tools and blasting powder. Although the shift usually began at seven in the morning, the miner was free to come and go as he pleased, and to quit when he had had enough. Miners usually worked by themselves, often working all day without seeing a boss. (As late as 1921 the editor of *Coal Age* magazine described mining as "a cottage industry, only the cottage is a room in the mines.")

To keep its work force happy, or at least quiescent, Stonega gradually added new amenities in its company towns. The company's records for 1912 show that the superintendent at the Stonega mine spent all of the town's rent receipts on house repair and sanitation. The first movie theater was built in the town of Stonega in 1914. A year later the company put up a new general store, including an ice cream parlor, which then was a novelty. Children's playgrounds and baseball fields were built in several company towns, provoking a response that took the company by surprise. It had assumed "that only children would participate in the sports afforded by these grounds," the company's annual report remarked. Instead "men well up in years and in many cases women came out, sometimes joining in the games[,] and every evening when the weather was favorable, large audiences were at these grounds participating in the amusements as spectators or in conversation with their friends." Soon the various collieries formed baseball teams that competed against each other, each team soliciting contributions from the mine superintendent to import semi-pro players, some from as far away as Cincinnati.

Now the company began building underground first-aid stations in the mines and organizing first-aid teams. In 1917 the company started funding welfare workers, welfare nurses, and missionaries to tend to residents' physical and spiritual needs. The value of these social welfare gestures, a coal trade magazine contended in 1915, was "demonstrated by the entire absence of strikes and other disaffection."

Nevertheless, in 1916 a report on the town of Stonega by a federal mine official criticized the company for neglecting to collect garbage and waste regularly. Some privies, the report noted, were located above houses on hillsides, so that their runoff might seep into the houses. Since there was no regular boarding house in Stonega, many of the town's homes were crowded with boarders, who represented a source of extra income for the tenant families but also a strain on community order. "With the yards themselves," wrote the Bureau of Mines inspector, Joseph White, "even where trash receptacles were provided, tin cans, rubbish and trash of all kinds were recklessly strewn about."

In a time characterized by high turnover and ethnic diversity within the company's work force, the company seems to have concluded that the most effective tool for maintaining peace in the company towns was not creature comforts but ethnic and racial segregation. In each town, each ethnic group was now given its own section and church, each with a colloquial name like "Hunktown," "Happy Hollow," or "Quality Row." In keeping with Virginia customs and laws, blacks were segregated from whites in the company's hospitals, schools, and bathhouses. Even the company's first-aid teams were segregated by race.

The county seat of Big Stone Gap, where the Stonega C&C maintained its headquarters and most of its officials lived, evolved into something of an urbane eastern oasis within the surrounding Appalachian hills. The company's officials, mostly eastern-bred and -educated, tended to assume that their presence was an uplifting influence on otherwise backward natives, "who a generation ago were at least two hundred years behind the civilization of the more densely populated sections of the United States," as Stonega's 1917 annual report described the challenge.

Appalled by the seeming absence of organized religion, these company managers—and especially their wives—hastened to form missionary societies for the salvation of the mountain folk. By 1917, in tandem with company physicians and officers, these societies began to attend the sick, teach cooking and sewing, and correct unsanitary conditions (although not in the mines themselves, where women weren't permitted). In the process these company wives also began to pressure the company itself, using their influence with their executive husbands. Perhaps as a result, around this time Stonega C&C enlarged the schools in most of its collieries and hired additional teachers.

All of these beneficent activities, of course, were designed with one unspoken purpose: to keep out the unions. The Wentzes felt just as threatened by unions as Rockefeller or Frick did, but they lacked Frick's stomach for violence. If the unions lacked fertile organizing ground in Wise County, the company reasoned, they would direct their limited resources elsewhere. Nevertheless, in June of 1917 the UMW's organizers appeared in the region for the first time in fifteen years, just across the state borders in Kentucky and Tennessee.

Although John L. Lewis recorded no great successes as a labor organizer, he made other substantial contributions: He served as Samuel Gompers's eyes and ears in UMW circles. In New Mexico, newly admitted to statehood, he lobbied for passage of a progressive state constitution in 1912. That same year he campaigned throughout the Midwest for Woodrow Wilson's progressive presidential campaign. Wherever he went, people remembered his jowly face, topped by a thick mane of hair and punctuated by unnaturally bushy eyebrows. But his most memorable characteristic, audiences found, was his voice. His speeches were not mere lectures but dramatic performances, delivered in a stentorian bass that moved slowly and ponderously, pausing for dramatic effect to build his listeners' anticipation. He studded his remarks with flowery phrases and archaic words deliberately calculated to remind listeners of the cadences of Shakespeare and the King James Bible. Years later Lewis confessed half-jokingly to a close friend that his pauses had less to do with dramatic emphasis than with his need for time to figure out what he would say next. But whether his presentation was genuine or calculated, Lewis left an impression that stuck in people's minds.

In 1916 Lewis accompanied Gompers on a speaking tour through Indiana and Illinois, introducing the AF of L chief to his audiences. This public proximity to the AF of L's legendary founder further compounded Lewis's appeal. At the same time he assiduously tended his UMW garden, campaigning for the re-election of his patron, John White, in the union's 1915 national election. This success earned him a prominent role at the UMW's 1916 national convention, where in the process of chairing the resolutions committee and presiding over debates he impressed even his critics with both his knowledge of parliamentary procedure and his tongue-lashings of White's opponents. In 1917 a grateful White appointed Lewis to two plum jobs: as the UMW's statistician (a role critical to the union's collective bargaining) and as business manager of the union newspaper, the *UMW Journal* (which enabled him to keep his name and image before the union's 350,000 members for the rest of his career).

His timing was again propitious, for with America's entry into World War I that year the coal industry became even more important in the life of the nation. Coal and coke were needed for iron and steel mills in the manufacture of arms and trucks, for railroad locomotives, for ships transporting troops, for electrical generators, and for home and commercial heating—not only for America but for her European allies. In these chaotic days the only thing limiting coal production was the number of railroad cars available to transport it. Bituminous coal alone accounted for nearly seven-tenths of the nation's wartime energy, a level never reached before or since. Coal prices shot upward, and so did coal production and profits for those who produced it. There had been 5,818 bituminous mines operating in the United States in 1910; by 1920 that figure had increased by more than half, to 8,921. In Washington a new federal Office of Fuel Administration was created to assure, by limiting prices and wages, that the disorderly coal industry would provide an orderly flow of coal to the war effort.

Stonega Coke & Coal, meanwhile, found itself in 1917 coping simultaneously with both the war and the UMW's first successful incursion into the area. In June 1917, the UMW hastily signed up ten thousand new members just across the state line in Kentucky and Tennessee. The subsequent wage settlements there didn't officially recognize the union, but as Stonega officials noted in their annual report, "it practically amounted to the same thing and conditions in that territory are going to be almost intolerable until this question is definitely settled, and intolerable if recognition is granted." The company had granted its workers a modest pay increase that April and another in May, both in anticipation of the UMW's arrival. But now, to forestall union activity on the Virginia side, Stonega officials hastened "to anticipate any demand that might be made by our employees" and tendered "substantially the same increase as given in April and May."

In November 1917, at the suggestion of the new U.S. Fuel Administration, Stonega officials reduced the workday for its hourly workers from ten hours to eight, and again wages and coal rates were raised so as not to reduce the miners' daily earnings. In the company's view, this fourth boost pushed its wages "so much higher than any previous scale and so far in excess of the increase in the cost of living" that company officials hoped it would take care of their labor shortage. Nothing of the sort happened. To their dismay, however, "corresponding increases were made throughout the entire coal mining industry," and competition for labor remained as fierce as ever.

This time, however, the Stonega company was precluded from engaging in a bidding war for labor against its smaller competitors, for the newly created federal Fuel Administration warned that further pay increases would be taken as evidence of excessive profits and would result in a reduction in the federally regulated selling price allowed the company. This meant that Stonega would have to find other ways to compete for workers and customers. One solution—the gambit attempted by Henry Clay Frick in the 1880s—was to organize the industry into bargaining cartels, and this time, thanks to the pressing needs of the war effort, the government was favorably inclined. When in September 1917 the National Defense Council asked operators to form themselves into trade organizations, Stonega's executives were eager to oblige. That year they helped organize both the Virginia Coal Operators Association and the National Coal Association, the fractious industry's first nationwide organization. Both groups quickly came to be dominated by the largest coal companies. Stonega's vice president, Otis Mouser, was elected to head the Virginia group. Soon after, Stonega's president Daniel B. Wentz, who had run Stonega's Virginia operations from John Taggart's death in 1896 until he was called back to the Leisenrings' Philadelphia headquarters in 1904, became the leader of the national association. For the first time, the coal industry could speak to government and labor with at least the semblance of a unified voice—and that voice, to a large extent, was the voice of the Wentzes and their widespread operations within the Leisenring group.

Yet it was at this very moment that Daniel Wentz resigned temporarily from Stonega and patriotically offered his services to the government for $1 a year. Daniel was a dapper man in his mid-forties with an eye for the ladies (he had conducted several semi-public romances). His education at Andover and Harvard had more than broadened him as his father had hoped: It had planted the seeds of yearning for the world beyond the coal industry. The war offered the perfect opportunity to satisfy that hunger.

In Washington Daniel served at first in the federal Fuel Administration as well as the Bureau of Explosives. Then in April of 1918, at the age of forty-five, he was commissioned a lieutenant colonel in the American Expeditionary Force. A month later he sailed for France. There he served as chief fuel supply officer under the American commander, General John ("Black Jack") Pershing, reorganizing fuel supplies to the French railroads that were transporting troops and munitions.

As his legman he brought along Ralph H. Knode, his bright and aggressive young protégé from Stonega's Philadelphia office. Knode had been born in Pittsburgh in 1893 into a family of modest circumstances but useful connections. As a young man he passed up college to enable his older brother to attend. His greatest asset was not his education but his gregarious personality, which endeared him to men and women alike. Knode grew into a short, bald, and jovial young man with an infectious smile and laugh and an uncanny ability to remember names, faces, and anecdotes about everyone who crossed his path. He was a classic hail-fellow-well-met who reveled in all social activities: golf, cocktail parties, nightclubs. Eventually his charm led to a marriage to Marian Head, who came from a wealthy Pittsburgh family, and his talents as a tournament golfer and social drinker generated a network of well-placed friends in Pittsburgh and throughout the country clubs of the East. One such golf buddy was Benjamin Fairless, later the head of U.S. Steel; another was his contemporary Ted Leisenring, just graduating from Yale as the United States entered World War I.

Knode actually arrived at Stonega's Philadelphia headquarters well before Ted Leisenring did. With his outgoing personality and his "can-do" approach to every problem he confronted, he seemed a natural salesman almost from his arrival. He was a born expediter with a knack for the logistics of moving coal and other materials from one point to another. So when Daniel Wentz was placed in charge of organizing fuel supplies for the war effort in France, it seemed only logical to him to bring Knode along. Since Knode in France would need to exercise the same authority as his civilian boss, Wentz persuaded General Pershing to commission Knode to his same rank—lieutenant colonel. In this manner Ralph Knode became, at twenty-six, the youngest soldier in the American Expeditionary Force to attain that rank.

Wentz and Knode spent the remaining six months of the war scouting Western Europe for coal to stoke Allied trains. Knode's magnetic personality was put to good use almost immediately on his arrival in France, when he struck up a friendship with the famous American aviator Eddie Rickenbacker, then twenty-seven. In his civilian life Rickenbacker had raced automobiles; as a wartime air ace he was credited with shooting down twenty-six German planes. But when Knode met him in France he had been grounded by an injury for about a month. In Rickenbacker's enforced idleness, Knode's pragmatic mind saw an opportunity: "I want to get Rickenbacker as my chauffeur," he told Wentz. "He knows this land from the air. He knows maps intimately. No one can move us around France as quickly and efficiently." Rickenbacker eagerly volunteered for the job, and with his unexpected help, Daniel Wentz and Ralph Knode eventually provided more than half a million tons of coal a month for the French railroads before the German government sought an armistice in November 1918. In honor of this feat, shortly before his discharge the following year Wentz was promoted to the rank of full colonel and made a chevalier of the French Legion of Honor. Thereafter he affected the title of "Colonel" for the rest of his life.

Daniel Wentz's cousin Ted Leisenring, meanwhile, graduated from Yale in 1917, married his childhood sweetheart, Margaret Pierce, and almost immediately joined the U.S. Naval Aviation Corps, eager for a taste of adventure with the new flying machines. To his dismay he was sent not to France but to the Panama Canal to fly in patrols searching for German U-boats, a largely fruitless exercise. (This diversion from the front in France left intact Ted's lifelong aversion toward all things French, presumably a product of his Midwestern upbringing; although Ted himself had been born in Nice, he often insisted as an adult that he had been born in Philadelphia.) With the end of the war he was ready to reclaim his more prosaic mining patrimony—to which, up to this point in his life, he had not yet been exposed.

The death in July 1918 of Daniel Wentz's father, the entrepreneurial Dr. John S. Wentz, meant that Daniel on his return assumed the presidency of all the Leisenring group's operations. He was thus ideally suited to introduce these companies to a new opportunity—the overseas export market—after the war ended. Ted Leisenring joined the group and was put to work in the family's Panther Valley Coal Company, in the Pennsylvania anthracite region, where he eventually became a mine superintendent. Ralph Knode, Daniel Wentz's young assistant in France, also remained with the Leisenring group, and he eventually became chairman of the executive committee of all of its companies. Knode's great value to his employers and to the coal industry had to do with his ingratiating personality, which he put to good use both as a salesman and a corporate diplomat. He delighted in the sort of industry politics that the quieter Ted Leisenring had little taste for. The same outgoing quality that had endeared Knode to the gregarious Eddie Rickenbacker during the war, now fostered a personal relationship with the labor leader who in theory should have been Knode's adversary but in fact became a close personal friend: John L. Lewis.

Lewis was thirty-seven and a rising star in the UMW when America entered the war in 1917. Now his forcefulness and astuteness played a critical role as the union accelerated its negotiations not only with coal operators but with government officials desperate to increase the supply of coal to the armed forces.

When federal officials appointed Lewis's patron John White that fall of 1917 as an adviser to the government's fuel administrator, White resigned the UMW presidency. White's automatic successor, the popular international vice president Frank Hayes, tapped Lewis as his own second in command. Although Lewis was barely eight years out of the mines of Panama, Illinois, he would be next in line to a president who suffered from numerous physical ailments as well as a drinking problem.

Lewis well understood the fortune that had befallen him. When the UMW's executive board unanimously approved his appointment on October 15, Lewis noted in his travel diary, "Our ship made port today."

By March of 1919, as Hayes sank into lethargy, Lewis was effectively running the union. He took control at a moment when the coal industry, having overexpanded during the war, was suffering from overproduction and decreased demand and feeling less expansive about meeting miners' demands. When the UMW's negotiations with the recently formed Coal Operators Association collapsed in the fall of that year, Lewis issued his first call for a nationwide strike, effective November 1. In the process he seemed to establish his reputation as a fearless champion of miners' rights. Yet anyone who bothered to probe beneath Lewis's militant facade would find a fundamentally cautious and conservative man, a lifelong Republican who believed in free markets and competition and who distrusted union radicals and insurgents almost as much as Henry Clay Frick and John D. Rockefeller did. Lewis perceived unions and corporations not as natural enemies but as natural partners: one organized production and sales, the other organized labor. A strike, to Lewis, was not a weapon of bloody class struggle but merely a viable negotiating tool in the give-and-take of the capitalist system.

Most coal operators were too myopic to perceive that Lewis might be the best friend they ever had. Ralph Knode would prove to be a rare exception.

Four weeks into the strike of 1919, the federal government issued an injunction against the UMW, and Lewis obligingly backed off, explaining, "I will not fight my government, the greatest government on earth." But to his embarrassment, most miners ignored him and refused to return to the pits, effectively curtailing three-quarters of the nation's coal production. The strike finally ended in mid-December, after six weeks, with a settlement supervised by America's first secretary of labor, William B. Wilson, himself a former miner and UMW member. The settlement produced a two-year contract that gave union "tonnage men"—those paid by the weight of the coal they mined—a 34 percent rate increase, and "day men"—the haulers, sorters, and processors who actually comprised two-thirds of a mine's work force—a 20 percent daily wage increase. Lewis had delivered tangible success for his miners, but he had done it by following the miners rather than by leading them.

Hayes, continuing to decline physically, officially resigned the UMW presidency as of the first of January 1920, and Lewis automatically succeeded him. Without a democratic mandate at the international level—or, on his way up, at virtually any other level of the union—Lewis had gained control of the nation's largest and perhaps most powerful labor union, representing more than half of the nation's seven hundred thousand coal miners.

Later that year Lewis was elected to the UMW presidency in his own right. Now he needed to convince his members that he was indeed on their side—that he possessed the soul of a miner rather than that of a politician or an entrepreneur. To make that case he would need to expand the union's effectiveness by expanding its reach and tightening his own control. He would need to score victories in places where no union had ever before shown its face.

The Leisenrings and Wentzes had fancied themselves enlightened employers, at least by comparison to their fellow coal operators. But by the time of Lewis's election their Stonega Coal & Co. employed three-fourths of all the miners in Virginia and was producing more coal than three-fourths of its Virginia competitors combined, and they could hardly expect to escape Lewis's notice.

In the same year that John L. Lewis was elected president of the UMW, Daniel Wentz was again elected president of the National Coal Association he had helped to found. To the extent that any operator could be said to represent the coal industry, the Leisenring group was it. The fourth generation of Leisenrings were about to experience their first confrontation with John L. Lewis.

CHAPTER 11

Utopia Goes Union

When Dr. John S. Wentz died in July of 1918, the national economy was on a war footing, the coal industry was expanding as never before, and Wentz's son Daniel was forty-six and in the prime of life. On his father's death he inherited the presidency not only of Stonega Coke & Coal but of all the family's various coal concerns in eastern Pennsylvania, Virginia, and West Virginia. He also inherited his father's private investments, including one of the largest single blocks of stock in Westmoreland Coal Company, the western Pennsylvania concern originally founded by Pennsylvania Railroad investors in 1854 in order to ship coal directly to eastern markets. (This was the legendary firm whose coal traffic provoked Robert E. Lee's 1863 incursion into Pennsylvania that ended at the Battle of Gettysburg.) Like John L. Lewis, Daniel Wentz had reached the summit of a booming industry only to find, now that the war was over, that there was little to celebrate.

The war effort had engendered a coal industry capable of producing one billion tons a year for a nation that now consumed barely half that amount. Americans still needed coal—it produced 68 percent of all fuel consumed in 1924—but thanks to their own ingenuity they no longer needed quite as much. The biggest coal consumers—the railroads, the iron and steel companies, and the electric power plants—had learned, while struggling with wartime coal shortages and postwar coal strikes, to use coal more efficiently. Thanks to their newly developed combustion techniques for squeezing energy units out of coal, by 1920 a kilowatt hour of electricity could be produced with only two-thirds of the amount of coal required in 1913.

The postwar coal strikes had also driven coal customers to other fuels that might be available on a less erratic basis. Americans were driving cars now, which meant that, for the first time since Colonel Edwin Drake first struck oil at Titusville, Pennsylvania, in 1859, the crude oil industry posed serious competition to coal. Between gasoline for automobiles and kerosene for illumination, cooking, and home heating, Americans in 1920 were using seven times as much crude oil as they had in 1900. Iron and steel works, meanwhile, were turning to natural gas to heat furnaces and generate steam.

From his headquarters in the Land Title Building in Philadelphia Daniel Wentz responded to this overproduction problem not by cutting back but by trying to drum up new customers. Exploiting his wartime contacts, he sniffed out new overseas markets for his companies in Europe and Brazil. During the 1920s several British officers he had known in France created the Brazilian Light & Traction Co. to build railroads in Brazil, and through this connection Stonega won contracts to supply coal to power plants in Rio de Janeiro, São Paulo, and Santos. Daniel Wentz also expanded Stonega's customer base from Midwest railroads and steel mills to southeastern industries hungering for steam coal for their power plants. He even launched several new anthracite operations in northeastern Pennsylvania at a time when anthracite's market was relatively depressed.

But Daniel Wentz's energy alone could not make up for some fundamental changes in his industry. Those thousands of smoke-churning beehive coke ovens that had formed the basis of Henry Clay Frick's fortune in the 1880s were now receding before the inevitable advance of time and technology. The much more modern "by-product" coke oven was first introduced in Europe in the 1880s and brought to the U.S. steel industry by Dr. Heinrich Koppers in 1906. It worked twice as fast as the beehive oven and was far more efficient as well: In the process of cooking coal into coke, it extracted valuable coal by-products. Chemicals that had formerly been wasted as noxious smoke could now be used to make dyes, inks, antiseptics, aspirin, saccharin, and many other products. Thanks to the by-product oven, just about any grade of bituminous coal anywhere would suffice to make coke—and unlike anthracite, the supply of this coal, as John L. Lewis explained to the U.S. House of Representatives in 1922, was virtually inexhaustible, and "so distributed over the entire country to be available with comparatively short hauls to every point that aspires to any degree of industrial activity." By the time Frick died in 1919, his trademark beehive ovens produced just two-fifths of the nation's coke; within a decade their share would decline to just 6 percent. And since the new by-product ovens were located at steel plants rather than coal fields, they eliminated the competitive edge that coking coal districts like Connellsville and Southwest Virginia had enjoyed for two generations.

An industry that employed 862,000 men in 1923 actually needed only about one third that many on a full-time basis. Some nine thousand coal mines competed with each other to serve this shrinking market. Thousands of these were tiny, thinly financed "wagon mines," so called because, as one observer put it, "almost any farmer can take a pick and shovel and go into a hillside on a coal seam and fill his wagon with coal and load into a flatcar at the nearest railroad switch." These drift-entry mines required no machinery to open and little cost to maintain. The biggest expense was labor, which accounted for at least 60 percent of most coal operators' costs (compared to 15 percent in most other industries). Thus a cheap wagon mine could be opened quickly if demand and price rose, and closed just as quickly if demand and prevailing tonnage rates declined. The lower the capital value of the mine and equipment, the less the operator stood to lose if the mine was idle.

In this environment, large operators like Stonega Coke & Coal were like elephants competing with pigeons: The haphazard wagon mines, with their low overhead and low capital costs, dictated prices to the rest of the industry. Most of Stonega's Virginia competitors were tiny mines averaging less than ten thousand tons of output annually—that is, less than one railway car per working day. In 1918, the last year of the war, a visitor driving through the Leisenrings' heartland in Wise County counted thirty-six wagon mines within sight of the road over a stretch of about thirty-five miles. "There were doubtless several times that many not in sight of the road," a local historian suggested.

Daniel Wentz expanded Stonega after the war by absorbing several of these pesky rival neighboring companies, and through the 1920s Stonega ranked among or near the nation's ten largest coal companies (its annual output exceeded three million tons by 1929). But that sort of heft wielded little impact on an industry in which no single company accounted for more than 5 percent of the market. When thousands of wagon mines shut down to save on labor costs, larger producers such as Stonega had no choice but to do likewise. The coal industry contained, in Lewis's mellifluous words, "Twice too many mines and twice too many miners."

Other splintered and cutthroat competitive industries had been consolidated and whipped into some sort of rational order by America's leading financiers, whose organizational skills were intended precisely for that purpose. In the late nineteenth century J. P. Morgan had imposed control upon America's tangle of railroads and steel companies from the outside in order to protect his clients' investments. But coal was not a capital-intensive industry like railroads, oil, or steel (most coal operators, the Leisenrings included, financed themselves from their own current revenues), and consequently coal attracted little attention from financiers and little benefit from their services. Anthracite coal—a specialty fuel that existed in finite quantities within one specific section of Pennsylvania—had indeed been organized by the railroads and a few large coal operators. But bituminous coal defied control: It was too widely scattered, and just about anyone could play.

It was a bitter joke for major companies like Stonega Coke & Coal to find themselves at the mercy of their smallest competitors, but not so bitter as the story then circulating among mine workers, in which a shivering miner's son asks his mother, "Why don't you light the fire?"

"Because your father is out of work," she replies, "and we have no money to buy coal."

"Why is he out of work?"

"Because there's too much coal."

Lewis had spent the previous decade preaching against the transgressions of greedy coal operators. Now his essentially capitalistic mind shrewdly perceived that the miners' enemy was not coal operators per se but the chaos caused by those thousands of marginal operators, whose fly-by-night tactics were undercutting the established operators' ability to pay stable wages. As Lewis saw it, by organizing virtually all the workers in the industry he would bring order out of chaos for the benefit of miners and operators alike.

This view was shared by, of all people, a group of relatively enlightened coal operators from the major unionized northern soft-coal states— Pennsylvania, Ohio, Indiana, and Illinois. They had formed a consortium called the "Central Competitive Field" for the sole purpose of bargaining collectively with the UMW. This organization cherished the hope that an industry-wide labor contract would stabilize their wages and their production costs. To these farsighted operators, the UMW was not an enemy but a potential friend that would enforce the contract, discipline miners who staged wildcat strikes (that is, strikes not authorized by the union or the contract), and unionize any mine that refused to honor the organization's agreements. In this vision, labor and management were almost on the same side, working toward the same goals.

Virtuous acts often produce unintended consequences, and at least one by-product of this labor-management alliance was a positive one. Electric undercutting machines for digging coal in the mines had been available to coal operators since the 1880s, and by the 1920s mechanical loading machines were on the market as well. But as long as labor in the mines was cheap, few operators had felt any need to invest in expensive machinery to dig, load, or haul coal. Miners with picks, and donkeys pulling carts were sufficient, just as before the Civil War. But now, as the UMW successfully negotiated higher wage rates for miners, many operators responded by reducing their work forces and turning instead to machines to do the work. By 1920, more than 60 percent of all coal was being mechanically undercut by electric machines, even while 99 percent of coal was still being loaded by hand. Now coal operators were increasingly installing mechanical loaders as well.

In virtually every industry since the industrial revolution began, workers and labor leaders have initially feared machines as a threat to their jobs. Lewis, to his credit, was an exception. He readily understood that, with the advent of machines, the number of men working the mines—and consequently the number of potential dues-paying UMW workers—inevitably would shrink. Nevertheless, he welcomed mechanization, on the theory that it would help drive out the smaller inefficient mines that were driving down coal prices and wages. He wanted *more* mechanical loaders, he said, not fewer. Only if the universe of mines and miners was sharply reduced, Lewis believed, could miners hope to gain higher wages, shorter hours, greater fringe benefits, and better working conditions. "We decided it is better to have a half-million men working in the industry at good wages ... than it is to have a million men working in the industry in poverty," he subsequently explained.

What neither Lewis nor the operators anticipated was the way machines would change the psychology of the mines. However much miners and operators may have despised each other before mechanization, their relationships were at least clear: Operators raised the capital and marketed the coal but left the miners free to make most of the day-to-day decisions underground.

But with the new technology, miners became subordinate to expensive machines that performed the most important work. Now there were fewer workers, organized into crews and clustered more closely around big machines—and consequently supervised more closely. "Supervision must be more intent," *Coal Age* magazine warned in an editorial in 1924. "Companies can no longer countenance superficial observance of processes by foremen." Machines inadvertently created rigid lines of authority that miners could resent. "In hand loading you were your own boss," a West Virginia miner later told an interviewer.

> For instance, I would get up at five o'clock. . . . Then I had two miles from the outside [the mine entrance] to where my place was [in the mine]. I would walk in there and bug-dust my cut and drill it and shoot it and be ready by the time the motor crew came at seven o'clock with my cars. I could load whatever I wanted to load, but I usually averaged six to eight cars—two-ton cars—and if I would clean my place up the second day—which I would do, take two days to clean my place up—. . . if I got done at two o'clock [the second day] I could go home.

But with the coming of the new Joy Manufacturing Company loading machines—named for their inventor, Joseph F. Joy—the miners necessarily found themselves working in tightly scheduled crews. "While the cutting machine was in one place," the West Virginia miner recalled, "the Joy would be in another. The timber men would be in another, and the drill crew would be in another. So we got it worked out in a cycle; everybody knew where he went next and such as this." And as miners grew more resentful, Lewis himself would necessarily grow less cooperative and more militant toward the operators.

Lewis's vision of a partnership with enlightened coal operators suffered from one other flaw. Most operators possessed neither the vision nor the resources to wait for him to organize the entire industry. They needed to survive in the face of cutthroat price competition, and for many of them the best short-term strategy was to avoid the United Mine Workers altogether. Thus operators in the northern states (which were heavily unionized) began hedging their bets by acquiring coal lands in southern and border states like Virginia, Kentucky, and Tennessee, where miners remained skeptical of unions and the UMW had made few inroads.

Here in the Southern Appalachian hills, both law enforcement and local sentiment sided with the operators. In this empowering climate, operators who wouldn't dare to defy the UMW in Illinois or Ohio eagerly stiffened their backbones and flexed their union-busting muscles. When the UMW tried to organize in Mingo County, West Virginia, some seventy miles from Stonega's Virginia operations, in late 1920 and early 1921, company guards and policemen forced striking miners and their families out of their company-owned houses. Dozens of miners, guards, and policemen died in shoot-outs and ambushes. In May 1921 angry West Virginia state police tore up a strikers' tent colony at Lick Creek and rounded up the men, women, and children who lived there. After threatening to burn their prisoners alive, the police pulled out one striker.

"Hold up your hands, God damn you," a policeman snarled at him, "and if you have got anything to say, say it fast."

"Lord have mercy," the striker mumbled just before he was gunned down.

Miners replied in kind, dynamiting railroad cars and powerhouses. In Matewan, West Virginia, seven company guards were killed while trying to evict striking miners from company-owned houses. One anti-union activist there was castrated and left to bleed to death. Following a three-day gun battle in Logan County, West Virginia, along the Kentucky border in May 1921, it took eight days to remove the dead from the woods and hollows. The union's effort soon flickered out, and by 1922 the UMW's presence had all but vanished from West Virginia.

The result of this "southern strategy" on the part of operators was a two-tier industry—one northern, one southern—that largely undermined the UMW's bargaining authority. The union's nationwide strike of 1922—in the midst of a national recession—meant instant prosperity for non-union areas like Virginia: With northern mines shut down, production at Stonega Coke & Coal suddenly jumped from less than half of capacity to 83 percent.

But of course this prosperity depended entirely on keeping the union away from Wise County. When 240 non-union miners in Virginia staged a strike that year in sympathy with the UMW, Stonega's ordinarily paternal managers panicked and fell back on a hated device that originated in the 1870s: the "individual contract," which required each worker to acknowledge that "so long as the relation of employer and employee exists between them, the employer will not knowingly employ, or keep in its employment, any member of the United Mine Workers of America, the I.W.W., or any other such mine labor organization, and the employee will not join or belong to any such organization." When these "yellow-dog" contracts—so called by miners because, as they saw it, only a "yellow dog" would sign one—were introduced in the late summer of 1922, they provoked more commotion at Stonega than the sympathy strike itself had caused. By that fall some men had been fired for refusing to sign; at Stonega's Exeter community, at least eighty quit rather than sign the despised pledge. Harry Wallace, editor of the nearby *Appalachia Independent,* boldly took the miners' side; the yellow-dog contract, he argued, "deprives a man of his liberty and freedom of speech . . . [and] that is what the country was founded for." But the miners' protest failed, and by December 1 every remaining Stonega C&C employee had signed.

Thanks entirely to the miseries of the rest of the industry as well as the yellow-dog contract, Wise County's future seemed bright. "The high quality of our coal," wrote the *Big Stone Gap Post,* "and the total absence of labor troubles which continually clog the wheels of Eastern industry will be leading factors in convincing buyers for the large industrial centers that our field is the logical source of supply."

Two years later, in 1924, the unorganized South again undercut the unionized North in much the same way. With the cooperation of the northern operators in the Central Competitive Field, Lewis pulled off a seemingly impressive coup: the "Jacksonville Agreement," in which the UMW and operators in the Central Competitive Field agreed to honor a fixed wage scale for three years—a year longer than the previous contract—even though coal prices were falling. "We got a three-year contract," jubilant UMW delegates chanted as they left the meeting in Jacksonville, Florida. "Next time we'll make it five."

But their celebration was premature. The Jacksonville Agreement did not apply to non-union southern operators, who remained free to reduce wages, so that by 1925 union workers received an average of $7.50 a day while southern workers were paid only $5. Veteran coal operators may have shrugged off this sort of differential as an inevitable annoyance in an irrational business, but newcomers like Andrew and Richard Mellon found it intolerable. These famous Pittsburgh banking brothers (whose father had bankrolled the young Henry Clay Frick in the 1870s) had just acquired a controlling interest in the Pittsburgh Coal Company, then the world's largest coal concern, operating more than one hundred mines employing some seventeen thousand union workers in Pennsylvania, Ohio, and Kentucky.

Like John D. Rockefeller at Ludlow, Colorado, in 1913, the Mellons were financial men who lacked the patience or interest to nurture long-term relationships with the union. Rather than honor what they saw as a suicidal union contract, the Mellons' managers peremptorily reduced the pay scale to $6 a day and announced that the mines would thereafter operate on a non-union basis. To avoid being sued by the UMW for breach of contract, Pittsburgh Coal technically went out of business and reorganized a few weeks later as a new entity. Its employees were now required, as a condition of employment, to sign a yellow-dog contract. When the company's chairman, William G. Warden, was asked by a U.S. Senate committee to justify these actions, he answered simply, "We were losing money. The firm had operated under a deficit for seven years in the decade after World War I."

Soon other northern operators were copying Pittsburgh Coal's tactics. By April of 1927, when the Jacksonville Agreement expired, few northern mines were honoring it, and even the Central Competitive Field's organizers refused to renew it or even to renegotiate on a regional basis.

Lewis and the UMW responded, predictably, with the only weapon at their disposal: a bitter strike that lasted eighteen months. But this time so much coal production had shifted to the non-unionized South that the Midwest operators weren't really suffering much by shutting down their mines. They could afford to bide their time until the desperate miners gave up. A U.S. Senate investigating committee touring the Pittsburgh area in 1928 reported finding "men, women and children, living in hovels which are more unsanitary than a modern swine pen."

Lewis effectively abandoned his dream of an industry-wide contract that summer, granting permission to individual districts to negotiate their own separate contracts with the Central Competitive Field. In effect, the union's hold on the Central Competitive Field which the UMW had celebrated at Jacksonville was broken, and northern operators were now free to slash wages and prices in order to compete with the South. To Lewis's embarrassment, union wages were actually lower after this strike than they had been before it. And without bargaining power, the UMW's membership declined rapidly. By the end of the 1920s its numbers had dwindled below one hundred thousand compared to four hundred thousand when Lewis assumed the presidency in 1919.

How Lewis survived as the UMW's president throughout this disaster, like Lewis's own rise to the union presidency, defies logical explanation. Many UMW officials distrusted Lewis as an unelected bureaucrat and social climber, more comfortable among operators than miners—as indeed he probably was. Miners tended to distrust bigness in unions as much as they distrusted big business, so much of the union's power rested in its locals and its "districts," which were usually defined by state boundaries. The union's powerful and independent-minded district leaders, for their part, resented Lewis's efforts to impose national contracts on them. Yet Lewis survived because he was more tenacious and ruthless than his adversaries, and because he skillfully employed the powers of his office.

Alex Howat, leader of the union's Kansas district and a hero to rank-and-file miners, led his followers on scores of unauthorized wildcat strikes and traveled through the UMW's western districts whipping up anti-Lewis sentiment with speeches that ran on for hours. After Lewis had Howat expelled from the union because of alleged irregularities in his Kansas district, Howat showed up anyway at the UMW's 1924 national convention, strode to the podium and, defying Lewis's gavel pounding, calmly poured himself a glass of water and launched into a speech. Lewis's bodyguards quickly threw Howat off the platform, setting off a wild demonstration in which Howat's supporters shouted "Mussolini!" at Lewis.

An even more tenacious—because he was more coherent—challenge came from John Brophy, head of the UMW's central Pennsylvania district. Brophy's fury at Lewis dated back to the UMW's nationwide strike of 1922. When Lewis settled that strike on the Central Competitive Field operators' terms, he pulled the rug out from Brophy's efforts to organize seventy-five thousand Pennsylvania bituminous miners and coke workers. To Brophy, Lewis's betrayal was a sign that the UMW wasn't serious about fighting for miners' rights. When Brophy's repeated criticisms of Lewis fell on deaf ears at Lewis's headquarters, Brophy announced his candidacy for the UMW presidency in 1926. Although he cobbled together an alliance of Lewis's traditional enemies (such as Howat) and even accepted support from American communists, Brophy's campaign received scant attention in the union's own *UMW Journal.* Lewis, firmly in control of the election machinery, won the election in a vote of 170,000 to 60,000.

Still Brophy refused to depart. He appeared at the UMW's 1927 national convention bearing countless examples of evidence that the election had been stolen. In one Kentucky local, for example, the vote was recorded as 2,686 for Lewis, none for Brophy, even though the largely defunct local had only one dues-paying (i.e., voting) member. Lewis responded by painting Brophy as a communist tool whose challenges threatened to undermine the union's united front. "In the days when people were besieged in a walled city," Lewis declared portentously at the convention, "and a soldier got upon the top of the wall and called to the enemy that the people were weak, they merely took his life and threw him off the wall to the dogs below."

A year later, Lewis made good on this implied threat. Brophy convened a "Save the Union Conference" in Pittsburgh, where eleven hundred men vowed to recapture the union from "its present incompetent and greedy leadership under President John L. Lewis." Lewis's international executive board responded the following month by expelling Brophy—without any hearing—on the ground that he was forming a rival union. Lewis thus disposed of his only democratic opposition in much the same manner that Joseph Stalin expelled Leon Trotsky from the Soviet Communist Party in 1927 and then banished Trotsky from Russia altogether in 1929.

Of course Lewis was no communist. But he was no democrat either. In this case, as throughout his career, he viewed his actions as a fitting reflection of the men he admired most: the corporate capitalists who ran the big coal companies. No corporation, he reasoned, would tolerate public attacks on its leadership from within—and the UMW needed to transform itself into a centralized operation as strong and efficient as the companies with which it was negotiating.

The irony, of course, was that the coal operators were neither particularly strong nor efficient, as Lewis well knew: He himself had derided coal companies as "incompetent, inefficient, backward, lazy and disunited." The similarity between the Stalin and Lewis approaches to dissent suggests the dictatorial mind-set that lurked beneath Lewis's empathetic rhetoric about the working man. In democracies and even corporations, leaders were answerable to their constituents. Yet Lewis had solidified his hold on the UMW even though, on his watch, the union lost three-fifths of its members during the 1920s. After Brophy's candidacy in 1926, Lewis never faced an election challenge again.

Instead he faced a secession revolt by the UMW's most solidly organized district, in Illinois, which in 1930 reinvented itself as the Reorganized United Mine Workers of America, or RUMW, under the leadership of old Lewis foes like Brophy and Alex Howat. Lewis ruthlessly suppressed that revolt during a year when RUMW speakers were beaten with brass knuckles and, in the words of one observer, "union man hunted union man." Beginning in 1928 Lewis was also challenged by a communist union, the National Miners' Union, which won the loyalty of mine workers in Ohio, Pennsylvania, West Virginia, and Kentucky who felt betrayed or abandoned by the UMW.

But ultimately no union could do anything about mine closures, which, in the words of two poetically inclined federal investigators, had left the countryside "dotted with industrial tombstones—burnt out slack piles, rotting tipples, here and there a smokestack standing alone in the midst of a pasture—to mark the graveyard of almost twenty thousand jobs."

By the end of the 1920s Lewis was the undisputed chief of a greatly weakened union in a greatly weakened industry. Even before the stock market crash of 1929, coal had been a depressed industry. To revive his organization, his course seemed clear: The nation would have to work its way out of the Depression. And Lewis would need to organize the South as well as the North. To Lewis's astute mind, the thought gradually dawned that the solution to both problems might lie neither within his own organization nor among the coal operators, but in Washington.

Just as Lewis's future success depended on organizing the South, Stonega's success depended on its ability to remain non-union. But was there any non-coercive, non-violent way for a coal company to resist unions? Christian charity, enlightened paternalism, "killing with kindness"—these strategies seemed incongruous in a fundamentally harsh industry such as coal mining (or perhaps any industry). On the other hand, military responses like the Rockefellers' Ludlow Massacre of 1913 in Colorado didn't work either. On the contrary, they intensified workers' anger and determination while rallying public opinion against the operators. By the 1920s John L. Lewis wasn't the only public figure attacking coal operators. The auto pioneer Henry Ford himself, hardly a friend of labor, remarked that even in times of prosperity the coal industry failed to meet its employees' welfare needs. "The conditions of coal mining," wrote Ford's newspaper, the *Dearborn Independent*, "are and always have been a disgrace to a civilized people." That Ford was a loose cannon and his newspaper an organ for racism and anti-Semitism did not mitigate Stonega's embarrassment when the *Independent*'s editorial was reprinted verbatim in the nearby *Crawford's Weekly*.

Stonega's sensitivity to this kind of criticism was reinforced in the early 1920s when Stonega's president, Daniel Wentz, became president of the fledgling National Coal Association, and his company, as a consequence, became the industry's spokesman. Since the company's creation in Virginia in the 1890s—and especially since the Ludlow massacre of 1913— Stonega had striven, wherever possible, to use carrots rather than sticks to influence and control its workers. Under Stonega's philosophy of "labor capitalism" (sometimes called "welfare capitalism"), all those Stonega company libraries, baseball fields, schools, churches, and soda fountains constructed before and during World War I were designed to keep workers happy, or at least quiescent. Every librarian, teacher, store clerk, and doctor in the company's towns was a Stonega employee, generating trust and goodwill for the company while simultaneously serving as the company's eyes and ears. In each Stonega company town only one person—the minister—was not a company employee. But even in church the company exerted controls: It owned the church building and collected the ostensibly voluntary contributions that paid the minister's salary, and ministerial assignments were usually cleared first with the local mine management.

Deep down, most people in and around Wise County knew, Henry Ford's point was inarguable. Even at a relatively enlightened coal company like Stonega, one-third of the residents of its largest colliery were still using surface privies in 1922 (a ratio, to be sure, that applied to most residents of the Southeast). The company was already reviled for imposing the yellow-dog contract that summer. Stonega needed some bold new strategy to improve its public image and cure its labor shortages—all while continuing, of course, to keep out the unions. Baseball fields and libraries alone would no longer suffice.

In a further effort to soften its image, the company began holding mandatory courses for its foremen in "giving the individual more attention, more implicit [sic] instructions and making the workman feel that his superiors have a real interest in his welfare." Instead of squelching miners' grievances, Stonega turned the monthly meetings of its mine safety committees into formal grievance channels. In theory, miners attended these meetings voluntarily; in practice, as one superintendent conceded years later, "We kinda insisted on 'em coming." These incremental changes provided the company with listening posts that enabled local mine superintendents to detect grievances early and act to prevent them from spreading. But something more was needed to dramatize the company's enlightened vision of its labor relations.

The answer, launched that same summer of the yellow-dog contract, was the construction of Stonega's ninth mining complex and coal town, a modern community like nothing the industry had previously seen. To the Wentzes and Leisenrings in Philadelphia, the happiest memories of their visits to Southwest Virginia were tied to their trips to the Kentucky Derby each May. And so their new model mining town got its name: Derby.

Like Stonega's other company towns, Derby was designed to reflect the hierarchy of work in the mines. The higher one's standing in the company, the larger the house and the better its location along the town's only road. As in virtually every coal town in America, Derby's houses were assigned in recognition of company status or as a reward for good performance. The most luxurious house, a two-story brick affair, belonged to the mine superintendent. The colliery doctor's house had six large rooms on a single story and was equipped with a hot-air furnace and indoor bathroom fixtures. A front porch, ten feet deep, extended across the front of the house. Six smaller replicas of this house stretched down the road for the assistant superintendent and certain lower-level straw bosses.

Twenty other double houses, each of two floors and ten rooms (that is, five rooms per family), were reserved for various "bridge" figures: lesser straw bosses, schoolteachers, store clerks, the local minister, and favored miners. The rest of the miners lived with their families in seventy two-story double houses of eight rooms, or four rooms per family—or, if the miners were single, in Derby's twenty-five-room boarding house.

But here the resemblance between Derby and other coal towns ended. At a time when 95 percent of all houses in American coal towns were built of wood, the Derby houses used hollow brick tile on their exteriors. All indoor walls were plastered and wired for electricity; every kitchen featured a sink with running water. The familiar outhouses disappeared, replaced in Derby by frost-proof water closets on the houses' rear porches, connected by sewer lines to septic tanks. Front porches with front lawns basked in the shade of 375 maple trees planted along the valley's only road. Even the lowliest miners' houses displayed a taste of elegance rarely found in mining communities.

Derby was different in another respect as well. In other coal towns, the superintendent and his managers lived apart from the workers, either on a rise overlooking the town or at one peaceful end of it. The "bridge people" usually had their own neighborhood, too, much like white-collar urban or suburban neighborhoods where professionals can mingle with neighbors of similar backgrounds and tastes. No such class segregation existed in Derby. Instead, the mine superintendent lived directly in the center of town. His assistants, the company doctor and the bridge people, instead of clustering among their own kind, were mostly scattered among the workers' houses. During the workday, the superintendent and the doctor similarly worked from accessible second-floor offices above the company store, the heart of the miners' community.

On paper this egalitarian layout had much to recommend it. It blurred the adversarial lines between managers and workers so that they tended to think of each other more as interdependent members of one big family. It promoted at least the appearance that management was available to its workers. It exposed hardened, uneducated miners to the softening influences of more cultured teachers and clerks. It also enabled managers and bridge people to intercept grievances and feuds before they could fester into violence or strikes.

To be sure, the benefits of this integration did not extend to Derby's black and foreign workers. At a time when racial segregation was required by Virginia law, black families occupied four-room houses in Derby's "colored camp," located directly underneath the tipple, where they suffered the constant noise and dust as coal was loaded and unloaded there. Black families' houses fronted directly on the road, so their porches opened to the side rather than to a front lawn. Their porches received no shade because their camp had no room for shade trees. They were barred from the Derby baseball team, although their own team occasionally played games against the whites. Their wages were kept lower by assignment to Derby's nearly worked-out No. 1 mine, where good coal was difficult to find.

The "furriners' camp" began at the bridge below the colored section. Here, too, the company neglected to build any ten-room houses. But the welfare of immigrant workers largely became a moot point after the U.S. Congress passed stiff immigration restrictions in 1924. In any case, Derby proved such an attractive community to native whites that the colliery never felt a pressing need for foreign workers.

As anyone who has lived at close quarters with neighbors in a city or a college dormitory can attest, the deliberate plunking together of bosses, clerks, and workers did not guarantee that they would develop friendships or even get along. It merely made such relationships possible. For the company's "Derby strategy" to work, other factors—such as a healthy economy and a healthy company—would be necessary.

From all surface indications, the Leisenring group was indeed growing and prospering. On Daniel Wentz's watch, the group's companies had reached the peak of their coal production (a level that they would not reach again until 1964). Colonel Wentz had spoken for the coal industry as a founder and leader of the National Coal Association. He and his family lived in palatial splendor in the northern Philadelphia suburb of Wyncote, and Wentz was often gallivanting off to Europe to see old friends from the war and reinforce his business ties. As if to signify its ascendance, in 1925 the group moved its headquarters into the newest and most prestigious office building in Philadelphia: the massive twenty-seven-story Fidelity Building on Broad Street, at the very heart of the city, just two blocks south of City Hall and three blocks south of the Pennsylvania Railroad's Broad Street Station, within easy access of coal men, executives, and politicians from New York, Washington, Pittsburgh, and throughout the Northeast. Here a staff of some one hundred employees occupied the Fidelity Building's entire twenty-fifth floor, just two floors below the penthouse and its Midday Club, which quickly became a leading lunch venue for local business executives.

Yet early in 1926, while Daniel Wentz was home preparing to go hunting for lions and big game on an African safari, he suddenly took ill and died. He was only fifty-three at the time; like Ned Leisenring and John Taggart before him, much had been expected of him. Now, like them, he was gone suddenly, almost a generation earlier than expected.

His death generated a chain of hurried promotions within the Leisenring group's domain. Otis Mouser, Wentz's longtime associate from Virginia, had been brought to the Philadelphia headquarters in 1923 as

Wentz's second in command. Now, after just three years in the home office, Mouser became head of all the Leisenring group's associated companies. Ralph Taggart, who had succeeded Mouser as the group's ranking Virginia official, now became a full vice president of the Virginia operation. At thirty-eight, after 16 years of working his way up through the ranks, he had grown into a serious, saturnine executive, with a domed forehead, a professorial manner, and a slow, stentorian speaking voice that endowed him with the air of an owner even though he wasn't one. No one questioned that he was eminently suited to take charge of the operation that his father had run thirty years earlier.

Ralph Knode, Daniel Wentz's gregarious assistant in France during the war, also rose within the Philadelphia inner circle. And young Ted Leisenring, the group's heir apparent by virtue of his father's 50 percent interest in the Leisenring-Wentz properties, now assumed executive status far earlier than anyone had expected.

After he got out of the Navy in 1919 Ted had been put to work in the engineering department; then in 1921 he had spent a year as superintendent of one of the family's collieries in the Pennsylvania anthracite region. But by 1922 he was back in the Philadelphia headquarters office. In the suburb of Ardmore, he and Margaret bought a large turn-of-the-century mansion with a garage, two working anthracite furnaces, a beautiful garden and one of the first private swimming pools on the Main Line. Here, attended by an Irish cook, four Irish maids, a chauffeur, and a gardener, they began to raise a family and Ted went through the paces of becoming a coal executive.

But despite his controlling share of the Leisenring and Wentz properties, by the time Daniel Wentz died in 1926 Ted held only one vice presidency, with the family's Hazle Brook Coal Co., in the Pennsylvania anthracite region. To the extent that he was noticed and admired among businessmen, it was for his remarkable athletic prowess, which if anything had expanded since he left Yale. Such was his physical coordination that he succeeded at virtually any sport he took up, and the list of these was endless: golf, tennis, court tennis, squash, fly fishing, fox hunting. Yet invariably he preferred individual sports over team sports—although he was a superb horseman, he never took up polo—and he similarly shied away from the nightclub scene that Ralph Knode so enjoyed. So there was reason to wonder how well suited he might be to the dynamics of running a business organization. Yet now, with Daniel Wentz's death and Otis Mouser's promotion, Ted became, at age thirty-one, vice president of most of the group's other companies. He also inherited Wentz's seat on the board of Westmoreland Coal.

Even in the traditionally hectic coal industry, it seemed the worst possible time for a change of management. Pittsburgh Coal Co. and other Midwest operators were in the process of abrogating their contract with the

United Mine Workers, undercutting the wage advantages that the Leisenrings' Stonega operations had counted on. And Stonega wasn't the only investment on the Leisenring group's plate: It also had the Wentz estate's significant chunk of Westmoreland Coal stock and a number of anthracite mines, a nineteenth-century legacy from John Leisenring Sr. and his son, Judge John ("The Boy Wonder") Leisenring.

But anthracite as a fuel for home heating had been largely replaced by oil, natural gas, or even bituminous coal. In a kind of industrial triage, shortly after Wentz's death Leisenring and Knode between them made their first major decision: They would liquidate the family's anthracite coal holdings, cutting their last link to the Mauch Chunk region where Ted's great-grandfather had launched the family's coal holdings a century earlier. In theory this would release money being sucked into the marginally profitable anthracite mines while it enabled the group's new managers to focus their attentions on Stonega and Westmoreland.

But events were spinning beyond their control. In 1928 one of Derby's mineshafts turned out to be nearly depleted, a surprise that instantly caused a 23 percent drop in production. At the same time, industrial demand for coal—and, consequently, the price of coal—collapsed throughout the nation. Although no one realized it then, this was a first omen of the Great Depression that lurked just around the corner. Smaller operators, desperate to protect their rapidly depreciating capital stock, offered coal on the market at almost any price, in some cases for as little as 10 cents a ton. The going price of coal dropped to one-third the level it had been when Derby opened in 1922.

In this environment, the utopian aspirations of a town like Derby became an unaffordable luxury. In 1924 and 1925 Derby's hourly workers had been making as much as 50 cents an hour. But that scale vanished by 1927. Stonega's managers, once so eager to communicate and consult with their workers, now simply announced the bad news on a take-it-or-leave-it basis, even in Derby. "When the company took a notion to cut your wages," one Derby motorman recalled years later, it never consulted the miners: "They'd post it up on the bulletin board: 'We regret to give you a reduction in wages.' "

In the midst of this crisis, two more unexpected deaths further depleted the Leisenring group's top management. In February of 1929, after witnessing a drop of 600,000 tons in his company's coal production the previous year, Westmoreland Coal's president, Pemberton Hutchinson, died suddenly of pneumonia at the age of 68. With the entire coal industry declining, Westmoreland's directors were hard put to find anyone eager to succeed him. Unlike Stonega C&C, which was still owned by a close family circle, Westmoreland's control was widely dispersed among Philadelphia bankers and merchants (Pemberton Hutchinson's father had been president of the Philadelphia Saving Fund Society in the late nineteenth century), and Westmoreland's managers were mostly salaried executives rather than owners who had a stake in the company's survival.

In just such a vacuum in the 1860s and '70s, Judge John Leisenring had picked up several small anthracite companies that no one else wanted. Now that process repeated itself: Westmoreland was leaderless and desperate, Ted Leisenring was already on its board, and the Leisenring group had a relatively stable management team in place. So rather than sell Westmoreland or shut it down, Westmoreland's board turned the company's management over to Stonega's officers, and Ted Leisenring became Westmoreland's president.

To Ted, who owed his job at Stonega solely to his inheritance, this appointment by people who were neither related to him nor beholden to him represented a confidence-building ratification of his merits. (Ted conveniently overlooked the fact that, since 1917, the estate of his late uncle John S. Wentz had owned the largest single block of Westmoreland stock.) The two separate companies—Westmoreland and Stonega—now operated under the same managements with very similar boards, in effect doubling the Leisenring group's coal properties (while recovering the acreage they had lost when they abandoned their anthracite mines in 1926). In this manner, some forty years after they were driven out by Henry Clay Frick and Andrew Carnegie, the Leisenrings quietly returned to the western Pennsylvania coal fields long after Frick and Carnegie themselves had departed. When Otis Mouser, the interim head of Stonega itself, also died in 1929, Ted Leisenring succeeded him as chairman of Stonega as well, and Ralph Taggart became Stonega's executive vice president.

Ted Leisenring was thirty-four years old, with less than ten years' business experience and a stock market about to crash. Only in retrospect does it appear that he may have been well suited to the moment. The stock market crash of October 1929 shattered many Americans' comfortable assumptions about the inevitability of prosperity and progress. Until then most Americans believed that the Federal Reserve System, created in 1913, had solved the vicious cycle of financial panics that had erupted every twenty years through the nineteenth century. This greatest of American depressions that began in 1930 was America's first in thirty-seven years— that is, the first within the memory of most working adults. The market value of stocks traded on the New York Stock Exchange dropped from a high of $90 billion in 1929 to less than $16 billion by 1932, and the market would not recover for another generation. By 1932 one out of every four Americans would be jobless. Even the Rockefeller family, which owned 38 percent of Consolidation Coal Co., was forced to take a complete loss on its investment and to retire from the coal business in 1932.

It was the sort of cultural crisis that sets generations against each other—the old (helplessly despairing at the passing of their world) against the middle-aged (fighting vigorously to restore the status quo) against the young (fighting just as vigorously to seize the opportunities presented by the new order). One of the men who seized this opportunity was Franklin Delano Roosevelt, who as the next president of the United States would radically expand the federal government's role in the nation's private economy. And in that change, John L. Lewis would find the opportunity to revive his deteriorating union.

Daniel Wentz, Otis Mouser, and Pemberton Hutchinson had been men of great experience in a world whose assumptions were no longer valid. Their untimely deaths spared them and their companies the frustration and expense of the hopeless railing against the New Deal that soon preoccupied the surviving captains of industry of their generation. Their successor Ted Leisenring, by contrast, had four generations of coal in his blood, as well as the counsel of older family retainers like Ralph Taggart (then forty-two) and Ralph Knode (then thirty-six), who were still young enough to adjust to radical change. In this bleak new world of diminished expectations, Ted Leisenring's very youth and inexperience were points in his favor.

Coal production had fallen sharply even before the stock market crashed. Now it dropped further still: by 42 percent nationwide and by 53 percent for Stonega C&C in the three years after 1929. In the process, Stonega was forced to close three of its eight remaining mines. As wage rates fell throughout the bituminous industry, Stonega struggled to stick to the pay scales that had last been reduced in 1927. Its reasons were less humanitarian than pragmatic: Recruiting mine workers was a major headache and expense even in good times, and Stonega hoped to avoid it by paying more than the going wage rate. Stonega's workers "thought we were much better off than the average mining field," the company's Virginia general manager, John D. Rogers, assured Ted Leisenring in December 1930.

Assuming that good times would return soon enough, as they usually did, throughout those three years Stonega fought desperately to maintain the entire work force at its five surviving mines. But the goals of steady wages and steady hours soon proved impossible. Wages throughout the bituminous industry had fallen by a quarter, and Stonega could not survive competitively without cutting its own wages. This it did in 1932 and again in 1933, by a total of 20 percent. But these were the last wage cuts made by any coal operator in Virginia, and also the smallest.

Maintaining a full-time work force in the midst of a Depression was no more realistic than maintaining high wage scales. "Several of the collieries have too many miners," the company's 1933 annual report noted. "In many instances places are doubled which should be worked by one miner only." Rather than lay anyone off, Stonega shut its mines for all but one or two days a week.

This combination of both reduced wages and reduced hours was of course devastating to Stonega's miners. Now long lines of miners formed outside the company's stores for scrip. But even the supply of these tokens—redeemable only at the company store—was limited, for the miners could draw only the equivalent of a month's advance pay. In 1932 the Derby superintendent, B. R. "Brownie" Polly, reported to Stonega's headquarters that many of his men had made less than $28 in a month's time—that is, less than $1 a day. "You can readily see," Polly wrote his superiors, "why I have so much O.K. scrip [on hand]."

Still, Derby was relatively lucky. In coal fields across the country as many as two-fifths of all miners were unemployed. Families evicted from company homes were camping in makeshift tents and abandoned beehive ovens. "Every aspect of their lives is ugly and anesthetic," the New York labor journalist Fannie Hurst wrote of her visits to coal towns. Unemployed miners roamed the countryside in search of work, only to return home to their familiar "patch" when they couldn't find it. "I've traveled three thousand miles in the past ten weeks trying to find a job," one miner told a reporter. "But it ain't no use. There ain't no job any place."

Those miners who found jobs were not much better off, for their employers had little cash to pay them. "We find we are starving even at our work," commented one, "as we can't get any food or money that we have sweated so hard to earn. . . . We have no doctor. We have no hospital for our sick, no graveyards for our dead. We have gotten nothing to eat at this job for ten days."

In southwestern Virginia, many of Stonega's smaller neighboring mines went bankrupt, whole towns disappeared, and thousands of Virginia miners were out of work altogether. By the end of 1932 the local relief chairman reported that in Wise County alone, 1,155 men were drawing relief "wages" of 20 cents an hour. To Fannie Hurst, the consequences of these conditions seemed clear: "Where human beings are living under conditions that generate hate," she wrote, "you can see that seeds of revolt are being sown."

But Stonega's managers, in their offices in Big Stone Gap and Philadelphia, contrived to draw one tiny ray of sunshine from this situation: Whatever the miners' condition, Stonega's workers didn't seem to blame the company. "The labor has been very well satisfied with working conditions and wages paid," a company report noted in 1932, "the only criticism being, of course, that it has been impossible to furnish continuous employment."

One obvious victim of Stonega's retrenchment was the company's whole program of "welfare capitalism," especially in the model town of Derby. That program's original purpose was to fend off the United Mine Workers, but in the depth of the Depression the UMW lacked the money, and apparently the interest, to organize Stonega. In the absence of a union threat, Stonega's social programs now withered away along with the miners' pay scales and working hours. The company's psychological training programs for new foremen were discontinued. House repairs grew infrequent. Movies at the local theater stopped. The coal-town baseball league folded for lack of company support. All that remained of Derby's amenities were the fine tile houses, the shade trees, and whatever congenial social patterns had managed to take root during Derby's first ten years.

Still, the miners rarely complained. They directed their anger against the national economy, not the company, which they continued to perceive as humanitarian, especially in contrast to rival operators.

But the company's withdrawal from its paternal role brought a subtle psychological change within its miners' families. Almost imperceptibly, they stopped relying on the company to solve their problems. Instead they looked to themselves, to other miners, and to sources of help outside the company. When husbands couldn't work, their wives found jobs cooking and cleaning. Sympathetic merchants nearby, unlike the company's stores, often extended credit to mine families even when they had little hope of repayment. Neighbors, including Stonega's mine officials, sometimes dipped into their own pockets to make small personal loans or provide occasional jobs.

The miners also turned to local charities—and it was here, for the first time, that they found the company actively aligned against them. Concerned about their undernourished children, a group of Derby residents resolved in 1930 and again in 1931 to raise funds to furnish a daily hot lunch at the local school. One of the first agencies they approached was the Virginia Tuberculosis Association, in part because the Derby superintendent's wife had long been an active fund-raiser for that group.

The local tubercular fund was administered by Dr. C. B. Bowyer, the physician-in-charge at Stonega's company hospital, and before approving the donation he consulted his superiors at the company. He found them embarrassed to have their employees perceived as objects of charity. Based on that perception, he wrote to Stonega's general manager, John Rogers, the Tuberculosis Association decided "that we would not appropriate any of this fund toward feeding the school children at the collieries of this Company, because we did not wish the publicity or a rumor to go out that the children of employees of the Stonega Coke & Coal Company were in such a destitute circumstance that they were being fed by funds created for the Virginia Tuberculosis Association." The next day Rogers apprised the company's vice president, Ralph Taggart in Philadelphia, of the situation: "We do not wish to deprive our poor children of a warm luncheon," Rogers explained in a memo, "but we're really afraid of the publicity."

Whether the miners understood that the company had actively obstructed their welfare in this case remains unclear. But they slowly came to perceive that the company had lost interest in their welfare. Struggling with the national economy and seemingly insulated from any union threat, the company had adjusted its priorities. But this adjustment had created unexpected side effects. Workers angry at faceless economic forces gradually began to focus upon a more tangible force—the company—as the villain in their midst. Families once dependent on the company's generosity were now of necessity becoming self-reliant. The docile model coal town of Derby had become ripe for the very thing Derby had been built to prevent: a labor union. Yet John L. Lewis and the United Mine Workers hadn't lifted a finger to create this condition. The economy, and the Stonega Coke & Coal Company, had done it for them.

Early in the Depression nearly one thousand non-union coal loaders, angered over a pay cut, struck the Consolidation Coal Company just across the state line in Harlan County, Kentucky—the same Harlan County where the United Mine Workers had quickly organized ten thousand miners in 1917. That first effort had soon collapsed, and so did this one: The striking miners received no assistance from the UMW and returned to work after a few days.

But in late 1931, a few remaining UMW members in Kentucky began signing up members and collecting dues. Then they notified the UMW's southern delegate, William Turnblazer, who responded by calling a mass meeting in Bell County, Kentucky, adjacent to Harlan County along the Virginia and Tennessee lines. More than two thousand people showed up, and before long the union had eleven thousand members in those Kentucky counties—enough to strike for official recognition as the miners' collective bargaining agent. But Turnblazer balked, fearing (probably correctly) that a strike would bankrupt the union.

"Everybody wanted action," one local organizer later recalled. "Nobody could wait much longer. And every time I'd go to Turnblazer and tell him so, he'd say, 'Now, Bill, wait just a little longer. This ain't no time for action, now.'"

Derby's workers seem to have taken little notice of this agitation just a few miles away from them. But coal operators in Virginia did. Sooner or later, they realized, declining wages would open the door to unions. In 1931 Virginia's operators gathered to discuss the wage problem, only to wind up dealing with another problem: the rampant price cutting among them. The solution promoted by the National Coal Association lay in so-called "regional marketing collectives," which was a euphemism for a price-fixing pool—the same device Henry Clay Frick had organized in the 1880s around Connellsville, Pennsylvania, with modest success. Virginia's first such collective, Appalachian Coals Inc., came to life at the end of 1931 with 147 member companies. Its most important member, of course, was Stonega C&C, which accounted for nearly three-quarters of Southwest Virginia's coal output.

In theory such combinations were illegal under various federal anti-trust laws enacted since 1890, and many observers presumed that the Appalachian Coals combine would be ruled unconstitutional. But at that point the U.S. Supreme Court was dominated by conservatives sympathetic to the plight of business in a depressed economy. When the government's anti-trust suit against Appalachian Coals finally reached the Court, the majority held in 1933 that this collective was not necessarily a "monopolistic menace" nor inimical to "fair competition." It suggested that the legality of Appalachian Coals could be answered definitively "only after thorough investigation of the economic effect of the decision."

Back in Wise County, Virginia, the reaction of the iconoclastic weekly newspaper editor Bruce Crawford was more succinct: "Appalachian Coals is a union of operators to get all they can for their coal," he wrote, "with no provision for miners to combine to get all they can for their labor."

In effect this union of coal operators to protect themselves had inadvertently revived the concept of a union to protect coal miners. But the United Mine Workers lacked funds to seize the moment. At this time it was hemorrhaging members in the face of the inroads made by rival unions alienated by John L. Lewis: John Brophy's Progressive Miners of America, Frank Keeney's West Virginia Coal Mine Workers' Union, and the communists' National Miners' Union, based just across from Virginia in Harlan County, Kentucky.

But Lewis had a new card to play. Franklin D. Roosevelt had been elected in November 1932 with a promise of economic relief but no specific plan for carrying it out. The new president would not take office until four months after the election. In this moment of uncertainty—with the public demanding action and competing interest groups promoting a cacophony of solutions—Lewis rushed into the vacuum. In congressional testimony and outside, he argued with his customary eloquence that the key to America's economic recovery was the revival of mass purchasing power—and the key to boosting purchasing power, he claimed, lay in the protection of workers' rights to join labor unions.

When Roosevelt took office on March 4, 1933, his administration immediately unveiled a decisive plan of action. FDR's National Industrial

Recovery Act (NIRA) was a recipe for more than five hundred mandatory business and labor provisions designed to stimulate industry and reduce unemployment. Its provisions included the specific amendment Lewis had sought, granting workers the right to organize themselves into unions.

Whether this amendment would survive Congress was still unclear in March of 1933; so were Roosevelt's feelings about it. But most industrialists saw only the terrifying prospect of a new administration that seemed determined to dismantle their prerogatives. Lewis, seizing the widely perceived belief that the government and FDR were on labor's side, gambled what remained of the UMW's treasury on a vast nationwide organizing campaign. Only by joining a union, workers would be told, could labor secure benefits under the NIRA. Lewis needed to get his agents into the field before workers and employers (not to mention his own organizers) realized otherwise.

Beginning on June 1, UMW organizers fanned out across the coal states, bursting with fresh confidence even though the recovery act would not be signed into law by Roosevelt for fifteen days. They went first to their most familiar coal fields, defying company guards and local sheriffs by invoking the authority of Roosevelt and the federal government. In effect their message to miners was, "The president wants you to join the union." This alleged interjection of federal authority into labor relations was a brilliant tactic: It sowed doubt and confusion among coal operators and prevented them from retaliating with their customary righteous muscularity.

The result was a stunning success for the union. In dozens of coal camps and towns through West Virginia, Pennsylvania, Tennessee, and Kentucky, miners organized themselves even before the organizers arrived, so the agents needed only to pass out membership cards by the hundreds. Like the French army a few years later in 1940, coal operators were completely caught off guard by this blitzkrieg. "The men flocked into the union so fast," reported an organizer in West Virginia, "it took away their [the companies'] breath." Within a few months of passage of the NIRA, membership in the United Mine Workers had quadrupled.

The UMW's agents didn't reach the previously untrod territory of Wise County, Virginia, until July 9. "For the first time in the history of the coalfields of Virginia," reported the *Coalfield Progress* in nearby Norton, "representatives of the United Mine Workers of America have held open meetings during the past week and openly solicited membership from among the ranks of Virginia miners."

In fact, a cadre of a half-dozen local miners had already been laying the groundwork on their own for the previous six months—"doin' what we knowed," as miner Joe Blanton recalled many years later. Within Derby's chummy confines, their activities—mostly casual conversations to promote unionization—invariably got back to the Stonega company's informants and to Brownie Polly, the superintendent. But the company didn't perceive these men as an appreciable threat. For all their efforts to open channels of communication, Stonega officials still mistakenly believed that their workers were grateful and loyal.

When the union claimed to have signed up four hundred members at its very first meeting, in Esserville, less than ten miles from Derby, even neutral observers scoffed. "Four hundred men couldn't be congregated at Esserville, even to see a hanging," remarked the weekly *Coalfield Progress.* But a second meeting the next day at Appalachia, just three miles from Derby, attracted four hundred men despite Stonega's efforts to prevent them from attending. On the day of the meeting, Derby's only road was barricaded below the camp. Miner Joe Blanton, moonlighting as a police officer, recalled years later that "they gave us orders not to let nobody out of the holler that day that that meeting was down there." But Stonega's officials themselves showed up at the meeting—to identify miners who joined the union and also to rebut the organizer's claims. Despite these efforts, between 150 and 200 men took the UMW oath that day.

Suddenly the scales fell from the company's eyes. Carrots were out; sticks were in. In the ensuing months virtually all of Stonega's union supporters were fired—twenty-two men at Derby alone. To be sure, the company took pains to find other excuses for these dismissals. "If they caught you talking union in Derby mines," Joe Blanton later recalled, "or associating with the union, or if they suspicioned you of doing that, they wouldn't haul off and discharge you just for union activity. They would try and catch you asleep, or catch you breaking some kinda rule, or put a rule so if you walked by it you had to break it, so they'd catch you, you see."

One Stonega miner, fired that August from the adjoining colliery at Arno, told his story to the local *Crawford's Weekly:*

> Saturday, July 15 C. W. Rothberry, my boss, and Charles Mylan, company sheriff, said to me they had been told I had been talking the union around the mines. I told them I had not done it. They asked me if I had joined, and I told them I had. They asked me why, and I said the union is like a church, I had heard. One joined the church to get a crown in heaven, and one joined the union to get a crown in life. They told me I could go to church, but I couldn't talk union around the company. Then August 10 they told me I had too much dirt in my coal, and told me to get my tools and get out of the mines. But the tipper sheet didn't show dirt in my coal.

Most of the fired workers were served immediately with eviction notices. In theory, tenants should have been able to remain long enough to appeal these notices in court. But as Stonega's general counsel, J. L. Camblos, pointed out in a memo, that would mean "we would have these hostile persons in the midst of our colliery villages for several months." So instead of involving the county sheriff, Stonega used private company guards to handle the evictions.

To the company's credit, its managers recognized that coercion alone wouldn't stop the union movement. Working in tandem with the local operators' association, Stonega officials now secured a uniform wage scale for all Virginia miners. That meant a 15 percent pay raise for the Derby miners, their first raise in nearly a decade. At a time when the UMW had yet to deliver anything but promises, this tangible benefit helped sow uncertainty among many mine families who still wanted to believe that the company had their best interests at heart.

Saylor Givens, who had left his father's farm in 1909 at age fourteen to work in Stonega's mines, was by now thirty-nine years old, a twenty-four-year veteran of the company with a wife and ten children. With a family and seniority came upgrading from a tiny two-room house to a more spacious four-room home at the Imboden mine. Everything considered, it wasn't a bad deal. Saylor's one-armed brother-in-law, Sam Warden, felt much the same way: Although he had lost an arm beneath a coal cart some twenty years earlier, he continued to work and was soon considered as capable in the mines as any mule driver with two arms. The Givens men couldn't help noticing that their few relatives who did not work for Stonega seemed far less fortunate: Saylor Givens's wife's brother-in-law Nip Barnett, desperate for work, had abandoned his family to work the mines in West Virginia; his wife and six children wouldn't see him again until years later, when the children were grown. Stonega, by contrast, looked out for its people and wasn't about to move.

"You didn't know what to do for the best," said the widow of a man who resisted joining the union for more than a year. "The company was *so* against the unions then. And the superintendent, of course—Brownie Polly was superintendent then, and he was really against the union."

These maneuverings took place, of course, in the shadow of the Roosevelt administration's recently passed National Industrial Recovery Act. Among other provisions, the act required each industry (through its trade association) to design a "code of fair competition" for presidential approval. In the absence of an acceptable code, the NIRA permitted the president to impose a code of his own. Many coal operators across the country, while terrified by the prospect of an imposed code, nevertheless yearned for some set of standards that would stabilize their industry and perhaps even provide some government support. This was especially true for the unionized northern operators suffering from price and wage competition with non-union southern mines like Stonega's.

In the predictable spirit of an industry that had never managed to speak with a single voice, regional coal associations presented twenty-seven different draft proposals for coal codes. The Appalachian operators' draft code stubbornly asserted the right of employers to bargain individually with their employees and to refuse to negotiate with anyone not on their payroll. It was the last anti-union gasp of a generation of managers who refused to accept Washington's new reality: No code could win the President's approval without endorsing the NIRA's Section 7a, which stipulated the right of all workers to organize and bargain collectively "through representatives of their own choosing."

When the coal code hearings began in Washington on August 9, the Appalachian operators were persuaded—in the euphemism of the newly created National Recovery Administration—to hold "voluntary" code negotiations with union representatives. The Appalachian operators agreed to meet the unions on August 22. But union recognition so repelled these operators that they continued to cling to the notion that they could somehow finesse or reinterpret Section 7a in their final code agreement. In this belief they were encouraged by the refusal of Roosevelt or his National Relief administrator, Hugh S. Johnson, to speak out forcefully on the matter one way or the other. Eventually the operators tried to interpret Section 7a so as to allow them to bargain collectively only with their own employees rather than with an independent union. That would have locked out the United Mine Workers, of course, and the union negotiators refused.

As the deadlocked negotiations dragged on for another month, miners across the country grew impatient. In his Indianapolis headquarters, John L. Lewis, convinced that an agreement was within his grasp, desperately counseled against any strikes, which he feared would antagonize the operators and stiffen their resistance to any code agreement. Daily he worked the long-distance phone lines, pleading with his field agents to keep the lid on the restless miners. But the Virginia miners understood from their own experience the depths of the operators' resistance to union recognition. Almost as soon as the negotiations broke down in Washington, the miners took matters into their own hands and went on strike.

The walkout started to the south of Derby, in Lee County, just across the state line from Harlan County, Kentucky. The UMW immediately designated the action a "wildcat"—that is, unauthorized—strike and refused to provide any benefits to participating strikers. Nevertheless, the strike soon spread north to Derby. "We . . . thought that maybe if we didn't get enough backing, we'd take care of it ourselves," the Derby miner Joe Blanton later recalled. More than forty years later, a Derby miner named Fred Sloan recalled this first formal workers' confrontation with the Stonega Coke & Coal Co.: "We went up here and we asked Brownie [Polly, the mine superintendent] to recognize us as the United Mine Workers, and the [union] president tacked up a notice on the driftmouth. We started inside and he [Polly] tore it down. He jerked it off, and about half of us came back, half of us went on."

Without union support, the miners nevertheless managed to sustain the strike with donations and credit from sympathetic merchants and farmers. Even when the union's own officials circulated a telegram from John L. Lewis urging the strikers to return to work, most refused to comply. But the miners also bent over backward to avoid confrontation. "They [the union] called us back to work several times," Blanton later recalled, "and we wouldn't do it. And we would put it off a day or two before we would tell 'em. Everything we done we had to play it safe."

In their effort to shut down Derby's mines, the strikers picketed the mines, the bathhouse, and the town's public buildings. When the mines continued to operate, the strikers adopted stronger tactics. After one meeting in Appalachia, some one hundred striking miners and sympathizers marched to Derby, where they confronted a roadblock set up by armed company guards. According to one striker, violence was averted only when state police persuaded Stonega's guards to remove the roadblock.

When the strikers weren't using the roads in and out of Derby themselves, they littered them with tacks and roofing nails, effectively preventing all motor traffic from reaching the mines. Even strikers' wives got involved by shouting insults at men who went to work or by pressuring their wives when their husbands were working in the mines. On one occasion about fifteen women tried to join the picketing around the Derby bathhouse. Two company guards stopped them, in the process tearing a woman's sleeve. When she swore out a complaint, a sympathetic local justice of the peace fined each guard $10.

This incident—like the state policemen's help in dismantling the company roadblock, the support of merchants and farmers, the sympathetic accounts in at least some newspapers, and the alacrity with which the strikers sabotaged the roads—suggests that Stonega Coke & Coal in 1933 was as much victim as victimizer (unlike, say, Rockefeller's Colorado Fuel & Iron Company at Ludlow in 1913). But in other cases that fall the authorities were less sympathetic. In September forty-five state policemen were dispatched to Wise County, ostensibly to preserve order but seemingly to harass the strikers into submission. "If we went down where we could come out of that holler, happened to walk out, we had to go through a line of state policemen, and they'd search you," Blanton later recalled. "But as long as you walked into that holler with a dinner bucket, you could have a high-powered rifle on your shoulder and they wouldn't even look at you. They'd let you go on—if they thought you was goin' to work."

Throughout the strike Stonega continued to operate its larger mines, including Derby's, though only on a 50 percent production basis. By drawing on the county's large pool of unemployed men as strikebreakers, and by preventing pickets from harassing men going to work, the company hoped to convince miners that it could stand a long strike better than they could. The key to settling the strike, the company perceived, lay with the large number of uninvolved miners who were reluctant to take sides. Some of these men would honor picket lines, but then go work when no picketers were around to confront them. Some would honor the strike but wouldn't join the union. Others joined the union but avoided union meetings or activities. "A large percentage of employees now belonging to the union joined to keep down dissension, discussion and possible harm to themselves," the company's annual report hopefully concluded. "Probably not more than twenty per cent of the members are paying dues to the locals to which they belong."

Both the company and the union believed the strike would end as soon as agreement on a coal code was announced from Washington. And a settlement seemed imminent, if only because Roosevelt was anxious to avoid an industrial war. On September 7, General Johnson tried to move things forward by offering a model coal code of his own. When a week passed and still no code emerged, Roosevelt threatened to impose his own if the operators and union negotiators couldn't reach an agreement within twenty-four hours. To show he meant business, the president cancelled a yachting expedition to await the result.

But even before this pressure was applied, the coal operators had decided to streamline their confused negotiation by reducing their Washington negotiating team to just four delegates. Two of these so-called "four horsemen" represented the non-union South. But one of them—James D. Francis, president of the Appalachian Coals price-fixing cooperative—had already seen his Island Creek Coal Company organized almost completely by the UMW's 1933 campaign, so the NIRA's collective bargaining provision was a moot point to him. That left the non-union operators' defense in the hands of the other southern delegate, Stonega's vice president Ralph E. Taggart. Now forty-six, having worked his way up Stonega's corporate hierarchy for thirty years, Taggart found himself thrust onto the national stage for the first time.

Taggart had been no friend to unions. Like his father before him, he believed in the company as a surrogate family. But Taggart's beliefs (and presumably those of his boss, Ted Leisenring) were not set in stone. The tenacity of the strikers in a model community like Derby, of all places, had convinced him that the unionization of Stonega's work force was inevitable. In that case, Taggart reasoned, managers would be better off recognizing unions and accommodating themselves to the new environment sooner rather than later.

For all the fears expressed by his fellow coal operators, Taggart had gradually come around to the heretical notion that a large company like Stonega could function just as well with a unionized work force, especially if most of its competitors operated on the same basis. On September 16 he and Francis signed their acceptance of Section 7a. And when Taggart backed away from his opposition, other operators were obliged to follow. If a southern company the size of Stonega could recognize the union, so could the entire industry.

Within days President Roosevelt approved the code, and southern operators, led by Taggart, signed the first Appalachian union contract on September 21. This agreement covered 340,000 miners, the largest number of workers covered by a single labor agreement in any industry in American history. It gave coal miners—generally the lowest-paid of all industrial workers in the United States—a far higher basic daily wage of $4.20, almost on a par with oil and gas workers. It also gave miners the right to join unions of their choosing and to hire a checkweighman to confirm the weight of the coal they dug. And it offered the prospect of a newly stabilized and modernized coal industry. But it failed to bring labor peace.

The new contract meant there could be no more neutral miners in Derby or anywhere else. Miners had to join the union or not, and in Derby only about half chose to join the union that had failed so notably to support their effort to organize. John L. Lewis, for his part, was so embarrassed by his inability to control the strike at Derby that in negotiating the contract with Taggart he made several concessions. Among other things, the contract failed to require operators like Stonega to rehire anyone who had been fired for union activity during that strike summer. Over this provision Stonega's miners voted to remain out of work.

This time the strike turned violent. In one camp near Derby, a company guard shot and killed a miner in an argument over the Coal Code. At another mine, a worker was kidnapped and forced to take the union oath at gunpoint. When a rumor spread that the company was importing trainloads of strikebreakers, hundreds of miners picked up rifles, shotguns, and pistols and mobilized near the town of Appalachia with the intent to stop all rail traffic. Only after Stonega officials agreed to renegotiate the discharged miners' reinstatement did the strike end, after a total of six weeks.

For better or worse, the rules of the game had been permanently changed. Miners now enjoyed union membership and a higher wage (which the company had previously insisted it couldn't afford), but their union seemed less attuned to their needs than the company had been. The company had a stable work force and a wage scale identical to its competitors, but it no longer enjoyed its workers' gratitude because every other union mine paid the same wage scale as well. The cost of labor had always dominated the coal industry, but now every issue was filtered through a union-management prism.

Less than a year after the Appalachian coal contract was signed, when seventeen Stonega miners were killed in an explosion of gas and dust, both parties seized on the tragedy as a way of scoring points against the other. The company falsely blamed the explosion on a union miner's cigarette. The UMW local won twenty-three new members by implying that a wrathful God had punished the company with seventeen deaths for refusing to rehire seventeen union members after the 1933 strike (even though the actual number not reinstated after the strike was twenty-one).

In later years the credit for unionizing the coal industry was generally given to Franklin D. Roosevelt (who had remained silent virtually throughout the strike) and to John L. Lewis (who had turned to Roosevelt rather than his miners). The strategic role played by Taggart and by the Derby miners went largely forgotten. Even Derby's miners themselves accepted the conventional interpretation. The first president of the UMW's Derby Local named his first son Franklin, after the president. "Abraham Lincoln freed the colored man and Henry Ford freed the mules," he declared in an emotional letter to a local paper, "and Franklin D. Roosevelt is trying to free us all." In effect the miners had transferred their loyalty from a paternal company to a paternal government, and no one could blame them. For ultimately it was only the government's intervention that had revived both the coal industry and John L. Lewis's moribund union.

CHAPTER 12

Be Careful What You Wish For

According to the theory embraced by John L. Lewis and Ralph Taggart alike, coal companies at last enjoyed a level playing field and uniform wages throughout most of the industry. Instead of undercutting each other in destructive price wars, now they could compete by mechanizing their mines to boost production. Mechanical loaders and other machines would offset labor costs even while individual miners' wages were rising. In 1910 a miner armed with pick, shovel, and black powder explosives could dig about three tons of coal on an average day; now, thanks to this modern equipment, he could produce fifteen tons or more. A partnership like the Leisenring group could economize and weather the Depression by merging the managements and sales forces of Stonega and Westmoreland Coal. John L. Lewis would guarantee coal operators labor peace from one union contract to the next. The only losers would be those pesky little wagon mines that lacked the capital to mechanize. Efficiency would triumph at last over inefficiency. That was the theory.

In practice, the coal operators weren't the only ones who needed to adjust to a union environment. America's miners had won the right to join a union, and within a decade more than 90 percent of all miners had joined the United Mine Workers. But they had not won any voice in that union. On the contrary, throughout the 1930s Lewis systematically transformed the UMW into a monolithic bureaucracy under his absolute control. He revoked the charters of rebellious districts and locals and replaced his opponents with his own loyal subordinates. He centralized the bargaining process into his own hands. Contracts that he negotiated were not even submitted to rank-and-file UMW members for their approval. Miners who were unhappy for any reason had only one weapon at their disposal: the unauthorized "wildcat" strike, in which workers protested grievances—against management or their union—by refusing to work, in violation of their contract.

In January of 1934, less than four months after Stonega's first UMW contract went into effect, workers at Stonega's mine in Roda walked off the job, ostensibly for medical reasons. From Stonega's headquarters in Big Stone Gap, the company's general manager, John D. Rogers, duly consulted the UMW contract and found that, in the absence of a doctor's certificate that they were too sick to work, the strikers could be legally discharged without any recourse. But he was reluctant to take that step.

"We realize that this so-called strike was sanctioned by the officers of the Roda local through ignorance," he wrote to the Roda superintendent, "therefore we do not wish to penalize the men for the ignorance of their leaders. Under these circumstances we will waive the discharge penalty, provided these men report to work tonight."

But the wildcat strikes continued off and on, often driven by animosity between Stonega's union and non-union members. Because only about half of Stonega's work force belonged to the UMW, Stonega's executives still cherished the hope that Lewis's dictatorial tactics might drive the miners back into the company's arms. At that point they could decertify Lewis's union, leaving the company the flexibility to outmaneuver its unionized rivals. Thus the company took care to screen its new hires for anything that smacked of radical union sentiments. Union members constantly complained that non-union men were given the best jobs and the most overtime work.

They also complained about lack of safety in the mines. New drilling and cutting machines were being introduced into American mines faster than most laws and labor contracts could keep up with them, and operators under pressure to make maximum use of expensive new equipment could be less than scrupulous about mine safety. On the other hand, Stonega's miners also complained about the union contract's compulsory safety rules, which they saw as a tool the company used to punish union men.

One recurring issue concerned smoking in the mines. In the volatile and unventilated depths of a coal mine, any spark, even from an unlit match, could set off an explosion. For this reason, the UMW contract and Stonega's company rules alike prohibited miners from carrying matches into a mine. But miners tended to be fatalistic about the risks of their job. They smoked for the same reason most everyone else did: because tobacco delivered a sublime (not to mention addictive) diversion from a grueling workday. Thus some miners smoked routinely underground, regardless of the rules or the consequences. To them, the burden was on the company to provide better ventilation through powerful fans placed at the mouth of each mine.

"You have no idea of the contention we have had," Rogers wrote in April to his boss, Ralph Taggart in Philadelphia, "not only regarding our No. 3 mine conditions [at Derby], but in regard to all conditions in the [union] agreement. . . . I might also add that Bill Minton [of the UMW] was very arbitrary, which, of course, was to be expected."

In June Rogers told Taggart of a new union problem: "I have just been informed that Mr. Martin Atkins, President of Roda Local and also check-weighman, absconded about two or three days ago with all the Local's money. They have not as yet completely checked the situation, but I am informed that the amount is well over a hundred dollars." Rogers managed to find a silver lining to this cloud: "Atkins," he explained, "has been a most undesirable tenant and checkweighman, and has caused us a lot of trouble. The man who will take his place will certainly be an improvement."

From Stonega's sister company, Westmoreland Coal in Irwin, Pennsylvania, Rogers's fellow vice president Levi Good wrote to assure Rogers that he wasn't alone: "I trust you are getting along with the Union better than we are. However, they tell us that after a year or two, in case we are able to live that long, the men will become accustomed to working under union regulations, and things will work out much more satisfactorily. Let us, therefore, live in hopes that there is a better day coming."

That day was still far in the future at 7 A.M. on the morning of Monday, August 6, 1934, when a crew of ninety-five workers entered Stonega's No. 3 mine at Derby. The mine had been idle since 7:30 Sunday morning—almost twenty-four hours earlier—because of inadequate ventilation due to a malfunctioning blower. One man preceded the others into the mine and restarted the two electrical fans that had been shut off on Sunday. The rest of the men assembled in what was known as the "safety room," just inside the mine entry, where the foreman routinely gave them their assignments. Four men left after being told to go home and report for the next shift or to retrieve their tools outside the mine. The remaining ninety-one men descended into the mine.

At 7:20, barely a few minutes after they arrived, two members of a nine-man crew were drilling a hole in the mine's "Main West" section when something combustible ignited the electrical equipment. What followed in that moment was a gas and dust explosion of such force that two of the men who had been told to go home, and had just reached the mouth of the mine on their way out, were picked up and carried about fifty to seventy-five feet by the impact.

At the moment of the blast, Derby's superintendent, Brownie Polly, was seated at a window in the mine office over the company store, about eight hundred feet from the mine entrance. He heard the noise and asked what had happened. When told of the explosion, he phoned the operator of the county-wide private telephone system and instructed her to make four calls for him: one to John Rogers, Stonega's general manager in Big Stone Gap; one to the company's safety engineer, also in Big Stone Gap; one to Dr. C. B. Bowyer, the company's chief surgeon, at the company town of Stonega; and one to Joseph F. Davies, the U.S. Bureau of Mines agent in Norton, about ten miles away. The operator reached Davies immediately, and Polly urged him to send help. At the government's Mine Rescue Station Davies rounded up two men, one car, and one truck, and together the three agents headed to Derby.

After making his calls, Polly went to an outcropping (or "crop") opening on a side of the mine that had not been affected. Here, to counter the carbon monoxide fumes from the explosion, he placed two electric blowers, capable of providing eight thousand cubic feet of air per minute. Then he entered the opening and helped gather men from the mine's undamaged section and led their escape through the crop opening. At least one miner Polly personally pulled out had nearly drowned in an accumulation of floodwater that reached to within a few inches of the roof of his mine section. One miner was quickly removed and given artificial respiration for an hour and a half, even though the company doctors had already pronounced him dead. Seven more dead were carried out as well.

By the next morning it was clear that the toll was much higher: seventeen men were dead, three injured. Within twenty-four hours an investigating party headed by the state mine inspector found, at the point of the blast, its apparent cause: the remains of a burnt match, as well as the remains of a partly burned cigarette, about three-quarters of an inch long. The most damning evidence of widespread smoking in the mine turned up in the cigarettes and cardboard matchboxes found on the bodies of many of the dead victims themselves. "Apparently no careful search was made for these before the men entered the mine," wrote Davies, the federal agent, in his official report, "as there was no apparent effort on the part of the men to hide them."

This disaster turned out to be the wake-up call the company needed to embark on an extensive modernization program. Over the next ten years almost thirteen thousand American miners would die on the job and 639,000 would suffer disabling accidents, but Stonega would not suffer another fatal explosion for nearly forty years.

The company's union problems—which were human rather than structural—were less easily solved. Less than a month after the Derby explosion a miners' committee complained to the government's Bituminous Coal Labor Board that "non-union men are getting all of the extra work" in the cleanup that followed the blast. Under Labor Board procedures, the company was required to draft a response to the complaint and also to hold meetings with the men. "Every day brings up something which cannot be overlooked," Rogers commiserated to Taggart, "and it requires time and attention."

John L. Lewis, meanwhile, had turned *his* attention to the national political stage. In May of 1935, the Supreme Court struck down Roosevelt's National Industrial Recovery Act—and with it, of course, the NIRA's Bituminous Coal Code. But by that time Lewis, with his customary prescience, had perceived the need to lobby Congress for other forms of legislation to stabilize the coal industry. He further perceived that the same whirlwind organizing tactics that had served him so well in 1933 had also attracted more than two million new union recruits in mass-production industries such as automobiles, steel, rubber, and textiles. In barely a year the labor movement's membership had mushroomed from less than three million to nearly five million. Here, Lewis saw, was a golden opportunity to move labor from the fringes to the center of American life—an opportunity that the aging leaders of the American Federation of Labor hesitated to seize.

It was also an opportunity to leverage the influence of the United Mine Workers at the very moment that coal itself was yielding the center of America's energy stage. In an age of cheap and convenient oil, gas, and eventually electric heat, homeowners were unlikely ever again to store coal in their basements and stoke furnaces each morning. The home-heating market was lost to the coal industry forever, and so was another major customer: the railroads. In 1897 a German engineer named Rudolf Diesel had invented an internal-combustion engine that was sufficiently rugged to power ships and industrial plants, using cheap low-grade oils that cost less than regular gasoline. When Diesel engines were first introduced to railroad locomotives in 1934, they proved so much faster and more efficient than coal-burning engines that more than one million horsepower's worth of Diesel engines were produced in the United States in the following year. By 1937 that figure had doubled; and by 1945 the coal-burning locomotive, which once consumed one-quarter of all the coal produced in the United States, would be all but obsolete.

For a century, every outmoded and discarded use for coal had been succeeded by some new use. That was still true: The rapid electrification of rural America being promoted by Roosevelt's New Deal opened up a whole new market for bituminous steam coal to power electric plants. But coal no longer held a monopoly on this or any other market. Electric utilities enjoyed other alternatives to coal power, such as oil and hydroelectric power (and eventually, of course, solar, wind, and nuclear power).

Lewis, to his credit, accepted the prospect that coal might no longer be king. Instead he sought new ways to lead rather than follow in the new industrial world he envisioned. In 1934 Lewis moved the UMW's headquarters from Indianapolis to Washington. Between 1935 and 1937, he pulled his union out of the AF of L and invested the UMW's staff and funds in a new labor movement, the Congress of Industrial Organizations. Under the CIO's nurturing—which is to say Lewis's nurturing—the fledgling United Automobile Workers and United Steel Workers eventually grew into unions even more powerful than Lewis's own UMW.

Lewis was no longer solely the leader of America's largest labor union; now he was the father of a whole confederation of large unions that depended heavily on his unique personal presence on picket lines, in negotiations with General Motors and U.S. Steel, in Congress, on the campaign trail for Roosevelt, and in newspaper headlines. At every turn in the CIO's subsequent organizing and negotiating successes, the journalist James Wechsler later observed, "it was his name that workers scrawled on the walls of corporate tyrannies" and "his poetry [that] lifted the CIO from the prosaic terms of another union drive to the level of a great crusade."

Lewis now had even less time and patience for the day-to-day grievances of miners, many of whom failed to appreciate the benefits of his broad new political alliances. Yet even as he expanded his operating field he refused to delegate authority within the UMW. The result was that once a coal contract was negotiated, Lewis often sided with the operators in seeing that it was enforced.

In April of 1937, for example, unionized coal operators signed a two-year labor contract with the UMW. Four days after the contract went into effect, workers at Derby and three other Stonega mines went on strike to demand closed shops—that is, a requirement that the mines hire only UMW members. Stonega responded by turning to two forces it had reason to believe were friendly. The first was the Virginia State Police, from whom Stonega sought protection "to protect life and property and make it possible for those to work who desired to do so." The other force was John L. Lewis in Washington. Stonega's president, Ralph Knode in Philadelphia, fired off a telegram to Lewis, noting that the strikes at the four Stonega mines violated the union's contract:

> OUR CONTRACT CUSTOMERS AND OUR COMPANY ARE BEING SUB-
> JECTED TO SERIOUS LOSS ON ACCOUNT OF BREACH OF CONTRACT
> ON THE PART OF YOUR UNION STOP NO CONTROVERSY BETWEEN
> COMPANY AND UNION INVOLVED STOP WHAT DIRECT ACTION
> WILL UNION OFFICIALS TAKE IN THIS EMERGENCY?

Lewis did not equivocate. Two days later he wired the four striking locals:

> I AM ADVISED BY YOUR DISTRICT OFFICERS THAT YOUR LOCAL
> UNION IS ON STRIKE IN VIOLATION OF THE PROVISIONS OF YOUR
> CONTRACT STOP YOU WILL PLEASE CONVENE YOUR LOCAL UNION

AND RESCIND YOUR ACTION IN CALLING THIS ILLEGAL STRIKE STOP PLACE THE MINE OVER WHICH YOUR UNION HAS JURISDICTION IN OPERATION AND THEN IN AN ORDERLY MANNER AS SET FORTH IN YOUR CONTRACT PROCEED TO TAKE UP ANY GRIEVANCES YOU MAY HAVE WITH THE COMPANY STOP PLEASE ADVISE BY WESTERN UNION THIS OFFICE IMMEDIATELY OF THE ACTION TAKEN BY YOUR LOCAL UNION.

Thus chastised by their international president, the strikers returned to work, each striker paying the $1-per-day fine stipulated in the contract. This intervention on Lewis's part was precisely the sort of stabilizing effect that Ralph Taggart had anticipated in 1933 when he signed the Appalachian Coal Code. Taggart, ironically, was no longer around to reap the benefits of his gamble: In 1935 he had left the Leisenring group to become president of the Reading Railroad's Philadelphia & Reading Coal & Iron Corporation, the largest of Pennsylvania's anthracite coal producers. So it was left to Ted Leisenring, the group's chairman, to express his gratitude at the result.

"I want to congratulate you on the fine way in which you and the rest of our official family handled the very serious strike situation which we have just gone through," he wrote to Rogers. "Needless to say, we are all delighted." This letter reflects one subtle change wrought by unionization: Stonega's once-paternal executives, feeling themselves besieged by the greater forces of labor and government, were beginning to apply the term "family" not to their employees but to their own management circle.

Yet the old notion of miners as family persisted among Stonega's managers. In 1938 Ted Leisenring's cousin Daniel Wentz Jr., a Stonega vice president based in Philadelphia, chastised a Stonega executive in Big Stone Gap for failing to notify Wentz of a miner's death. "You can readily understand why I am so interested in being notified," he wrote, "for as you know, my father worked with these old employees a great many years and I know they will think it strange if they do not receive some remembrance from me at the time of the funeral." When the Depression took its inevitable toll on the Warden and Givens families, Stonega's managers in 1938 found a job for James Warden as a night watchman, even though he was seventy-six by then and long since finished as a miner. James's son Sam, who had lost his arm in a Stonega mine as a teenager in 1910, continued to work for the company and live in his rent-free company house, the company's compensation to him for the accident. When the UMW's national contract expired in April 1939 and a strike (over the closed-shop issue) shut down the entire coal industry for weeks, Stonega's general manager John Rogers notified his mine superintendents to extend credit to strikers, "to eliminate suffering in the families of our employees, and to help tide over this suspension period those living at our collieries needing credit at our commissaries."

By that April—six years after Roosevelt's inauguration and John L. Lewis's organizing triumph—American workers and businesses alike were still struggling to reverse the nation's economic decline. Both Roosevelt, who had triumphed despite his polio, and Lewis, who had triumphed over poverty, had conveyed a refreshing confidence that things could improve. But men remained out of work, and business executives felt resentful and hamstrung by all the New Deal's controls. Despair and frustration persisted as management, labor, and government blamed each other for their problems. The New Deal had laid the groundwork for recovery, but it had not cured the economy, nor had it unified Americans with a positive sense of common purpose. Those tasks would be accomplished not by Roosevelt or Lewis, but by Adolf Hitler.

Hitler's invasion of Poland in September 1939 launched the most destructive war ever known to humanity, but it also endowed Roosevelt's administration with both the determination and the legislative clout to spend public money on a previously inconceivable scale. Hitler's war also answered Americans' deep-seated psychological need to feel good about themselves again after so many years of selfish squabbling. Nazi aggression was both more dangerous and less complicated than an economic depression, and Americans responded to World War II as they had always responded to external threats: by joining hands and rolling up their sleeves.

Even before the United States entered the war, America's industrial economy became, in FDR's felicitous lend-lease metaphor, a good neighbor lending "a length of garden hose" to a friend whose home has caught fire—in this case, Great Britain. After Japan drew the United States into the war by attacking Pearl Harbor in December 1941, the government itself became the economy's prime customer, quadrupling defense spending and military production within a year. In the process, the government did what it had been unable to do throughout the 1930s: restore full employment for the first time since 1929.

Suddenly the nation's problem was not oversupply but scarcity. Now America needed more of everything—not only planes, tanks, and guns, but jeeps, trucks, trains, uniforms, shoes. It needed steel, iron, rubber, aluminum, tin, and electric power for everything. After a decade as the nation's designated villains, America's industrialists found themselves revered again as saviors.

Coal was in demand again like everything else, and managers and workers competed now only to demonstrate their common patriotism. After Pearl Harbor, Lewis and other labor leaders signed a "No Strike Pledge" as their contribution to the war effort. Barely two months after Pearl Harbor, the same members of the UMW Local at Stonega's Derby mine who had been consumed by bitter strikes through the 1930s now voluntarily contributed a day's earnings to the war effort.

"I know you will agree with me that it is men of this type who are winning the war," Ted Leisenring wrote in congratulating the UMW's district president. "I hope that this action of the Derby Local will act as an inspiration to all our employees, not so much as to contribute their earnings to the cause, but that they will want to work every day possible."

This warm patriotic harmony unraveled soon after Pearl Harbor. The war did indeed prove an economic bonanza for industry, for coal operators, and for workers in America's defense plants. But coal miners again found themselves excluded from the party. Coal operators reaped large wartime profits while miners' wages, regulated by wartime government scales, remained stagnant. Amid the pressure for greater production, accident rates in the mines soared. The UMW had surrendered its only weapon—the strike—but the government had dragged its feet on its promise to create a fair and effective system for resolving disputes without strikes. Nor could miners seize other opportunities: Those high-paying defense plants were located mostly in big cities, geographically and psychologically inaccessible to rural coal miners.

These seething dissatisfactions came to a head in 1943 when Lewis abrogated his no-strike pledge and led a half-million soft-coal miners in a series of nationwide strikes. Now Lewis was called "the most hated man in America" by newspaper editorial writers. *Stars and Stripes* published a cartoon portraying Lewis throwing dirt on a soldier's grave. "Speaking for the American soldier," the army newspaper declared, "John Lewis, damn your coal-black soul!"

Yet as Lewis well understood, these strikes never jeopardized the nation's coal supply. The hostility toward him and the UMW sprang instead from the fear that coal strikes would provoke walkouts in other industries (which did in fact occur). In wartime, Stonega and other large coal companies had further mechanized their mines so that they needed fewer miners to produce greater quantities. As a result, the coal industry was always able to recover its lost production capacity whenever the mines reopened. By the war's end in 1945 Stonega's production had climbed almost to the level of its previous high prior to the Depression, in 1929.

At the outset of the war, the fifth generation of Leisenrings was growing up on his parents' place along Glenn Road in Ardmore, outside Philadelphia. Edward B. Leisenring Jr.—called "Teddy" as a boy to distinguish him from Ted, his father—was his parents' fourth child but their only son. He spent much of his boyhood free time swimming in the family's large pool, a rarity among suburban homes at that time. Before he reached adolescence he was a proficient swimmer. Like his father, Ted Jr. found his intended spouse at an early age and within his own close circle (indeed, Ted Jr. and his bride-to-be were both delivered by the same Bryn Mawr obstetrician, although they didn't meet until some ten years later). Julia Bissell was a daughter of the Delaware establishment whose mother was a du Pont and whose father and grandfather were prominent stockbrokers in Wilmington. Teddy met her at Saranac Lake in upstate New York, where the Bissells and Leisenrings had summered together for years.

Teddy was fourteen in 1940 when his parents sent him off, in his father's footsteps, to the Hotchkiss School in Connecticut. Teddy's father, who as a Cleveland banker's stepson had received no exposure to the coal industry as a boy, was determined to groom Teddy for the family business; and in an age when nepotism was widely accepted and youthful rebellion universally frowned upon, it never occurred to Teddy that he would do anything else. When he was fifteen his father and Ralph Knode took him along on a business trip to Wise County, and it was here that Teddy entered a mine for the first time. To the children of a James Warden or a Saylor Givens, the mines were a constant and inescapable presence. But to a fifteen-year-old from Philadelphia's Main Line, nothing could quite match the masculine excitement of going underground in wartime, or the sight of miners with their soot-covered faces, hard hats, and cap lamps. Young Leisenring was permanently smitten.

Many of the younger miners he met would depart shortly for the Army and Navy. Saylor Givens, who was forty-seven when Pearl Harbor was attacked, saw several of his sons and nephews enlist and head overseas. Yet the family's only fatality during the war occurred not in combat but in the mines. Saylor's nephew Worley Givens had followed his father, Albert, to work in Stonega's mines in 1924, when Worley was fifteen. Although he was tall, for most of the next nineteen years he worked eight hours a day on his knees, drilling and cutting coal in dark, narrow, cramped spaces.

With the advent of World War II came a sharp increase in European demands for coal, and with it increased competition among U.S. mines for workers. Around this time Worley and his brothers Robert and Ed left Stonega to work for the Pocahontas Fuel Company at Amonate, a company town (named for the daughter of Pocahontas, the seventeenth-century Indian princess from eastern Virginia) that straddled the Virginia–West Virginia state line, some sixty-five miles northeast of Big Stone Gap. Pocahontas Fuel was one of the region's larger mining companies, controlling at its peak more than one hundred square miles of land and mining rights.

Worley and his wife, Clara, and their three small children settled along the Dry Fork stream on the Virginia side, about a mile from the mine. On the morning of July 26, 1943, Clara Givens, then twenty-eight and pregnant with their fourth child, watched Worley go off to work for the last time. Later that day, while he was working alone, a slab of slate was jolted from the mine roof just above him, falling and crushing him to death at the age of thirty-four. His widow and children were left to subsist on their Social Security survivors benefits.

Worley's brothers reacted to this tragedy in the customary stoic miners' fashion: They remained at Amonate, and both subsequently worked their way up to jobs as supervisors. But Worley's father-in-law, Jonah Brown Linkous, was not a miner. He was a carpenter, and when his daughter's husband died he vowed to move his daughter and her children so far away that his two grandsons would never have to work in a mine.

After saving up for three years, Linkous took off one day after the war was over to visit friends in central Ohio. There he found more jobs and opportunities than he had ever imagined possible back in Virginia. On the spot he bought a farm near Sparta and soon settled his wife, his daughter, and his daughter's four children there. From this community, many years later, one of Worley Givens's grandsons would grow up to achieve a level of affluence beyond the wildest dreams of his Givens relatives at Stonega—and even, for that matter, a level of fame beyond the imaginings of the Leisenrings.

But that chapter in the Givens family saga was still far in the future when World War II ended in 1945. Some of the Givens brothers and cousins remained in the service after the war, but most drifted back to the mines, to be followed by their younger brothers. The youngest of Saylor's six sons, Don Givens, grew up in Imboden, dropped out of school after seventh grade, and went to work for Stonega in 1947, at age seventeen. At first he built houses for a new company town at Madison, West Virginia. Then in January 1949 Don was assigned underground as a coal loader at Stonega's Imboden mine, where his father worked. (Wherever possible, the company tried to assign new men to the same mines with their relatives.) Here Don Givens was placed in a section called "the kindergarten," where the newest (and usually youngest) miners were taught to load coal with a shovel. He was paid by the weight of the coal he loaded, which usually worked out to the equivalent of between $2 and $3 an hour. Even taking into account the unpredictable hours and long idle stretches, that was the best laboring money in the area at a time when the federal minimum wage was 75 cents an hour.

But much of what Don Givens learned there was about to become irrelevant. In 1948 the Joy Manufacturing Company introduced its revolutionary new "continuous mining machine," a long, low-slung crane-like machine designed to combine the separate functions of undercutting, drilling, blasting, and loading within a single process. Although this machine required only a single miner to drive it, it was capable of ripping two tons of coal per minute out of a solid seam of coal. By 1960 the continuous mining machine would double coal production per miner and, in the process, bring the price of coal down to the level where it could still compete with oil and gas.

Teddy Leisenring, meanwhile, followed his father's footsteps into the Navy and then to Yale, where he paid little attention to his studies. Instead, on a lark, he put $3,000 in a Broadway musical—*Small Wonder,* starring Tom Ewell—and frequently skipped classes to run down to New York to keep tabs on his investment, which was essentially an excuse to hang out backstage with actors and chorus girls. The show ran for a year and a half, and when Teddy took his final comprehensive exam he had been partying in New York for days and consequently flunked. He qualified for graduation only after passing a makeup exam in July of 1949.

Young Ted's cavalier approach to his studies infuriated his father, for a reason that Teddy may have been unaware of. In 1948, when he was fifty-three, Ted Sr. noticed a tremor in his hands. A doctor diagnosed the problem as early-stage Parkinson's disease, a degenerative disorder of the central nervous system. Without a counteractive drug such as dopamine—which wasn't developed until the 1970s—the disease was ultimately fatal. Both Ted Sr.'s father and his cousin Daniel Wentz had died by this age, but nature had an especially cruel fate in store for Ted Sr.: This superb athlete, who had continued to play tournament golf into his fifties and who prided himself on his physical coordination, would now be forced to watch helplessly as his motor skills deteriorated. Shortly after the diagnosis, when he was asked to serve a two-year term as chairman of the U.S. Professional Golfers Association, he turned it down; although the post would have been the capstone of his career, he feared what lay ahead during that two-year term. And in the summer of 1949, when Yale mailed young Ted's belated diploma to his parents' home, his dismayed father refused to forward it to him for months. (Ultimately Ted Sr. left it on Teddy's doorstep as a Christmas gift.)

By that time Teddy had arrived in Big Stone Gap in September 1949 to pursue his claim to the family legacy. His father wanted him to learn the business from the ground up, and quickly: Lacking an antidote for his Parkinson's disease, Ted Sr. had no idea how much time he had left.

At first young Ted was assigned to Stonega's engineering department, which occupied a sprawling room on the fifth floor of the Minor Building, the company's headquarters in Big Stone Gap. The room consisted mostly of long rows of tables on which engineers constantly updated large maps of the company's mines. This was presumed to be the best way to familiarize a young trainee on an executive track with the layout of the mines and how and where the entry to each mine was placed. Each morning at 7:30 Teddy would assemble with a survey crew, proceeding to the mines in two- or three-man teams that used a transit—that is, a telescope on a tripod—to peer through sights both aboveground and underground in order to measure the depth of coal seams and confirm that the mining machines were "on line"—that is, proceeding exactly in the right direction. When they returned to the Minor Building each afternoon their findings were duly recorded on the engineers' maps.

Like many another boss's son, Teddy was eager to prove himself on his own. The miners, for their part, were happy to test him. One day at the bottom of a mine, Teddy and a group of miners were sitting in a mine car being pulled by an electric locomotive to their work section. As the mine train carried them along its track, a gnarled old miner smiled at Teddy and casually proceeded to pull out a pack of cigarettes and a lighter.

Teddy was shocked: Although the mine was well ventilated—which meant that the danger of an explosion was more theoretical than real—smoking underground nevertheless remained a dischargeable offense. As a management man, he was duty-bound to report it.

"You're not going to smoke, are you?" he asked.

"Ah, honey," the miner replied, as he lit up, "you're not going to tell on me, are ye?" He had calculated that this young greenhorn wouldn't say anything, and he was right. A manager thinks of the greater good of the enterprise; a miner thinks of himself and his immediate comrades; and Ted, determined to experience the life of a miner, was not yet ready to think as a manager.

After a year or so he approached Stonega's general manager, Harry W. Meador, with a request. Young Ted wanted to be transferred to the underground work force, which would require him to join the United Mine Workers. (The closed union shop, which the UMW had finally achieved at Stonega Coke & Coal after battling through the 1930s, was technically overturned by the Taft-Hartley Labor Act of 1947. But Taft-Hartley permitted a union shop if a majority of employees voted for one, so union membership effectively remained a requirement of employment.)

Ted's request was an audacious one for a chief executive's son. Few major coal operators in the twentieth century had ever put a son—much less an only son, as Teddy was—at risk in the depth of mine. And virtually no coal operator's son had ever joined a miner's union anywhere.

Meador was a kindly, conservative man of medium build whose unprepossessing appearance masked his skills as a coal executive. He was unfailingly gentlemanly in his dealings with women and children, but he turned stiff when the subject turned to unions. He had come to Stonega from Hinton, West Virginia, as vice president and general manager in the early 1930s, only to see the mines unionized in 1933 against his will. He had been an implacable foe of the UMW ever since.

"What do you mean, you want to belong to the union?" Meador asked incredulously.

"I think this is what I'd like to do," young Ted replied. "I really want to know what it's like to be a coal miner."

"Well, I'm going to have to ask your daddy."

"I think my father will say it's a good idea."

"I doubt it," Meador replied.

It turned out that Teddy knew his father. "Your daddy says if you're determined to do this, it's all right," Meador later reported. "But we gotta make sure you're as safe as you can be in this work."

Thus in the fall of 1950 Ted Leisenring at age twenty-four went to work as a hand loader in Stonega's Imboden mine. Here the coal was drilled in its seam, then exploded with a charge of black powder, then shoveled by hand by the loaders. He started in Imboden's "kindergarten," where Don Givens had started, but Don Givens was no longer there. Although he was only twenty, Don Givens had already advanced from a coal loader to a brakeman—one who rode the coal cars, detaching individual cars for loading by miners or mechanical loaders at intervals along the track, then later reattaching the loaded coals cars when the mine train returned from the working sections to the mine mouth. The mine roof was barely higher than the tops of the coal cars, and for safety reasons miners were not supposed to ride them; instead miners customarily entered and left the Imboden mine on foot, through a drift or slope entrance. But one day as the empty coal train was returning into the mine, Don Givens noticed a miner's headlamp shining from the floor of one of the cars. When he investigated, he found a young miner inside who had hopped aboard for a free trip to the bottom.

"You're not authorized to ride that car," Givens lectured him, halting the train. "You could get killed." That was Givens's introduction to young Ted Leisenring.

Eventually Don and Ted Jr. became mining buddies. With Don's brother-in-law Hagy Barnett, they would go carousing in bars on Saturday nights, ride horses at Hagy's sister's farm in Lee County, or just drive the countryside in Teddy's Buick station wagon. Teddy was, Don later recalled, "a good one to get on with."

For the most part young Ted's fellow miners didn't know what to make of him. During the thirty-minute lunch break one day, Teddy found himself sitting against the rib—that is, the side of the entry—with an old toothless miner named Pap Lee (old by miners' standards: He was in his sixties). As they opened their lunch buckets and ate their soot-covered sandwiches and hard-boiled eggs, Pap Lee struck up a conversation.

"Kid," he asked, "isn't your daddy some kind of a big rabbit in this coal company? Isn't he the daddy rabbit? What's his title?"

"He's chairman of the board," Teddy replied.

Pap Lee shook his head. "Chairman of the board," he repeated, "and he sent you down here to do *this*. Your daddy sure must hate the shit out of you to git you a lousy job like this."

Young Ted's isolation from home and family were mitigated by frequent visits from his fiancée, Julie Bissell. By this time Julie had grown into a beautiful girl with reddish brown pageboy and a decidedly independent mind. She was the oldest of three attractive sisters and the apple of her father's eye, but she was not sent to college because her father didn't believe college was necessary for women. Instead she was sent to Foxcroft, a girls' finishing school in the heart of the Virginia hunt country. Here she was put off by the cliquishness and vapidity of some of her classmates, many of whom arrived at school with their own horses or ponies. She hungered for intellectual challenge, and the cultural shock of moving to Big Stone Gap from the Delaware of her du Pont relatives inadvertently provided it.

Julie was still in her teens when she first came to the Gap, and because she and Ted were not yet married she stayed with the Meadors on her visits there. Both she and Ted knew they wouldn't live in Wise County forever, and both were determined to immerse themselves in this new environment.

They were back East in the fall of 1950 for their wedding, at which young Ted asked his father to serve as best man. But Ted Sr.'s Parkinson's was now in an advanced state, and he declined because of his impairment. Because he could no longer walk the grounds or swim in his pool, in 1950 Ted Sr. and Margaret sold their Ardmore mansion and moved to a smaller house not far away in Bryn Mawr. As his disease advanced, Ted Sr.'s main concern was the preparation, from a distance, of Ted Jr.

Back in Big Stone Gap after the wedding, young Ted would leave for work each morning and Julie would walk into town and explore. To her fresh outsider's eyes, Big Stone Gap was a beautiful valley town, with a small reservoir and its own little golf club. She shopped at the local Kroger supermarket, a few drugstores, and a variety of other shops (even though, by this time, most miners owned cars and did much of their shopping thirty-five miles away in Kingsport, Tennessee). Here people socialized through the rituals of garden parties, quilting parties, spring cleaning (necessitated by the pervasive coal dust), and large Christmas parties for which the women cooked for three days in advance. There were, she found, plenty of bright people in Wise County, as well as colorful hillbillies who sold moonshine liquor or made a few dollars sharpening knives along the roadside. To make friends, Julie sharpened her bridge game, volunteered in a pre-school program for mothers and young children, and attended funerals. Eventually she struck up friendships with the Carroll Knights, who owned the weekly *Big Stone Gap Post*, and with the local railroad agent, who grew a variety of apples with his horticulturalist wife.

With Ted, she drove the countryside, visiting other mines and discovering that Stonega's were relatively civilized by comparison (one at Pardee, owned by an absentee Philadelphian, was especially terrible, with houses neglected and an uncertain water supply). With Harry Meador's connivance, she smuggled herself into a mine, dressed in miners' work clothes because a woman in the hypermasculine subculture of the mines was considered bad luck. She investigated technical problems, such as how to prevent moisture in the coal hoppers. And she followed with fascination the seemingly inexplicable wildcat strikes. Once when Julie asked Teddy why his mine had gone on strike, he replied, "Because cap lamps were not handed to the miners. The miners had to pick up the cap lamps themselves." Yet to Ted and Julie the cup was always half-full rather than half-empty, and the strikes were relatively rare punctuations of otherwise friendly relations between Stonega's miners and managers. They lived in Big Stone Gap almost two years, and Julie's first baby was delivered at the hospital in Kingsport.

After his first month at the Imboden mine Ted was transferred to a mine at Derby, where he worked on his knees, wearing rubber kneepads, eight hours a day in a space no more than four feet high, shoveling coal around his side and loading it onto conveyor belts which in turn dumped it into the mine cars. Here he learned to keep his helmet on and his head down when riding the mine car, because a man could be hurt or even killed if his head hit the mine roof in a moving train.

During this time, Ted attended meetings of the UMW local at the union hall every second Saturday, when the mine was idle. (Because the contract now provided overtime pay above forty hours a week, the mines generally shut down on weekends except when demand was high.) In this respect he was not a typical miner at all: Most miners had little time for or interest in union meetings, which were typically attended by only about twenty or thirty of the mine's 150-odd workers.

A month later he was moved again, across the state line to Harlan County, Kentucky—the infamous "bloody Harlan" of violent labor-management confrontations—and the company's much newer Glenbrook Mine, named for his family's home on Glenn Road in the Philadelphia suburbs. Here, in a union-grade job especially created for him—"loading machine helper's helper"—Ted received his first exposure to the state-of-the-art loading machinery. He also received his first exposure to militant union leaders.

When Don Givens returned in 1954 from a two-year army hitch in Korea, his father, Saylor Givens, was still working for Stonega at Imboden, forty-five years after joining the company. Saylor's older brother, Albert Givens, had retired, only to die soon after, in 1951, of black lung disease contracted in the course of his many years in the mines. Their brother-in-law Sam Warden had also finally retired; he had mined more than forty years, and had done so almost all of that time with one arm as a result of a mule's momentary panic in 1910.

An army recruiter, well aware of the gruesome details of the Givens family's mining history, tried to persuade Don to re-enlist. "You never had it so good [as] in the Army," the recruiter said. Don just shook his head. "It was better in the mines," he replied. From his perspective, mining wasn't just a job; it was inseparable from his family and community. Don and his wife-to-be, Shelby Jean Stanley, had grown up together in the company town of Imboden, in a row of nine homes called Dude's Row; Shelby's father, John Arthur Stanley, had put in eighteen years with Stonega (and would accumulate more than forty-three before he retired in 1979). At the age of twenty-four Don instinctively perceived that what the nineteenth-century Pennsylvania coal baron Eckley Coxe had said about miners and operators was almost literally true of Stonega's Virginia operations: They quarreled like husbands and wives, but they learned to live together and look out for each other. In this climate a miner (and even a union official) could be loyal to both his union and his company simultaneously.

But Glenbrook, as young Ted Leisenring discovered, was a different story. The UMW local there was run by a nasty pair of midwestern cousins named Jack and Wilmore Deaton. Jack, a rough character with closely cropped hair and an intimidating military bearing, had served time in a federal prison for second-degree murder. Wilmore was shaggier and less articulate. Ostensibly they were the local's elected leaders; in practice they were John L. Lewis's designated representatives, and between them they dominated the local. As Ted sat quietly in the back row at the Deatons' biweekly union meetings, a recurring theme emerged: their hostility to management and to anyone who cooperated with management.

This combativeness represented a sharp departure from common practice. In the mines, cooperation had always been a basic survival instinct. If a machine broke down, it was only natural for the section foreman to pitch in and assist with the repairs. Technically this was a violation of the union contract, which forbade managers to engage in such work, but in the urgency of the moment miners often welcomed such assistance. The Deatons spent much of their time hectoring foremen and miners alike for allowing this kind of unauthorized work. If management failed to learn this lesson, the Deatons could call a wildcat strike that would shut down a mine for a shift, a day, a week, or longer.

Or they would find other ways to deliver their message. During one dispute the Deatons and their cohorts tore three telephones out of the phone stations at the Glenbrook mine entrance. On another occasion, shortly after Ted Leisenring finished his month's apprenticeship at Glennbrook, the Deatons learned that two section foremen had been in the mine on a Saturday with the fire boss, measuring gas levels and making repairs that should have been left to union men. When the foremen returned the following Saturday to finish this work, the Deatons and four or five union men waited for them in the locker room. They beat one foreman as he was getting undressed; the other was dragged naked from the shower, stretched on a table and beaten with a leather mine belt that left huge welts on his back. The union men involved were subsequently suspended without pay "with intent to discharge." But when the Deatons threatened to shut down the mine with another wildcat strike, the company backed off and restored the men's jobs.

Ted's memories of what he had seen and heard of the Deatons at Glenbrook did not subside when his Stonega apprenticeship ended abruptly at the end of 1951, after his father, no longer able to walk, resigned as chairman of both Stonega and Westmoreland Coal. Now Ted was moved to Westmoreland's operations at Irwin, in western Pennsylvania. He was learning the ropes there in June of 1952, when his father, in a final assertion of control over his deteriorating body, took an overdose of sleeping pills and died at the age of fifty-seven.

Now Ted Sr.'s longtime colleague, the loquacious super-salesman Ralph Knode, succeeded to the presidency of both companies, and Ted Jr. as heir apparent was summoned to Philadelphia as Knode's assistant. In short order he would assume his father's board seats.

Much of the company's future ambitions hinged on the success of the Glenbrook mine where Ted had last worked. Glenbrook was a model mining town, designed by a Philadelphia architect. Its mine yielded a rich grade of bituminous coal that was almost equal in quality to anthracite. When it operated, the Glenbrook mine made more money than any other Stonega mine. Yet this showplace was constantly shut down by the Deatons' seemingly endless harassments. Like a drop of ink in a glass of water, these two cousins seemed to possess the power to poison the company.

As Ted vented his frustration to Knode one day in 1954, a novel solution occurred to the older man. Although Knode was and always had been a management man, he and John L. Lewis had developed a friendship over the years. He could not be described as a close friend of Lewis's, for Lewis had few close friends; but he was close enough that they called each other "John" and "Ralph" in private (while maintaining the formal "Mr." in public). Knode knew that Lewis often felt more comfortable among business executives than among miners; while he publicly thundered against bosses and politicians, in private he was warm and even obsequious toward them. Why not, Knode suggested, make an appointment to see Lewis in Washington and discuss the Deaton situation directly?

Ted Leisenring had never met Lewis before. He was twenty-eight and still a tenderfoot; Lewis at this point was seventy-four and, after thirty-four years as UMW president, a figure of mythical proportions on the American scene. He had defied the American Federation of Labor as well as two U.S.

presidents. In the five years after the war ended in 1945, he had paralyzed much of the nation by calling four strikes, including a strike in 1946 that led President Harry Truman to nationalize America's coal mines and operate them for more than a year.

Yet Lewis had also seized that moment to negotiate—with the government, not the operators—one of the most humane and farsighted innovations in labor history. In May of 1946 the government accepted his demand for a miners' Welfare and Retirement Fund, to be financed with a nickel royalty on every ton of coal mined. The fund—jointly administered by representatives of the union, the operators, and the public—essentially replaced the old and abandoned system of "welfare capitalism" once practiced by the more paternalistic coal companies like Stonega. Now the union, not the operators, would care for miners in sickness, disability, retirement, and death. After an explosion in Centralia, Illinois, killed 111 miners in 1947, the new fund issued its first payments: death-benefit checks of $1,000 each sent to ninety-nine families. By 1952, when royalty payments had grown to 40 cents a ton, the union was building its own hospitals, leading Stonega to close its company hospital and dissolve its medical department by 1957.

The flaw in the fund's conception was its reliance on a royalty per ton of coal mined. Like the cycles of the stock market, coal production was often inconsistent and unpredictable. The new fund was created at a time when coal production was dropping from 688 million tons in 1947 to barely half that amount by 1954, even as droves of miners were retiring and expecting pensions. For that reason, eventually the fund would become a major actuarial burden. But in its first years this UMW fund set a shining example that unions, employers, and even (eventually) the Social Security Administration yearned to emulate.

Even more remarkable than this achievement was Lewis's apparent capacity for personal growth. He had fought the coal operators bitterly when he needed to—for union recognition in the '30s and the Welfare and Retirement Fund in the '40s. But once he got what he wanted from them, he had virtually reinvented himself as their partner. Now the United Mine Workers created a Research and Marketing Department to do what the fragmented coal operators could not: promote their product and encourage research into new uses for coal.

Just as Lewis and his minions (like the Deatons in Kentucky) had tried to indoctrinate miners to distrust operators, now Lewis impressed on his colleagues that the rules of the game were changing. A young union employee named Joe Brennan, who went to work in the UMW's new Research and Marketing Department in the 1950s, found himself summoned to Lewis's office during his first week. After walking the intimidating length of Lewis's long, deep-carpeted office, Brennan was seated opposite Lewis's desk in a chair somewhat lower than Lewis's own.

Lewis looked down at Brennan and began with a simple question: "Do you like coal operators?"

Brennan's father had long served as president of a UMW District in the Pennsylvania anthracite region; his grandfather had been active in the UMW since the union was formed in 1890. Brennan assumed he knew the proper reply. "Hell, no," he said.

"That's the wrong answer," Lewis replied. "I want you to listen. If the coal companies lose markets, especially the big ones, they'll take their money and they'll invest it someplace else. But what do you do with the fifty-year-old coal miner who loses his job in McDowell County, West Virginia?" As Lewis saw it, he was still fighting for the miners, but now the fight involved turning enemies into friends. In the process, the former "most hated man in America" had evolved into a genuine statesman.

Like the most skilled political survivors of history—Queen Elizabeth I, say, or Franklin D. Roosevelt himself—Lewis was an enigma who kept most people at arm's length and outlasted his rivals by skillfully avoiding commitments and confrontations on all but the most major issues. How, or if, he would deal with Stonega's complaint about the Deatons was unknown to Ralph Knode and Ted Leisenring when they walked into Lewis's office about a week after Knode suggested the visit.

With them was Westmoreland's executive vice president, Bill Gallagher, a mine safety expert who did possess some direct familiarity with Lewis's operating style. In 1951, Lewis was exhorting Congress to pass stricter laws on health and safety in the mines. Large operators resisted mine safety laws for the same reason they had resisted unions: not because they objected to mine safety per se, but because the added cost would put them at a price disadvantage against the small mines that eluded regulation. Because the states and the federal Bureau of Mines had only so many mine inspectors, they tended to focus on the large coal companies and ignore the thousands of smaller ones. As Ted Leisenring himself observed later, it was much easier for a police officer to direct traffic in Times Square than to track down a criminal in a tiny apartment in the slums.

In December 1951, while Lewis was filling the pages of the *UMW Journal* with searing photos of mine disasters, a mine explosion in West Frankfort, Illinois, killed 119 men, and Lewis rushed to the scene, both to investigate and to dramatize his cause. As it happened, Westmoreland Coal owned a 12 percent interest in the mine's parent company, the Chicago, Wilmington & Franklin Coal Company; and as it further happened, Bill Gallagher of Westmoreland had formerly supervised mine safety for the U.S. Bureau of Mines. For both these reasons, Gallagher's employers dispatched him to Illinois to join the Bureau of Mines group investigating the tragedy.

Gallagher and Lewis arrived at the stricken mine almost simultaneously. They descended separately but arrived underground in adjacent holding stations at the bottom of the shaft. When Lewis saw Gallagher and learned that Gallagher was representing the mine's owner, he refused to proceed unless Gallagher left. Lewis apparently feared, with good reason, that Gallagher would discover that the explosion had been caused not by unsafe conditions but by miners smoking cigarettes. Gallagher left the mine rather than create a disturbance, but he subsequently learned from others on the scene that cigarettes and lighters had been found on the bodies of some miners. Yet the official report described the cause as "uncertain." Lewis had needed an indictment of management—or at least an exculpation of labor—to help pass his proposed safety laws, and in this case he got what he needed. It was a classic example of Lewis's ability to make the right things happen for the wrong reasons.

When Gallagher, Knode, and Ted Leisenring walked into Lewis's office more than two years later, Lewis betrayed no recollection of that mine episode. He rose from his desk and warmly ushered his visitors to a couch and easy chairs at the other end of the room. He greeted Knode as "Ralph" but addressed Ted as "Mr. Leisenring," even though he was Ted's senior by forty-six years. Once seated, Lewis took out a long black cigar and rolled it between his lips, but without lighting it.

"Ralph," he said to Knode, "I want to tell you, I've got a doctor who tells me that these cigars are going to kill me, and he forbids me to smoke them anymore. But he did give me permission to chew them up."

Over the next hour and a quarter, Ted noticed, Lewis periodically bit off a piece of tobacco, chewed it, and replaced it in the ashtray, so that by the end of the meeting the cigar had been reduced to a pile of wet, black tobacco.

After the initial pleasantries, Lewis turned to the matter at hand. "You say you want to talk about some problems you have at your Glenbrook mine down in Harlan County, Kentucky." Knode recounted Stonega's troubles with the Deatons and their henchmen and asked if Lewis could help. Lewis acknowledged some familiarity with Jack and Wilmore Deaton: On three occasions he had appointed them to one of the commissions that Lewis and his assistant, W. A. (Tony) Boyle, periodically created to investigate union issues. Ted Leisenring and Gallagher were already familiar with Boyle, who had been Lewis's assistant since 1948; they knew him as a loyal if unimaginative lieutenant who often functioned as Lewis's eyes and ears in the field, just as Lewis himself had once served for Samuel Gompers. Boyle had been born in Colorado of Irish immigrant parents, and Lewis, who was of Welsh descent, was said to feel affinity for other people who had suffered at the hands of the English.

"I have to tell you," Lewis said, "in my dealings with Jack Deaton, I've found him to be a very reasonable man and very helpful and no problem. As a matter of fact, I've found him to be just an extremely personable fellow."

Knode persisted. "Well, you know," he said, "young Ted Leisenring here worked down in those mines, and he worked in that local union, and he got to know Jack Deaton, and he has related to me—and so has Bill Gallagher, who travels down to the mine—how Deaton operates."

Lewis turned and considered Ted for the first time said. "Well, now, then, Mr. Leisenring," he said to his young visitor, "tell me what you know about Mr. Deaton and how these problems came about."

"Well, Mr. Lewis," Ted said, "I can see that when Jack Deaton came up here, it would certainly be in his interest to do everything he could to charm you."

"Let's just hold on a minute here, Mr. Leisenring," Lewis interrupted. "He may or he may not have charmed me. But in my generation, only the *ladies* were considered charming."

Momentarily thrown off balance by Lewis's own charming non sequitur, Ted nevertheless doggedly plunged ahead with an earnest recounting of the whipping of the foremen in the shower, the yanking out of the telephones, and his own outraged observations of how Jack Deaton controlled the local union meetings. Gallagher followed by giving Lewis a summary of the number of wildcat strikes the Deatons had called and their cost to Stonega.

Lewis listened but expressed no opinion. "I can see that there are two sides to this story," he said. "Maybe what you say is right, and maybe what Deaton tells me is right. We'll have to look into it."

The meeting broke up on that note, and as Knode and Gallagher rose to leave, Ted found himself a few steps behind them. In that moment Lewis fell in with him and placed a fatherly hand on his shoulder.

"Now, young man," he said, as Ted later recalled, "you know, I can remember, believe it or not, when I was your age. I went through a great many experiences with people in the coal industry, people of all kinds, and I can see that you are emotionally disturbed by this situation. I can understand how that could be, because you worked in the mine and you knew the people and you knew the foremen who were beaten. But I'm telling you, if you are fortunate to live to my age, I think that you will look back upon this episode as though it were a mosquito trying to bite the ass of an elephant."

The three Stonega executives never heard another word from Lewis about the Deatons. But two months later they received a call from Harry Meador, their general manager in Big Stone Gap, with a piece of news: Jack Deaton had been offered a job as a UMW International representative in a Midwest district. His cousin Wilmore Deaton would be leaving with him. John L. Lewis had solved Stonega's Deaton problem in his own way.

CHAPTER 13

Prelude to Murder

Stonega's union problems at the Glenbrook mine failed to evaporate when Lewis removed the Deaton cousins to the Midwest. Lewis had never enjoyed the absolute power in the UMW that the world (not to mention young coal executives like Ted Leisenring) believed he possessed. Now that Lewis was in his mid-seventies, his health was failing—he suffered a heart attack in 1956—and the troops below him were jockeying for position in anticipation of his retirement. Within a few years Jack Deaton himself returned to the UMW's District 19, which covered Kentucky and Tennessee. Here, in a district that was poised to play a pivotal role in the coming post-Lewis power struggle, Deaton hooked up with the district's secretary-treasurer and (because he controlled the district's funds) de facto boss, Albert E. Pass—a stocky, cold-eyed little man with close ties to Tony Boyle, Lewis's right-hand man in Washington.

Ted Leisenring never met Pass but knew him by reputation. Within the UMW Pass was widely referred to as "Little Hitler." He was said to have instigated and funded the "Jones boys," a gang of some hundred miners who terrorized coal operators in eastern Kentucky and Tennessee. When operators refused to sign union contracts, the Jones boys dynamited and burned their tipples, forced their drivers to dump their coal in the middle of the road, and in one case buried a Tennessee operator named John Van Huss alive in a ditch. Pass was hated and feared by operators and union officials alike—including District 19's mild-mannered figurehead president, a lawyer named William Turnblazer, the son of the William Turnblazer who had organized Harlan County for the UMW in the 1920s and '30s. "That son of a bitch Pass would cut my throat if he had the chance," a District 19 official once remarked to Ted Leisenring.

In the summer of 1958 Stonega attempted to reduce the number of loading machine helpers at Glennbrook. The miners retaliated with a strike that closed the Glennbrook mine for five weeks. Like all wildcat strikes, this stoppage violated the UMW's contract with Stonega. But as both sides well knew, such a technicality was trumped by the realities of the situation—specifically, that it was more expedient for Stonega to settle the strike than to lose time and money enforcing the contract through the courts. John L. Lewis, at seventy-eight, and his Stonega counterpart Ralph H. Knode, at sixty-five, had been worn down by years of such expediencies; both had long since buried their respective hatchets and were now winding down their respective presidencies. But Ted Leisenring burned with youthful indignation toward the UMW's leadership at Glennbrook; at thirty-two he was preparing to succeed Knode as Stonega's president, and he was eager to take a stand.

The Glennbrook strike wasn't settled until Lewis invited Stonega's officials to Washington to meet with him and his advisers, including Deaton and Pass—a meeting that, for all practical purposes, was almost the equivalent of a contract negotiation. Under the terms of the settlement, Stonega agreed to keep its existing workers on the payroll. In exchange, Deaton and Pass promised to end the strike and to guarantee that Glennbrook would produce at least 1,800 tons of coal every day. As Ted Leisenring stressed to Stonega's general manager Walter Schott in Big Stone Gap, "It was emphasized and reiterated that a *minimum* of 1,800 tons per day was necessary, and not an average of 1,800 tons."

Six months later, having succeeded to the presidency of Stonega, Ted observed to Schott that Glennbrook had averaged only 1,407 tons since the agreement and had never exceeded even 1,700 tons in any single day. "It is no secret," Ted wrote,

> that the management of the working force, as stipulated in the Contract, has been usurped by the Local Union more at the Glennbrook Mine than at any other mine operated by Stonega Coke & Coal Co., that absenteeism, with many promises given and no results shown, has been flagrant, and that overtime, partly due to absenteeism, has been excessive.
>
> More important, little, if any, cooperation by the working force has been shown, in that workmen will not pitch in to help with another man's task unless directly ordered to do so by the foreman. . . . Since the agreement entered into in September 1958 has not been honored, there are two alternatives: to make such changes in the management of the mines, under the Contract, as may be necessary to reach an efficient level of operation, or to close the mine until some other solution presents itself.

Those were the combative words of a fifth-generation inheritor struggling to preserve his shrinking birthright. From the moment he went to work in Big Stone Gap in 1949 Ted was clearly the heir apparent to the Stonega and Westmoreland operations, but throughout the 1950s there was reason to wonder precisely what he would inherit. As late as 1946, coal had provided nearly half of all fuel energy used by Americans, but just eleven years later it was supplying barely one-fourth. Imagine newspapers after the arrival of television, or the typewriter industry after the advent of computers, or the record industry after the birth of compact discs, or the U.S. Mail after the introduction of facsimile machines and e-mail: Some demand for the product persisted, but clearly it was yesterday's product.

The smell of death pervaded the industry even as the rest of the U.S. economy grew by 50 percent in the fifteen years after World War II. Unsuccessful coal companies closed their mines or vanished into mergers with larger ones. Only diehard coal operators like the Leisenrings persisted—convinced that, as in the past, some new use could be found for coal and that the survivors of the shakeout would reap the benefits at some future time. Surviving companies shut down their older mines to focus on newer, more efficient ones. Rather than replace its aging rail facilities, after 1948 Stonega gradually closed many of its Virginia mines opened by John Taggart and the Wentzes before World War I. Westmoreland Coal likewise phased out its Pennsylvania mines in the early 1950s, either because the mines were depleted or the cost of operating them was excessive; its last Pennsylvania leasehold was sold to the Pittsburgh-Consolidation Coal Company in January of 1957. In their place, both Stonega and Westmoreland opened new, fully mechanized mines in Virginia and West Virginia.

When the Imboden mine closed in 1955, Saylor Givens chose, after forty-six years in the mines, to retire once and for all at the age of sixty-one on a union pension of $100 a month. But his six sons continued to work at Stonega's newer mines. By the time Don, the youngest, retired in 1985, the six Givens brothers had logged a total of 164 years with Stonega; including the forty-six years worked by their father, the total exceeded two hundred years.

As mines opened and closed and cut back and expanded, that seniority meant reasonably regular work as well as promotions. After returning from Korea in 1954 Don married his childhood neighbor Shelby Stanley—he was twenty-four, she eighteen—and the newlyweds moved into a company house in Imboden, which they rented for $5 a month. Don spent three years at Stonega's Crossbrook mine, running the shuttle car that hauled coal to the surface. When Cross Brook's operations were cut back, he was assigned to the same job at Stonega's Pine Branch mine, where he spent a dozen years and rose to a section foreman. Later he recalled Pine Branch as "one of the easiest mines I ever worked in," because the coal was so soft and easy to load that "everything just went right." In 1969, when Pine Branch too was cut back, he was promoted to foreman at Stonega's Wentz mine. He was not yet forty years old.

The same automobile that liberated postwar Americans from the fixed routes and timetables of railroads also liberated miners from their coal towns and from their thrall to the company store. After the Imboden mine closed—and, with it, the company town of Imboden where he had lived since childhood—Don Givens's main worldly assets consisted of a car, a pickup truck, and some cash savings. At this juncture, when land prices were depressed because of the mine closings, he traded his car plus $500 for a house and a fifteen-acre spread in Powell Valley, about five miles from Big Stone Gap. At the same time, at virtually no cost, Don bought two of the abandoned company houses at Imboden, tore them down for the lumber—good hemlock lumber, rarely found in those parts—and built a house next door to his for his father, Saylor.

Yet this new mobility for miners carried a price. For the first half of the century, mining companies like Stonega had given erstwhile drifters like the Givens family a stability and close-knit coherence they hadn't enjoyed before. Now, thanks to the automobile, the urge to drift was reasserting itself, for better or worse. After less than ten years in Stonega's mines—a short career by Givens family standards—Don's older brother Charles Givens moved to Chicago, where the family lost touch with him. After the Imboden mine closed, Don's brother Ray moved with his family to eastern Indiana, where he took a job in a glass factory.

Thanks to the UMW's higher wage scales, miners like the Givens brothers could afford to own cars, and consequently they could live and shop anywhere. As a result, many coal towns were emptying even before the mines closed. As early as 1947 a survey found only three-fifths of American mining families still living in company-owned housing.

In October of 1953, Stonega's officials at Big Stone Gap received permission from their Philadelphia headquarters to sell off the mostly empty dwellings at the company towns of Osaka, Roda, and Arno, as well as their very first Virginia town of Stonega, which John Taggart had plotted in the 1890s and where he had been killed in 1896. Miners who had rented their houses for $5 a month ($10 for the larger ones) now had the opportunity to buy houses for a few hundred dollars; the company, which had never made money on the houses and had built them merely to accommodate its miners, was content to divest them at bargain prices.

The company's official inventory of buildings recalled a bygone age of horses, wagons, and ethnic and class divisions. At Osaka, for example, the company inventoried five three-room houses in "Slabtown," a ten-room boarding house for "colored," the servant's house for the superintendent, another servant's house, an amusement hall, a garage near the company store, a barn and garage at the "white" church, a chicken coop, and another garage. Above the coke ovens, where black miners had once been housed, the company now offered to sell four four-room structures, four six-room structures, and one two-room structure. At Roda the locations were still identified in the inventory as "Hunktown" (for Hungarians) and "Happy Hollow" (for supervisors). The selling and dismantling continued through 1958. At Dunbar, where the empty houses had deteriorated, Stonega set a book value of $230 per dwelling.

To John L. Lewis, this inevitable shutdown process represented progress, and he welcomed it. Once again he repeated his mantra: A shakeout of inefficient mines would produce better leadership within the coal industry. "When we had eleven thousand producing entities in the coal industry," he remarked in 1959, "no one operator could speak for the industry as a whole, on legislation, on wages, on anything. They were all competitive enemies, and acted accordingly. Now the big companies give national leadership to the industry side."

Lewis was no doubt thinking of Ivy League–educated executives like the tall and genial George Hutchinson Love, a golfing friend of Ted Leisenring's father whose coal career had not flourished until his horizons expanded. After graduating from Princeton in 1922, Love had entered the coal industry as the operator of Union Collieries Company in Pennsylvania in 1926; when that company was acquired by the huge (albeit unprofitable) Consolidation Coal Company in 1943, Love joined his new corporate master. Two years later, when Consolidation merged with the equally huge and distressed Pittsburgh Coal Company, Love became president of the newly created Pittsburgh Consolidation Coal Company, whose domain now encompassed forty-eight mines in Pennsylvania, West Virginia, and Kentucky. The merger, Love explained at the time, would "build a company that could afford to close down the poorer properties and concentrate on the better ones." Only companies of that size, for example, could afford the revolutionary continuous mining machine introduced by Joy Manufacturing Company in 1948. Love, Ted Leisenring Sr., and Ralph Knode all belonged to a circle of Pittsburgh and Philadelphia golf-playing executives who cemented their relationships on the links and in the men's bar at clubs such as Richard King Mellon's Rolling Rock Club in Ligonier, Pennsylvania, outside Pittsburgh. In 1961 Love was elected chairman of Chrysler Corporation, thus becoming perhaps the first coal executive since Henry Clay Frick to apply his executive skills to another industry.

Ted Leisenring, whose company ranked about fifteenth in size among some five thousand coal operators, also seemed to represent the sort of new leadership Lewis had in mind. Yet Ted at that stage of his life harbored no similar sanguine feelings about Lewis's union. By the early 1950s the UMW had negotiated provisions for a closed union shop as well as for a guaranteed hourly wage (in addition to the tonnage rates already paid to miners and loaders). In theory the continuous mining machine had removed the need to motivate men to produce, but in practice the men were still as important as the machines, at least at first. Stonega's Derby colliery was shut down in 1954, Stonega's general manager Walter Schott explained to Ralph Knode and Ted Leisenring in Philadelphia, because of "the high cost of operation," which Schott blamed on "the continuous decline in the tons of coal per man production." This decline was caused, he added, "by the men taking full advantage of the hourly payments in addition to the contract provisions and also due to the method of paying the guarantee wage." The contract at that time guaranteed men a daily wage of $20.25 even if they loaded no coal at all. "We have found," Schott observed, "that this daily wage guarantee works an undue hardship on the company and tends to guarantee to the workers wages which they do not earn."

Utilizing the closing as a bargaining tactic, Stonega reopened the Derby mine two months later after reaching an understanding with the miners and UMW officials that "the men would put out all efforts possible and cut the absenteeism to a minimum," as Schott put it. Although the UMW's district officials and most of the miners had been "very helpful in this entire matter," Schott reported a few years later, "since May 1957, we do not seem to get the required cooperation from all the men."

From Philadelphia, Ted Leisenring responded by instructing Schott to remove all unqualified men from the Derby payroll, "and the sooner it is done the less difficult it will be to accomplish." He added: "We are not anticipating a work stoppage over these replacements, but if one should result, it would be more desirable now than after working the mine a longer period."

John L. Lewis suffered a second heart attack in 1959 and began to think seriously about retiring. He was seventy-nine and had run the UMW for nearly forty years, but he had devoted virtually no time or attention to cultivating a new generation of union leaders. His loyal vice president, Thomas Kennedy, was the default choice to succeed him when Lewis finally did retire in January 1960. But Kennedy was himself in his seventies and would clearly be only an interim president. In any case, Lewis did not really retire; he merely moved from his second-floor office in Washington to the sixth floor of the UMW building, where he assumed the new title of president emeritus and continued to serve as chairman of the UMW's Welfare and Retirement Fund, drawing his old salary of $50,000 a year. His true successor remained to be chosen.

Ralph Knode was one of the few people Lewis took into his confidence, albeit rarely. At Stonega's Philadelphia headquarters one day in the early '60s Knode called Ted Leisenring and Bill Gallagher into his office. "Tell me again what you think about this fellow Tony Boyle," he asked them, "because John Lewis told me he's thinking seriously of designating Tony Boyle as his successor."

Leisenring and Gallagher knew Boyle better than Knode did, in part because Boyle was their counterpart at the UMW. Like them, Boyle had spent much time coping with labor issues out in the districts. The two Stonega executives perceived Boyle as a dedicated, hardworking assistant to Lewis, loyal to his boss and his union, willing to shoulder responsibility and capable of demonstrating initiative. But they also found him devoid of imagination or charisma. Boyle struck them as a "black Irishman"—a humorless, narrow-minded man devoid of the warmth and charm conventionally associated with sons of the Old Sod.

"I would hate to find him running that union," Ted remarked. "He'll never have the intelligence, or a fraction of it, that John Lewis has."

Knode subsequently conveyed their misgivings to Lewis. "Are you sure this is the man you want?" he asked.

"Well, I'm thinking about it," Lewis replied in his customary noncommittal fashion. But when Kennedy died in January 1963, Boyle was indeed chosen to replace him. "He's the best I've got," Lewis explained to Knode. "There are other people in this union, but I don't know of anybody that I think would do a better job for the union than Boyle."

At that moment, everything that Lewis had neglected for years—the funding of the Welfare and Retirement Fund, outdated labor contracts, the consolidation of coal companies, the transformation of the work force from old docile miners to young rebellious ones, the bitter district and local leaders whom Lewis had repressed for decades—all had been dumped in Boyle's lap. Lewis himself had lost the ability to cope with these problems. And Boyle, as he quickly demonstrated, was no John L. Lewis.

After Ted Leisenring assumed Stonega's presidency in 1959, Knode stayed on for two more years as president of Westmoreland Coal before turning that presidency over to Ted as well. At thirty-five, Ted was the youngest president of Westmoreland in a history that stretched back to 1854.

His first order of business was to consolidate his two companies in order to beef up their clout within the industry. Since 1929 Westmoreland and Stonega had functioned independently, even though they shared a sales force, an executive office in Philadelphia, and numerous overlapping shareholders. In 1962 Ted engineered a complicated financial transaction through which Stonega's parent holding company—the old Virginia Coal

& Iron Co.—acquired 52 percent of Westmoreland's stock, including all the Westmoreland stock formerly held by Dr. John S. Wentz's estate. Then in April 1964 the two operating companies, Stonega and Westmoreland, were consolidated into a single company with about $30 million in assets. This consolidated company was more than four times Stonega's size at the end of 1949, and large enough to compete for markets with industry giants like Peabody, George Love's Pittsburgh Consolidation, and Amax Coal, each of which had assets exceeding $100 million.

The name of the newly merged company was an awkward matter. The Leisenrings had operated Stonega since 1896 and Westmoreland only since 1929. Stonega was also financially stronger than Westmoreland, and so it technically became the surviving company. But the Westmoreland name was older and more prestigious: It was America's oldest surviving coal company, and its legacy and goodwill were valuable assets. So the Westmoreland name was chosen for the merged company. From that point forward Stonega would be known as Westmoreland Coal, and the Stonega name vanished from public use.

In the nineteenth century the Scottish essayist Thomas Carlyle observed, "Every new idea, at its starting, is precisely in a minority of one." He might similarly have noted that every great political and sociological phenomenon begins as a tiny and barely noticed blip on the distant horizon. For the coal industry and indeed for all of American society, one such blip surfaced in the spring of 1962 with the publication of *Silent Spring*. In this book, the marine biologist Rachel Carson documented for a general audience the insidious dangers of pesticides and other chemicals to the natural world. Carson's target was not the obvious hazard of coal smoke but something far more dangerous precisely because it was invisible. The "silent spring" was Carson's metaphor for the unseen destruction that humans were wreaking on the environment and themselves through their virtually unregulated dumping, spraying, and dusting of harmful chemicals into the environment. Although coal escaped direct condemnation in *Silent Spring,* certainly Carson could have been thinking of the nineteenth-century coal and transportation barons when she declared, "The 'control of nature' is a phrase conceived in arrogance, born of the Neanderthal age of biology and philosophy, when it was supposed that nature exists for the convenience of man."

The chemical industry, government scientists, and the news media attacked Carson as a hysterical woman, but her book became a best-seller and eventually won even her harshest critics' grudging acceptance. (Nearly forty years after questioning Carson's accuracy and the validity of her conclusions, *Time* magazine identified her as one of the hundred most influential Americans of the twentieth century.) What *Uncle Tom's Cabin* did for abolitionism in the nineteenth century, *Silent Spring* now did for environmentalism in the twentieth: Carson's book aroused an entire generation to a previously overlooked social issue. In the process, *Silent Spring* launched the modern environmental movement.

Silent Spring inspired a new wave of legislation and litigation aimed at protecting the environment. The Wilderness Act, passed by Congress in 1964, established a process for permanently protecting some lands from development. In 1965 the Sierra Club successfully sued to protect Storm King Mountain in New York from a proposed power project. In 1969, public outrage against the Santa Barbara oil spill—a massive fouling of southern California beaches by offshore oil wells operated by Union Oil Company—was largely responsible for passage of the National Environmental Policy Act, which created the federal Environmental Protection Agency. The new government policy, the act declared, was "to create and maintain conditions under which man and nature can exist in productive harmony."

This was an issue that few coal men had given much thought to, at least since the discovery of anthracite coal more than 200 years earlier. Earlier Americans and Europeans alike had instinctively recoiled from the foul odors and dark fumes of bituminous coal, which was why coal burning had been widely banned in medieval Europe and why the cleaner anthracite coal had been welcomed with such enthusiasm in cities like Josiah White's Philadelphia. It also explained why coal mines usually operated far from population centers. But until Rachel Carson focused the public's attention on the environment, most people had assumed that the benefits of coal justified its myriad social and environmental costs. If the price of warm homes, iron, railroads, steel, and electric power was polluted skies and disfigured hillsides, most people were happy to pay that price. But now, as the environmental movement gathered strength in the wake of Carson's book, coal operators would be subjected to a radically different cost/benefit calculus.

Previously public opinion had presumed that the coal industry's sole victims were coal miners, in which case the solution seemed finite and feasible: Improve miners' wages and working conditions, and/or help them to find other lines of work. But now coal operators were vulnerable for two newly perceived and far broader offenses: for digging up pristine countrysides, and for polluting the air with the product they extracted. These transgressions affected not merely miners but virtually everybody everywhere. As the '60s proceeded, Ted Leisenring would find himself grappling with philosophical questions that his ancestors had never encountered. Eventually his attempts to respond to these questions would set in motion a chain of events that would threaten Westmoreland's very survival.

The appearance of *Silent Spring* coincided with two other seemingly in-consequential blips on the world's horizon. In 1955 the U.S. government agreed to help train the army of South Vietnam against aggression from Communist North Vietnam. Seven years later, as South Vietnam struggled unsuccessfully to defend itself, President John F. Kennedy announced that the several thousand U.S. military advisers in Vietnam would "fire if fired upon," effectively engaging the United States in a war against the North. But for most Americans at that point the war in Vietnam was only a minor annoyance. Following Kennedy's assassination in November of 1963, his successor, Lyndon Johnson, won a landslide election victory in 1964, which he interpreted as a mandate to launch another kind of war: an ambitious crusade to eliminate poverty in America through a new array of gov-ernment-sponsored social programs like Medicare, Medicaid, Head Start, Upward Bound, housing assistance, and student loans.

The cost of these two "wars"—against North Vietnam and poverty—at first seemed scarcely relevant. Johnson's election victory over the conserva-tive Republican Barry Goldwater, who advocated reducing the govern-ment's role in American life, reflected a widespread exuberant confidence in the government's ability to control the U.S. economy and to tackle and solve any social and economic problems that might arise. Economic growth was indeed strong, and America's technological pre-eminence in the world was unquestioned. "As 1965 begins," exulted the president's *Economic Report* that year with only slight hyperbole, "most Americans are enjoying a degree of prosperity unmatched in their experience, or indeed in the history of their nation."

Yet the strength of the economy depended on factors that most Ameri-cans foolishly took for granted: stable price levels and an ample and cheap supply of energy. In July 1965, six months after the publication of his *Economic Report,* Johnson authorized U.S. military commanders to send twenty-three thousand combat troops to Vietnam; within four years that number had grown to 543,000. Yet the government, deluded about its abil-ity to control events, increasingly chose to finance both its foreign military war and its domestic social war not only by borrowing and taxing but by the more politically expedient tactic of expanding the nation's money supply. The consequence would be a nationwide (and subsequently global) inflation that would cause interest rates to soar and would soon be-come America's chief preoccupation. From December 1947 through De-cember 1964, the nation's average annual rate of inflation had been less than 2 percent—barely 1 percent if the Korean War years of 1950 and 1951 were excluded. But by 1969, the national inflation rate had risen to more than 6 percent, its highest level since World War II. And that experience of-fered just a taste of the sharply steeper inflation that lay ahead.

Throughout the '60s Ted Leisenring's strategy as CEO of Westmoreland was to nail down long-term contracts with major customers while simultaneously acquiring new production sources in order to satisfy those customers' demands. This balancing act required an ability—transmitted almost innately over three previous generations of Leisenrings—to anticipate future demand as well as future prices. Since the 1920s under Ted's father's cousin Colonel Daniel Wentz, Stonega and Westmoreland had offset the wild swings of the U.S. coal market by shipping a small but steady amount of steam coal to electric power plants in England, France, and especially the Brazilian Light & Traction Company in Brazil. Long-term sales contracts, of course, could be disastrous in a period of high inflation. Nevertheless, in the mid-1960s, blissfully ignorant of the inflationary storm that was about to break, Ted was puzzling how to expand Westmoreland's export business when an opportunity to do so fell serendipitously into his lap.

At that time America's most profitable coal exporter was C. H. Sprague & Son Company, an old-line, privately held Boston firm that traced its roots to southern West Virginia in the 1890s. Sprague was an attractive operation in the mid-1960s not only for its overseas connections but for its vast holdings of so-called "smokeless coal"—that is, low-sulfur bituminous coal that was less effective but also less polluting than the coals mined by Westmoreland in Appalachia.

Phineas Shaw Sprague, the company's egotistical chief executive, was a son-in-law of Thomas Carnegie, whose brother Andrew Carnegie, in tandem with Henry Clay Frick, had caused the Leisenrings such aggravation in western Pennsylvania in the 1880s. Shaw Sprague, as he was known, had concluded that his son Phineas Sprague lacked the makings of a business executive, and he had made no secret of his feelings. Phinny Sprague, in turn, had developed a close friendship with his contemporary Ted Leisenring as part of a small circle of thirty-ish coal executives—many of them sons and nephews of domineering chief executives—who began meeting twice a year in the mid-1950s to provide mutual support and to discuss how their companies could jointly increase their market clout. These meetings of what they called the Interstate Coal Conference alternated between two elegant West Virginia resort hotels—the Greenbrier and the Homestead—and in the course of these gatherings Julie Leisenring developed a lifelong friendship with Phinny Sprague's wife, Mary Lou. By 1968 Shaw Sprague was nearing retirement age; his opinionated ways had alienated many acquaintances, including his wife, the former Lucy Carnegie, who had divorced him, and his son-in-law, Robert Monks of Boston, who was pressing Shaw Sprague to "monetize" the company—that is, provide some liquidity so his heirs could cash in their shares if they chose. Shaw

Sprague believed no one could run his company as well as he could, but under these conditions he reluctantly agreed to sell his family's company.

But to whom? Shaw Sprague was determined to control the choice of buyer. He would not sell the company, he said, to anyone who was not a gentleman. Eastern Gas & Fuel Company failed to qualify, he told his son Phinny, because its owner—a capable Harvard graduate named Eli Goldston—was Jewish. The giant Consolidation Coal Company also failed Shaw Sprague's test because, he said, its owners were strong-arm thugs who had treated him badly in the past. Island Creek Coal was run by Raymond Salvati, who was, he said, "let's face it, a talkative wop." Westmoreland, on the other hand, measured up to Shaw Sprague's crusty criteria because he had known Ted Leisenring's father at their mutual alma mater, the Hotchkiss School, and Ted had attended Hotchkiss as well.

Thus at this critical moment Phinny Sprague found that his father had a critical use for him after all. "Ted Leisenring is your friend," Shaw Sprague told Phinny. "Ask him what he thinks our company's worth, and we'll trust his judgment."

In this manner, Westmoreland found itself the leading, and perhaps only, candidate to acquire a company that sold roughly $100 million worth of coal annually, much of it overseas, at a time when Ted and his sales staff believed the price of export coal in the global market was poised to rise sharply. The acquisition would also bring Westmoreland nearly ninety thousand acres of West Virginia coal lands, much of it producing low-sulfur coking coals at a time when environmentalists, lawmakers, and customers were beginning to look askance at some of Westmoreland's high-sulfur coals from Virginia, West Virginia, and Kentucky. If Ted could pull it off, the Sprague acquisition would effectively double Westmoreland's size; it would become the nation's largest coal exporter and the largest company in the industry still dealing exclusively in coal.

But Ted saw another benefit in the merger as well. Coal could be, ultimately, a narrow and monotonous business. The excitement of closing and digesting the deal in itself would recharge his managers' batteries as well as his own. In a letter to a Sprague vice president Ted wrote of "the challenge and the requirement for hard work and imagination, which Westmoreland was almost beginning to lack. Things were getting too easy for us."

Over the next two years he got more challenge than he had bargained for. Ted had assessed the Sprague coal properties at $21 million, a price Shaw Sprague accepted with little negotiation. But Westmoreland was able to raise only $13.6 million of the purchase price in cash; it would pay the rest by issuing Westmoreland stock—300,000 shares, or fully 20 percent of Westmoreland's stock, as the figures ultimately worked out. The deal also gave the Sprague people two seats on Westmoreland's board—one to

Sprague's son-in-law Robert Monks, and the other to a Sprague vice president. Shaw Sprague may have trusted the Leisenrings as gentlemen, but Ted Leisenring feared that Monks could well sell the stock to the highest bidder, giving a potentially unfriendly shareholder (or even a rival coal company) a 20 percent foothold in Westmoreland, not to mention seats on Westmoreland's board and access to its management.

To prevent this outcome, Ted negotiated into the deal a consent decree that precluded the Sprague shareholders from selling their Westmoreland stock for at least two years after the deal was consummated in September of 1968. In effect this agreement gave Ted two years to get the Sprague heirs out of his hair and find an acceptable buyer for their stock.

His strategy was to begin shipping the Sprague coking coal overseas almost from the very moment the acquisition went through. The merger enabled Westmoreland to offer a much broader combination of coal products at the very moment that the expanding Japanese steel industry was happy to pay higher prices to satisfy its rapidly accelerating demand for coking coal of low and medium volatility. Ted's acquisition of C. H. Sprague & Son had been a calculated gamble based on his hunch that overseas coal prices would rise; in fact prices rose beyond his expectations, doubling within three years.

Meanwhile Monks, as Ted had expected, scouted for bids on the Sprague family's 20 percent share in Westmoreland stock, which he planned to sell as soon as the two-year time limit expired. But as Westmoreland's profits rose during those two years, Ted was able to co-opt Monks's effort by arranging a public offering of a block of Westmoreland stock at a higher price than anything Monks could obtain privately. In December 1970 Westmoreland offered, through the investment house of Kidder, Peabody & Co., four hundred thousand shares of stock, including the block of three hundred thousand held by Monks and his Sprague relatives. The offering raised $17.6 million for Westmoreland stock that had been valued two years earlier at $11 million. At the same time, it absorbed the Spragues' minority shares among a broad shareholder base that was unlikely to threaten Ted's control.

In effect Ted got to eat his cake and have it too: He had consummated the Sprague deal and catapulted Westmoreland into the top ranks of the industry without giving up effective control. The coal industry, notably deficient in heroes, suddenly embraced Ted as a corporate miracle worker who seemed to have found the key to making money from an outmoded product. His name and face appeared in *Forbes* and *Fortune* and *Business Week*. In 1970 Ted was elected president of the National Coal Association, the industry organization that his father's cousin Daniel Wentz had helped to found in 1917.

Some of Ted's plaudits were deserved, of course. But some he owed to his family's reputation and to his Hotchkiss School tie. And some of it had been luck. Ted had caught the Sprague company just as the coal market was approaching one of its periodic peaks. That crest, as his family well knew from more than a century's experience, would not last forever. Nor would Ted's ability to foresee the future.

The newest age of uncertainty for coal operators began in 1968 when seventy-eight miners were killed in an explosion at a Consolidation Coal Company mine in Farmington, West Virginia. In the wake of that tragedy, Congress passed a strict new mine safety law. Then in February and March of 1969 the UMW went on strike for protection from black lung disease, which led Congress to pass the federal Coal Mine Health & Safety Act of December 1969. Unlike previous mine safety legislation, this act transcended mere safety precautions and instead addressed the very methods of production, forcing operators to rethink the whole way they went about producing coal. Like many virtuous acts, these new laws promised unintended side consequences whose costs the operators couldn't hope to gauge. Their old fear of the '50s and early '60s—that coal would lose its hard-won ability to compete with oil, gas, and nuclear power—began to haunt them again. But the greatest uncertainty faced by the coal industry at the end of the 1960s was the same uncertainty it faced at the beginning of the decade: how to deal with the United Mine Workers after John L. Lewis.

Tony Boyle was, by most accounts, a loyal union functionary. As International president after 1963 he tried to run the UMW as Lewis had. Like Lewis, Boyle preferred the halls of Congress and the company of coal operators to the grit and grime of the mines; even more so than Lewis, he brutally suppressed dissent within his union in order to maintain his power. But the similarities between the two men ended there.

Lewis was overwhelming physically as well as psychologically; no one who saw him ever forgot him. Boyle, by contrast, was the sort of slight, forgettable figure that people rarely noticed on the street. Lewis, whatever his personal ambitions, instinctively empathized with the miners' cause and never forgot that ultimately his power depended on the miners' welfare. On their behalf he was eloquent, dramatic, and charismatic—he had, after all, once been an actor.

John L. Lewis would have been a tough act for any successor to follow, but especially for the introverted Boyle, who was suspicious by nature and incapable of hiding it. There was little about Boyle that could be called charming or endearing. Where Lewis instinctively reached out to people, Boyle instinctively constructed walls around himself. Although the UMW had always held its conventions in or near the coal fields in order to be accessible to miners, Boyle moved the 1964 convention to Bal Harbour, Florida, and extended its length to eleven days, so that only delegates whose expenses were paid by union headquarters could afford to attend. He surrounded himself with gun-toting goons, some of whom had done time in prison for violent felonies. The *UMW Journal,* which under Lewis had at least striven for the appearance of objectivity, now became a heavy-handed promotional organ. "In our book," the *Journal* wrote in 1969 in a typically clumsy rhetorical swipe, "persons who accuse the United Mine Workers and its dedicated international president, W.A. Boyle, of not doing their jobs are finks. Don't listen to them! Don't support them!" One issue of the *Journal* carried twenty-eight photos of Boyle in sixteen pages.

During those years Boyle was a co-chairman of the National Coal Policy Conference, a joint industry-labor lobbying group. Ted Leisenring got to know him there, as well as in Ted's capacity as a member of the operators' bargaining committee. Boyle, he found, spoke in a stilted manner that struck Ted as a ludicrous attempt to mimic Lewis. "Now, Ted," he would say (in contrast to Lewis, who addressed nearly everyone as "Mr."), "you and Bill Gallagher know exactly what's been going on down here in the Glennbrook mine. Those bastards"—meaning his own UMW officials—"are hard to deal with, and I've got to put up with them."

The gulf between Boyle and his miners was most glaringly displayed after the Farmington explosion in 1968. Lewis, during his reign, skillfully—and, to be sure, sincerely—exploited mine disasters as a way to cultivate public sympathy for his miners and to pry concessions from operators and lawmakers. Frequently he would rush to the scene of the tragedy, don a hard hat, and command the rapt attention of reporters and photographers with his denunciations of callous operators and lax federal mine inspectors. Lewis instinctively rose to such occasions; Boyle seemed to shrink from them. He did not come to Farmington until four days after the explosion, and when he finally arrived he chose to lecture not the operators but the dead miners' relatives.

"I share the grief," he said unconvincingly. "I know what it's like to be in an explosion. I've gone through several of them. But as long as we mine coal, there is always this inherent danger." Seemingly unaware that the mine's owner, Consolidation Coal Company, had violated federal safety regulations, he defended the company as "one of the best companies to work with as far as cooperation and safety are concerned." Then he returned to union headquarters in Washington.

Under Boyle the slipshod handling of the UMW's Welfare and Retirement Fund—Lewis's fault, to be sure—continued unabated. When some operators, in protest, withheld full royalty payments to the fund, Boyle's union stopped paying benefits to members who worked for those operators—in effect punishing union members for their employers' transgressions. To destroy the patronage power of UMW district and local officials who might challenge him, in the late 1960s Boyle negotiated a system of posting for mine jobs that assigned work according to seniority—a system that managed to anger miners and operators alike. In effect seniority obliterated the old cohesiveness of a mine crew, in which crew members would bring in their sons or relatives and look out for them. Herbert Jones, the president of Amherst Fuel in West Virginia and one of Ted Leisenring's fellow "young Turk" executives, told Boyle that this destruction of a unit's cohesion would devastate productivity. Boyle listened, nodded, and did nothing.

Lewis had lost touch with his miners but managed to maintain the illusion that he still cared about them; Boyle lost even the appearance. He blamed everyone but himself for his problems, and his paranoia ultimately became a self-fulfilling prophecy. Like Richard Nixon, his counterpart in the White House after 1968, Boyle believed himself besieged by enemies and withdrew within a small circle of loyalists whose isolation merely reinforced his suspicions. His closest confidant appeared to be his devoted assistant, an unsmiling lawyer named Suzanne Richards who acted as Boyle's secretary and gatekeeper and was known among union officials and operators alike as the "cobra woman."

But Miss Richards, as Boyle called her, possessed no weapon more lethal than her cold eyes. Albert Pass, Boyle's lieutenant from Kentucky, was another story. Beginning with his "Jones boys" mob of the mid-1950s, Pass was always ready to instigate violence in the union's behalf, even though he appeared never to engage in it himself and took care to distance himself from the perpetrators. At least as early as 1968, Pass gave $1,500 to a union subordinate named Bill Prater to find someone to murder Ted Q. Wilson, the chief counsel of the rival Southern Labor Union. Prater in turn had offered the $1,500 to his friend Silous Huddleston, a loyal UMW local president in Tennessee who had spent two years in prison for grand larceny. Huddleston and Prater both believed, as Pass did, that outside enemies were bent on destroying the union and, consequently, their pensions. Huddleston was eager to demonstrate his loyalty. "If you don't believe your leaders' word," Huddleston explained later, "you're in pretty bad shape." He turned down the Wilson job only because the price was too low.

As Boyle plunged deeper into paranoid schemes to destroy his perceived enemies, rank-and-file miners as well as public social critics urged John L. Lewis—by now in his mid-eighties—to speak out against Boyle. Privately Lewis was said to have remarked that promoting Boyle as his assistant had been "the worst mistake I ever made." But publicly he said nothing.

The inevitable challenge to Boyle's presidency came in 1969 from a UMW official whom Boyle had good reason to fear. Joseph "Jock" Yablonski, at fifty-nine, was one of the union's few officials who were elected rather than appointed by Boyle—his Pittsburgh-area district was one of the few remaining UMW districts that retained the right to vote for its officers—and as such he was the rare UMW official who could claim a popular mandate. Yablonski was his district's representative on the international executive board. He had the outgoing personality, as well as the barrel chest and muscular build, that Boyle lacked. He was combative and seemingly fearless. He was psychologically secure in a way that Boyle was not. He was well known and well liked within the UMW. Despite his misgivings, in 1968 he had demonstrated his fealty to Boyle by delivering the keynote address at the UMW's convention. He enjoyed the support of popular public reformers like Ralph Nader. And he had two lawyer sons, one of whom, Chip Yablonski, worked for the National Labor Relations Board.

Following the 1968 UMW convention, Boyle had rewarded Yablonski for his keynote address by appointing him to direct the UMW's lobbying effort in Washington. But after Yablonski announced that he was running for president in May of 1969, Boyle fired him. Yablonski refused to accept his dismissal; he filed suit to enjoin the discharge, on the ground that it was an obvious reprisal for Yablonski's political activity in the union. Yablonski also courted John L. Lewis's endorsement and received some indications that he might secure it. The two men spoke several times on the phone but never met. On June 8, Lewis entered a Washington hospital, suffering from internal bleeding; three days later he was dead at the age of eighty-nine.

Lewis's death seemed to remove whatever restraints Boyle felt about moving against his perceived enemies. During a meeting at UMW headquarters shortly afterward, according to later testimony, Boyle told two colleagues, "Yablonski ought to be killed or done away with." The two colleagues to whom he spoke were the two top officials of the UMW's District 19 in Kentucky and Tennessee: the district's secretary-treasurer and de facto boss, Albert Pass, and the district president, William J. Turnblazer. As Turnblazer later reconstructed the conversation, Pass peered intently at Boyle and replied, "If no one else will do it, District 19 will. District 19 will kill him." Boyle answered, "Yeah," and turned away.

Yablonski's defeat by Boyle later that year settled nothing for either man. Yablonski claimed the election had been stolen and vowed to challenge the results. He threatened to expose the union's dirty linen as well. "If I don't get the ability to speak my mind and have a free election here, I'm going to spill the beans," Yablonski told Ted Leisenring and Bill Gallagher. "I'm going to go public with some of this stuff." There were plenty of beans to spill: A U.S. Labor Department investigation reported in November that some UMW officials had claimed hotel and travel expenses for practically every day of the year, and that the UMW had created a $1.5 million retirement fund for its top officers that had barely been disclosed in the union's annual reports.

Yablonski was equally forthright in public. "I've taken Boyle to court before," he declared to his supporters, "and I'll take him there again. And next time he'll go to jail, where the hell he belongs. Maybe we lost the skirmish, but we're still going to win the war! We haven't given up. We're going to fight this thing all the way." He asked the U.S. Department of Labor to impound the ballots and to investigate corruption in the union, but in both cases the government declined to act until Yablonski exhausted all his grievances through the union's machinery.

Despite these setbacks to Yablonski's cause, the perpetually embattled Boyle felt as threatened as ever, and Yablonski appears to have understood that he might be in danger. At his three-story nineteenth-century fieldstone house in the small borough of Clarksville, fifty miles southwest of Pittsburgh, the driveway was lined with stately evergreens and firs that rendered the house almost invisible to the nearest neighbor, four hundred yards away. Yablonski took the precaution of installing floodlights around the house. Late in December he noticed a car with three people driving slowly past his property; they seemed to be casing the place. Yablonski recorded the license plate on a yellow pad but didn't notify police.

A few days later, on the next-to-last night of 1969, Yablonski, his wife, Margaret, and their twenty-five-year-old daughter, Charlotte, were asleep when three men slipped into the house. The intruders were not professional killers but oafish small-time criminals from Cleveland. One of them was Silous Huddleston's son-in-law, Paul Gilly, who had recruited the other two. The three men had fortified themselves for their task by juicing up on beer and whiskey in Yablonski's barn. Yablonski's floodlights, far from protecting his family, cast a dim glow inside and guided the intruders as they took off their shoes and ascended the stairs to the second-floor bedrooms.

The Yablonskis' friendly dog, Rascal, was in the house but did not bark. The men first entered Charlotte's bedroom and shot her twice, fatally, in her bed. In the master bedroom, Margaret awoke and screamed; she too was shot and killed. Yablonski himself tried to grab a shotgun that he kept near his bed, but he was cut down by five bullets before he could reach it.

In the silence that now settled over the house, the killers lumbered down the stairs and disappeared. The three decomposing bodies were discovered five days later by Yablonski's eldest son, Kenneth, a lawyer twenty-five miles away in Washington, Pennsylvania, who became concerned when he was unable to reach his family by phone.

The killers were not hard to find: They had left a trail of beer and whiskey bottles, with fingerprints, along the road to Yablonski's house and near his barn. One of the killers, Claude Vealey, got drunk afterward and talked about the shootings in Cleveland bars. And police found the license plate number on Yablonski's desk. Three weeks after the murders, all three men were arrested, and Vealey quickly confessed. That was the first of a chain of arrests and confessions that eventually led investigators into the UMW hierarchy and up to the union's very highest level.

In Philadelphia, Ted Leisenring was astonished when he heard of the killings. That Boyle and his confederates wanted to silence Yablonski did not surprise him. But that Tony Boyle, who had unburdened himself to Ted on more than one occasions, might have sanctioned a murder seemed inconceivable. Like most coal operators and UMW officials who had dealt with Boyle, Ted had found him a congenitally bitter and vindictive man, but not so bloodthirsty or foolish as to have ordered someone's murder. Surely, Ted presumed, the Yablonski deaths had been commissioned by some deluded union henchman acting in Boyle's behalf but without Boyle's specific knowledge or direction.

Yet in fact Boyle's obsession with control had set off a chain of events beyond his control. The killers had been paid with $20,000 from the UMW's treasury. Prater and Pass, it later developed, had drawn this money, ostensibly for organizing expenses. In the end the trail would lead to Boyle's office. It would destroy Boyle and many other people and almost destroy the United Mine Workers itself. In the process, it would come close to destroying Ted Leisenring's company as well.

Mauch Chunk (pictured here about 1845) was at first an idyllic company town carved out of the wilderness by Josiah White's Lehigh Coal & Navigation Co. The peak of Mount Pisgah, where the coal was mined, can be seen in the distance. The gravity-based "switchback railroad," America's first railroad, descends from the mountain to the Lehigh River, where coal was loaded on canal barges bound for Philadelphia. *Free Library of Philadelphia.*

The coal-loading facilities at Mauch Chunk in the 1840s. The system of rail tracks, bridges, and canals was considered one of the engineering wonders of its time. Note the steep angle of descent of the tracks at left and right. *Free Library of Philadelphia.*

A water-level view of canal barges being loaded with coal on the Lehigh River at Mauch Chunk. It was in the engineering of these and similar facilities at other coal works that John Leisenring Jr. got his start. *Free Library of Philadelphia.*

Judge John Leisenring Jr. joined the Lehigh Coal & Navigation Co.'s engineering corps in 1836, when he was 17. Eventually the "Boy Wonder of the Anthracite" formed a series of coal mining partnerships throughout Pennsylvania's anthracite region. Before he died in 1884 he had expanded to the coke fields of western Pennsylvania and Virginia, and had also become a judge. *Hagley Museum and Library.*

Edward B. ("Ned") Leisenring, son of Judge John Leisenring Jr., inherited control of most of his father's companies in 1884. Throughout the 1880s he battled Henry Clay Frick in the Connellsville coke fields, only to find himself outmaneuvered. *Hagley Museum and Library.*

Henry Clay Frick, the "Coke King of Pennsylvania," seemed to his contemporaries "more like a machine, without emotion or impulses." *Free Library of Philadelphia.*

The steel titan Andrew Carnegie resented his dependence on Frick's coke supply for steel fabricating. He thought he could control Frick by making Frick his partner, but Carnegie learned otherwise. *Free Library of Philadelphia.*

At Frick's 5,000 beehive coke ovens (shown above in this poster from the 1880s), bituminous coal from Pennsylvania's Connellsville region was cooked into coke and then shipped by rail to the steel mills of Pittsburgh. Each of the windows in the long line stretching across this poster represents a separate oven, each tended by a worker who poked and prodded and cooked the coal into coke and then pulled the coke from the oven. *Carnegie Library of Pittsburgh.*

The "Leisenring group" was a circle of relatives and close friends who created and controlled some dozen coal companies from their headquarters in Mauch Chunk and (after the early 1890s) Philadelphia. They are shown here in 1890, just after they abandoned Connellsville and launched operations in southwest Virginia. Edward ("Ned") Leisenring, president of most of the group's companies, sits in the front row at right. Next to him is his uncle, Daniel Bertsch Jr. Ned's brother-in-law and successor, Dr. John S. Wentz, who developed the Virginia mines after Ned's sudden death in 1894, is second from the left in the rear. Wentz's brother Dr. George Wentz, physician at the mining camp in Eckley, Pa., sits in the front row at left. Standing at right is Ned Leisenring's uncle, Alexander W. Leisenring. *Hagley Museum and Library.*

The dapper Daniel B. Wentz coordinated Allied coal resources in France during World War I and succeeded his father as head of the Leisenring Group in 1918. He opened up overseas coal markets and became the coal industry's spokesman before his untimely death in 1926. *Hagley Museum and Library.*

E. B. ("Ted") Leisenring Sr., born after his father Ned's death in 1894, was raised far from coal mining. As a young man he was admired mostly for his athletic prowess—he was a champion golfer, among other things—and social charm. But when a series of deaths in the 1920s propelled him to the head of the Leisenring group at age 34, he rose to the occasion. *Hagley Museum and Library.*

The utopian mining town of Derby, shown here about 1925, sought to soften the Stonega company's image in the face of union organizing efforts. In an age of wood walls and outhouses, its houses used hollow brick tile, not wood, and featured plastered walls, electricity, running water, indoor plumbing, and spacious porches. Even the lowliest miners' houses displayed a taste of elegance rarely found in mining communities. *Hagley Museum and Library.*

A locomotive hauls a loaded coal car from a working face in the Stonega company's Roda, Va., mine, about 1920. In the "room and pillar" arrangement typical of many mines, the pillars functioned not to support the mine roof but to provide advance warning of impending collapse. *Hagley Museum and Library.*

Despite their elegant pretensions, the homes of Derby (shown here in Oct. 1934) remained dominated by the mine's entrance, tipple (where coal was sorted) and loading area. *Hagley Museum and Library.*

During his 40 years as president of the United Mine Workers, John L. Lewis became "the most hated man in America" as his combination of vision, charisma, and demagoguery transformed the union into a powerful force in American life. But he often seemed more comfortable in the company of coal executives than among his union's coal miners. *Hagley Museum and Library.*

Ralph Knode, Daniel Wentz's resourceful assistant in France during World War I, was an affable salesman whose magnetic personality served Westmoreland Coal well when he became its chief in 1952, after the death of E. B. Leisenring Sr. He was said to be one of the few men in whom John L. Lewis confided. *Hagley Museum and Library.*

Saylor Givens, left, patriarch of the Givens clan, started mining for the Stonega company in 1909, when he was 14, and worked the mines for 47 years. All six of his sons and numerous other relatives became miners as well, and two of them died in the mines. He is shown above about 1952 with his nephew, Hagy Barnett. *Courtesy of Donald Givens.*

Don Givens, youngest son of Saylor Givens, went to work at Stonega's Imboden mine in the late 1940s and soon found himself supervising young Ted Leisenring, son of the company's chief executive. He is shown here in the early 1950s, preparing to go to work at Imoden, where he and his wife grew up. *Courtesy of Donald Givens.*

By the 1970s mechanization and increased safety measures had sharply reduced the number of accidental mine deaths. But that was cold comfort to the family of Robert "Rome" Givens, above, victim of a freak accident at Stonega's Pine Branch mine in 1979. *Courtesy of Donald Givens.*

The UMW's four-month nationwide coal strike of 1977–78 reached such an impasse that the antagonists were summoned to the White House. Above, President Jimmy Carter exhorts Westmoreland's E. B. (Ted) Leisenring Jr. (left), representing the coal operators, and the United Mine Workers president Arnold Miller, facing Leisenring across the table. *Jimmy Carter Library.*

Tony Boyle, John L. Lewis's longtime aide, tried to run the United Mine Workers with Lewis's iron hand but went to prison for ordering the murder of his rival, Jock Yablonski, in 1969. *United Mine Workers of America.*

Wave of the future: In this aerial view, a drag line extracts coal from surrounding Montana farm land. Surface mining is cleaner, safer, and less expensive than underground mining. But Westmoreland's arrival infuriated farmers, environmentalists, and members of the Crow Indian tribe alike. *Westmoreland Coal Company.*

CHAPTER **14**

The Age of Uncertainty

If Yablonski's murder was intended to eliminate Boyle's opposition, it failed. Almost immediately, two more formidable adversaries sprang up in Yablonski's place.

The first was the Inspector Javert of his day, Richard A. Sprague, a short, jowly Philadelphia prosecutor who rarely smiled and who seemed to derive his sole pleasure from the relentless pursuit and prosecution of suspected killers. Sprague—no relation to the family that owned C. H. Sprague & Son—had arrived in the Philadelphia District Attorney's Office in 1958 as a thirty-three-year-old former defense lawyer; by 1970 he was Philadelphia's first assistant district attorney, with nearly ten thousand prosecutions under his belt. He had won convictions in nearly all of the four hundred murder cases he had tried. He was already a local legend, if only for his successful prosecution of a 1961 case in which the alleged victim, Marie Coleman, had never been found.

In that case, Sprague had devoted the better part of a year to an intensive study of Marie Coleman's habits, which led him to conclude that she had been murdered by her common-law husband, Thomas Burns. At the subsequent trial, the defense attorney, in his closing statement, stunned the courtroom by announcing that Marie Coleman had been found and was standing behind the courtroom door at that very moment. As every head in the jury box turned to the doorway, the lawyer explained that in fact Marie Coleman was not there—but he added, "What I did was show each and every one of you that you had a reasonable doubt, because you looked."

This was the sort of prosecutorial test that Sprague relished. In his closing statement, Sprague acknowledged that the jurors had indeed looked toward the door. But he added: "There were two people in this courtroom who didn't. One was me and the other was the man I was looking at, and that was Thomas Burns. He didn't look because he knew she wasn't coming through that open door." The jury found Burns guilty—only the second time in U.S. history that a first-degree murder verdict was rendered without clear evidence that the alleged victim was dead.

Connecting the Yablonski killings to Boyle and his lieutenants posed a similar inferential challenge. The murders had been committed more than three hundred miles west of Sprague's base in Philadelphia, but the unraveling of the crime clearly required someone of Sprague's skill and persistence. For Sprague, the Yablonski case offered a test of his talents as well as temporary liberation from his day-to-day administrative chores. When Sprague took the Yablonski murder case on special assignment to Washington County in February 1970, four suspects—Gilly, Vealey, Martin, and Gilly's wife, Annette—had already been arrested in Ohio, and Annette's father-in-law, the Tennessee UMW loyalist Silous Huddleston, was about to be indicted for financing the conspiracy. Sprague presumed that the case did not end with them. He announced that he would start at the end and work back to the beginning.

The second Boyle nemesis created by the Yablonski murders was in some respects an even more unlikely adversary than Richard Sprague. Arnold Miller was a soft-spoken, lantern-jawed, curly-haired second-generation coal miner who had worked in mines in his native West Virginia since 1938, when he was sixteen. After serving in the U.S. Army during World War II and suffering severe wounds during the Normandy invasion, he had returned to the mines in 1951 as a repairman and electrician. But in 1970 a combination of arthritis and black lung disease forced him into a disability retirement at the age of forty-eight.

Thus liberated from the mines, Miller began to dabble in union politics. He became head of his UMW local in West Virginia and soon was also serving as president of the Black Lung Association, a miners' group organized in 1969 to prod union officials to fight for relief from that pervasive miners' disease. As Boyle's control over the UMW gradually eroded in the wake of the Yablonski killings, UMW reformers settled on Miller as an ideal candidate to challenge Boyle for the national presidency in 1972. Union members and operators alike perceived Miller as a decent and honorable man, an earnest idealist devoid of any of the past baggage of the Lewis or Boyle eras. But Miller was also weak and untested in leadership roles. Unlike Yablonski, Miller lacked significant experience in either union administration or union politics; his only real power base was his Black Lung Association. Nevertheless, during the bizarre hiatus between the Yablonski murders at the end of 1969 and the UMW national election in

1972, Miller's qualifications, or his lack of them, mattered less than Boyle's growing vulnerability.

Boyle's public responses to the Yablonski killings were predictable and unconvincing. His first formal statement expressed shock at the "tragedy" and pledged full cooperation with authorities. He posted a $50,000 reward for information about the murder (although his assistant Suzanne Richards later testified that Boyle had actually cut her suggested $100,000 reward in half). He appointed a UMW fact-finding commission which, after a single public hearing in March of 1970, concluded that the murders were unrelated to the UMW election.

He also hired a New York public relations firm which, shortly after Sprague took charge of prosecuting the Yablonski case, arranged a press conference at the National Press Club in Washington, D.C. There Boyle told the assembled reporters, "I hereby swear to Almighty God to tell the truth, the whole truth and nothing but the truth." He spent the next fifty minutes reading a fourteen-page prepared statement that accused his critics of spreading "outrageous lies" and "dastardly, outrageously scandalous and inaccurate charges." Yet his answers to specific questions were vague and evasive. When a reporter asked a member of the public relations team why Boyle needed a high-priced PR firm when he already had publicists on his headquarters staff, the PR man snapped, "That's perfectly obvious."

After the initial arrests in the Yablonski case, Vealey quickly offered to confess in order to avoid the death penalty, but Sprague insisted he would make no deals. Only as his investigation moved methodically up the UMW ladder toward Tony Boyle did Sprague agree to concessions for several key witnesses. In mid-1971, a year and a half after his arrest, Vealey formally confessed and, in the hope of winning leniency, agreed to testify against his two confederates. Gilly was convicted in 1971 and Martin the following year. Both were sentenced to die in the electric chair. Soon after Gilly's conviction, his wife, Annette, confessed to conspiracy in helping to arrange the killings. It was her cooperation that put Sprague's investigators on the first rung of their climb through the UMW ranks.

In a lengthy statement to the FBI, Annette Gilly implicated her father, Silous Huddleston, the former Tennessee UMW local president, as the man who had paid union money to the killers. She added that her father had told her that Yablonski's killing had been approved by "the big man"—that is, Boyle. Annette Gilly and her father both pleaded guilty to conspiracy charges and were put in the federal witness-protection program in exchange for implicating two higher UMW officials: William Prater and Albert Pass, who they said had supplied $20,000 to Huddleston for the killings.

Boyle, meanwhile, was under legal attack on two other fronts. In March of 1971 he was indicted by a federal grand jury for illegally channeling $49,250 in union funds to political candidates, and for embezzling $5,000 from the union himself. Five days later he suffered a heart attack. A year later he was found guilty on all thirteen counts of the indictment, and in May of 1972 U.S. District Court Judge William B. Bryant in Washington overturned Boyle's 1969 UMW election victory over Yablonski on the grounds that massive vote fraud and financial manipulation had been committed. When the court ordered a new election to be held that December, Boyle declared that he would run again, even as he appealed his sentence to concurrent five-year prison terms and a $130,000 fine on the embezzlement charges.

But Boyle's campaign never got off the ground. Instead of visiting mines and bathhouses to defend himself against the corruption charges, he delivered a single speech over and over again, denouncing his challengers as "outsiders" and "stinking hippies." When the votes were counted in December, Arnold Miller had 70,373 votes to 56,334 for Boyle.

It was the first successful rank-and-file challenge to the UMW hierarchy in the union's eighty-two-year history. Indeed, the *New York Times* remarked, "Nothing like it had ever happened in the labor movement before." In a sense this was true: A movement of idealistic but inexperienced rank-and-file coal miners had wrested leadership of one of America's most powerful and important labor organizations from the union's incumbent hierarchy. In another sense, the closeness of the vote in spite of all his transgressions suggests that Boyle's paranoid tactics had destroyed his own presidency. That Boyle's vindictiveness had created new enemies for him, few observers could deny. But had Boyle gone so far as to order Yablonski's murder? That question remained to be pieced together by Sprague's prosecution team.

William Prater was convicted of murder in March 1973. The day after his conviction, he made a full confession to the FBI and agreed to testify against Pass. Pass refused to crack, but he too was convicted in June of 1973 (unrepentant to the end, he was subsequently sentenced to three consecutive life prison terms). At the same time, William Turnblazer, the UMW's lawyerly president of District 19 in Kentucky and Tennessee, agreed to cooperate because, he said, his conscience was bothering him. In exchange for a deal that allowed him to plead guilty to a lesser charge of violating Yablonski's civil rights, Turnblazer reconstructed the conversation at which Boyle supposedly approved Yablonski's killing. "This is the end of the line," Sprague announced when Boyle, now seventy-two years old, was arrested for the Yablonski murders in September 1973. "I do not expect any more arrests."

In the heady aftermath of his election triumph, Arnold Miller moved quickly to correct the wrongs of the past. He was determined to bring democracy to the UMW as soon as possible. From now on, he promised, power would flow from the bottom up: from the miners through their locals and districts to the national officials, rather than vice versa. Under Lewis and Boyle in the past, the UMW's leadership had simply assumed the authority to bargain collectively on behalf of the union, even though no such authorization existed in the union's constitution. Under Miller, workers' delegates would be consulted at every step of any negotiation. As a symbol of the new order, Miller reduced his own presidential salary from $50,000 plus an automatic $25 per day for expenses to $35,000 with no automatic per diem.

It was a well-intentioned gesture, but nothing more. After generations of dictatorship, democratic processes could not be imposed overnight on the UMW any more than they could be imposed instantly on the Soviet Union—and even if they could, the instinctively reticent Miller was not the man to accomplish the task. Like Francisco Madero, the idealistic reform president of Mexico in the early twentieth century, Miller would discover that entrenched power bases are not easily uprooted by strong leaders, much less by weak ones. For coal operators like Ted Leisenring, the Holy Grail had always been personified by union leaders capable of enforcing their contracts. Now the coal operators came to feel that the fall of Boyle's vindictive regime, far from ending the chaos within the UMW, may have actually exacerbated it.

This upheaval in the UMW occurred when Ted Leisenring was serving as head of both the National Coal Association (the industry's lobbying and promotional group) and the Bituminous Coal Operators Association (which negotiated national contracts with the UMW). It was no easy time to become the coal industry's symbol and spokesman. Aside from the Yablonski murders and Boyle's downfall, and aside from mine disasters, labor strikes, black lung disease, environmental rape, and air pollution, the actual commodity known as coal had largely vanished from America's public consciousness. By the 1970s most Americans no longer felt a direct personal relationship to it. They no longer used it to heat their homes, nor did they observe it being delivered to shops or stoked in railroad engines. Nor were Americans likely to know anybody who worked for a coal company, since most Americans now lived in and around cities and fewer men were mining coal, most of them having been replaced by machines. In the public mind, coal was a relic of another age, much like, say, firewood or kerosene or whale blubber. The fuel of the future—beginning in 1963, when the first large-scale commercial atomic power plant was ordered—was presumed to be nuclear power.

Although they didn't realize it, Americans still used more coal than oil. In the age of automobiles, oil was essential for getting from one place to another; but electricity was essential once they got wherever they were going. Bituminous steam coal still provided some two-thirds of the nation's electric power, and coal still represented 88 percent of America's known fuel reserves. For that matter, America's supply of the uranium necessary for nuclear plants was a small fraction of America's coal supply. But few Americans perceived these connections. Instead they worried about filling their cars with gas at the service station, and they reveled in the exploits of colorful oil billionaires like J. Paul Getty, H. L. Hunt, Daniel Ludwig, and the Bass family. In their celebration of oil and their contempt for coal, Americans in 1970 were not unlike the myopic rabbi in the Sholom Aleichem story who, after much reflection, concludes that the moon is more important than the sun "because the moon shines at night and gives us light with which to see, whereas the sun shines during the day, when there is no use for it whatsoever."

There were indeed still places in the world that venerated American coal operators the way the boosters of Connellsville and Big Stone Gap had once done, but they were oceans away. In the spring of 1970 Ted Leisenring had the heady experience of leading ten U.S. coal operators on a tour of the Soviet Union, the first ever undertaken by an American coal delegation. Here they found commissars desperate for their advice about improving mine production and safety. The U.S. delegation members visited some fifteen mines, where they were staggered by the contrasts between the Russian operations and their own. The Soviets, they found, were using more than ten times as many workers to produce only about 10 percent more coal than American mines did. Where the Soviet Communists measured their success in terms of gross tonnage, the capitalistic American yardstick was total cost per ton. As for mine safety, the Soviet record was so appalling that no statistics were disclosed to the visiting Americans, nor were they allowed to visit any but the very safest mines.

In all, Ted concluded, the Russians were at least forty years behind the United States in terms of production and safety. Yet coal meant so much to the Soviet leader Leonid Brezhnev that he had exposed the industry's shortcomings to the visiting Americans in exchange for their advice. The following year, the twenty-fourth Soviet Party Congress approved two of the delegation's recommendations: the use of much larger trucks and excavating equipment, and the restructuring of industrial ministries along the lines of U.S. business corporations.

Back home, however, Ted and his colleagues withdrew once again into a defensive siege mentality not terribly different from that of the embattled Tony Boyle. Nineteenth-century Americans had cheered the smokestacks of coke ovens and factories as symbols of industrial progress; now, with the passage of the first air pollution control laws of the late 1960s, electric utilities in big cities grew hesitant to muddy the skies with high-sulfur coal, no matter how efficient it might be. Mining was no longer a matter of hiring a few hundred men and handing them picks and shovels; now it was a highly mechanized process that required many millions of dollars up front to open a new mine. Yet the added cost of complying with the government's new health, safety, and environmental laws had resulted in a rate of return so low that coal operators were hard-pressed to attract the capital they needed to open new mines, even as their electric utility customers demanded ever greater (and cleaner) quantities of steam coal.

Nor could coal operators raise their rates to reflect the added costs: In August 1971 the Nixon administration, in an effort to fight inflation, had imposed a general freeze on wages and prices. Many coal companies were losing money in more years than they made profits. To be a coal operator at the dawn of the 1970s meant shouldering all the burdens of providing a vital commodity while seemingly enjoying few of the benefits.

It was in such moment of self-pity throughout the coal industry that Ted Leisenring received a phone call in his Philadelphia office one day in the late summer of 1970, not long after his return from the Soviet Union. On the line was William Wikoff Smith, a fellow Main Line Philadelphian who was then president of Kewanee Oil Company.

"Would you like to get in the coal business in Montana?" Smith asked.

On first reaction it was a puzzling question, somewhat like asking a sheep farmer if he'd like to try raising cattle. Coal operators had known for years about the coal deposits beneath the Powder River Basin—a huge, oblong, football-shaped field, the largest in the world, stretching from Montana through Wyoming, Utah, and Colorado. The coal beneath the Powder River Basin represented more energy units than all the known oil fields in the world. But Powder River coal was of a lower grade—lignite or subbituminous, less efficient per ton than the high-quality coals the Leisenrings had always mined around Mauch Chunk, Connellsville, and Big Stone Gap. It was usually mined not underground but on the surface. Instead of burrowing caves into the sides of mountains, as Westmoreland and Stonega had been doing for a century, it involved constructing and deploying a dragline—a massive contraption the height of a six-story building and costing $25 million to $35 million—to cut the tops and scrape the sides of mountains. Most daunting of all, the Powder River Basin lay literally thousands of miles from most of Westmoreland's potential customers.

On the other hand, western coal was unusually clean coal, with very low sulfur content—just the thing to mollify environmentalists and regulators. Western coal was also much easier and cheaper to mine: Unlike the thinner eastern coal seams, which extended vertically to depths hundreds of feet beneath the surface, the Powder River seams were flat and horizontal, ranging in thickness from ten feet to as thick as 250 feet. So the coal could usually be mined from the surface, using high-capacity draglines and trucks, with none of the expense and danger of tunnels, support beams, cave-ins, and explosions. "The problem," as the coal association executive Joe Brennan later put it, "is not can you get it out, but what do you do when you get it out?"

The cost of shipping western coal back East by rail would be more than offset by the savings from the cost of mining it. And if Montana was far removed from Westmoreland's customers, it was also far removed from the United Mine Workers union. Ted could not help but be tempted by the prospect of doing business in a non-union environment blissfully devoid of the UMW's interminable chaos and high costs. Besides, Ted's ancestors in an earlier and more primitive century had made the difficult shift from anthracite coal to bituminous, and from Pennsylvania to Virginia; why, in an age of modern transportation and communication, couldn't he make a similar move? At the very least, Ted figured, western coal might offer Westmoreland a hedge against the headaches and uncertainties of eastern coal.

"Yes," Ted answered instinctively. "But I don't know anything about it."

Three weeks later Ted and two of his top executives, Pemberton Hutchinson and Howard Frey, flew to Montana. Three days later, at an auction in Big Horn County, just north of the Crow Indian reservation, Westmoreland made the winning bid of $480,000 for initial prospecting rights to thirty thousand acres of coal lands which were said to contain roughly 900 million tons of low-sulfur coal. That was the beginning of Westmoreland Resources, a joint venture between Westmoreland Coal, its affiliate Penn Virginia Corporation, Kewanee Oil (later Kewanee Industries), and a fourth partner, Morrison-Knudsen Company, which Westmoreland enlisted to operate the project because, as Ted readily acknowledged, "We knew nothing about surface mining in the West."

Not everyone in Westmoreland's orbit was as sanguine about the Montana venture. John Shober had just been brought in by Ted Leisenring as vice president for operations at Penn Virginia Corp., Westmoreland's landholding affiliate. He was a coal man by blood but not experience: His father was a Philadelphian who got involved in coal mines in the early twentieth century after the family brokerage firm went under, ultimately developing a friendship with Ted Leisenring's father. Shober was also a distant cousin of Westmoreland's president, Pem Hutchinson. But Shober himself was a computer man with a law degree who had launched new ventures for Smith Kline & French, the Philadelphia-based pharmaceuticals firm. His only personal connection to coal was a thesis on John L. Lewis he had written as a Yale senior in 1955. Ted Leisenring, in his customarily instinctive fashion, had hired Shober based on his knowledge of Shober's shrewd and industrious father.

Shober was uncomfortable about virtually every aspect of the Montana venture. The risks, he pointed out, were huge—a rail spur and a dragline might cost $30 million, at a time when Westmoreland's annual profit was less than $5 million. And Westmoreland controlled none of the variables: not the land, not the railroad, not the water, not the utilities, not the work force, not the state's taxing and environmental authorities, and certainly not the price Westmoreland could ultimately charge for the coal it extracted there. The sections of the dragline would have to be shipped to Montana in hundreds of railroad cars and assembled on the plains in the dead of winter by untrained local crews. Merely providing water for surface dust control, so that trucks could reach the site, involved drilling a well eighty-six hundred feet deep and praying that they didn't hit oil—because Westmoreland had leased only the coal and water rights to the land, not the oil and gas rights. In Appalachia and even at the UMW headquarters, Shober argued, at least they were dealing with coal people. Montana, by contrast, struck him as a massive earth-moving operation—a totally different business in a strange and distant land.

But Ted Leisenring's gut told him otherwise. In the future, he told anyone who would listen, Powder River coal would be shipped both to the West Coast and as far east as Ohio. Even a coal-rich state like West Virginia might buy its coal from Montana, Wyoming, and Colorado. On that hunch, Westmoreland proceeded to borrow so heavily that through the early 1970s the interest cost alone far exceeded the company's annual operating profits of between $4 million and $5 million.

That hunch conveniently overlooked a reality that Ned Leisenring and John Taggart had discovered in Virginia eighty years earlier: that every piece of potential coal land is occupied by somebody—and most likely somebody who is deeply attached to that land. In his zeal to meet the nation's demand for cleaner coal and his own enthusiasm for this new solution, Ted failed to remember two basic truths of his industry: First, nobody wants a colliery in his neighborhood; and second, most people would rather not think about coal at all. In securing the necessary mining rights to the Montana land, Westmoreland would necessarily disrupt people's lives. And here, unlike Virginia in the 1890s, the deedholders were not Appalachian farmers and squatters but an even more recalcitrant landlord: the collective members of the Crow Indian tribe, whose reservation stood just twenty-five miles away.

The Crows, for good historical reasons, were suspicious of white men bearing offerings. Before the Civil War, while other Indian tribes had resisted the advance of white civilization, these warriors of the Great Plains had acted as scouts for the U.S. Cavalry. In 1851 they were rewarded with millions of acres of rolling grassland in what subsequently became eastern Montana. But the Crows, as Ted Leisenring and his executives would discover, operated under a tribal system that was so unstructured that it put them almost entirely at the mercy of the good faith of the U.S. government, as well as other tribes. The Crows would elect a chief and follow his orders, but only for so long as they chose, which often was not long at all. They were tolerant to a fault of the social and sexual taboos that characterized most social systems, European as well as Native American. They were, by most accounts, a pleasant, happy society, and a non-violent one as well: They were the only Plains tribe genuinely friendly to whites, and they were said to have gone eighty years in one stretch without having killed a white man.

Thanks in part to this easygoing cultural mind-set, by 1868 the Crows' vast preserve had been reduced by encroaching white settlement to 8 million acres; by the twentieth century it had shrunk to 2.3 million acres; and by 1970 the poverty-stricken Crows had sold or leased more than three-quarters of that inheritance to non-Indians. About four thousand Crows remained, in squalid poverty, on the reservation, where the unemployment rate hovered around 40 percent and average family income was about one-third that of a typical American family. Their chief town—Crow Agency, Montana, about sixty miles east of Billings—was little more than a cluster of houses divided by highway and punctuated by the sorry clichés of twentieth-century Native American life: a forlorn curio shop, a grocery stocked mostly with junk food, a Baptist mission, and a building housing the local offices of the federal Bureau of Indian Affairs.

To nail down the necessary leases, Ted Leisenring dispatched his most trusted lieutenant, a man with impeccable coal credentials and Philadelphia lineage but no experience whatsoever at dealing with either cowboys or Indians.

S. Pemberton Hutchinson III, or "Pem," as he was known, was descended from a long line of Philadelphia merchants and bankers. He was the fifth generation of male Hutchinsons associated with Westmoreland Coal; his grandfather had presided over Westmoreland for ten years until his death in 1929 precipitated the company's merger with Stonega, and his father was executive vice president of Westmoreland's General Coal sales arm until he retired in 1965. As a young man, Pem Hutchinson had been determined to break out of this mold. After the University of Virginia and military service in Korea, he tried teaching in Charlottesville, but he found he needed to make more money before he could get married. Back in Philadelphia he went to work for a manufacturer of forklift trucks, where he demonstrated a talent for salesmanship. When the company was sold and Pem was at loose ends, Ted Leisenring was ready with an offer Pem couldn't refuse.

"You've got a heritage in the business," Ted told him. "You're educated in it; we need a salesman in Richmond, Virginia."

The job involved a pay cut, but little more needed to be said. Pem Hutchinson started in Westmoreland's sales division in 1961 and was on track for the division presidency in 1972 when Ted Leisenring tapped him for the presidency of the newly created Westmoreland Resources out West. In short order this proper Philadelphian, whose manner was as precise as the Crows were imprecise, found himself removed from the genteel stone homes and shaded streets of Philadelphia's Chestnut Hill section to the forlorn village of Crow Agency, Montana.

To a new, more activist generation of Crows, the coal beneath their lands represented the ticket to prosperity that would compensate them for all the exploitation they had suffered in the past. "We've been ripped off royally," one twenty-eight-year-old tribe member told the *New York Times.* "We don't want to see this place ripped off any more." The tribe's six thousand members still collectively owned the rights to roughly one-fifth of the 30 billion tons of strippable coal in the western United States. But negotiating with them would be no easy matter. The Crow tribe was run as a "pure democracy," which meant that every decision needed approval by a majority vote in tribal council. And consensus was a foreign concept to them. "The Crows are the Armenians of Indian country," one observer told a reporter. "For every four points of view, there are five factions." The Crows, in short, would be no easier to deal with than Arnold Miller's newly democratized United Mine Workers Union.

Even if the Crows could agree to lease the coal rights at a given price, yet another problem remained: Although the tribe owned the coal beneath the ground, in most cases it did not own the surface lands. These had long been homesteaded by white farmers determined to defend their way of life, with rifles if necessary. They were families like the Reddings, who had arrived in southeastern Montana in 1916, spending their first winter in a tent, without irrigation water, surviving by eating porcupines and feeding thistles to their cows. By the time Pem Hutchinson and the Westmoreland agents arrived, the second-generation Redding, John (Bud) Sr., was in his early seventies and prepared to chase coal men off his land with a short length of chain with an iron ring on one end. "That's sufficient, I expect," he grinned to a writer.

One Westmoreland coal survey crew encountered John Redding Jr., age thirty-eight, and a 30.30 rifle. When the crew chief refused to leave, John Jr. fired at the ground about four inches in front of the crew chief's toes. "It kicked up a little dirt," he later explained laconically. "They left." When asked what he would have done if they hadn't left, young Redding paused and replied, "Well, they were trespassing, you see."

In opposition to the Crows (who perceived coal mining as their potential windfall), these farmers now formed an alliance with passionate environmentalists who feared, as one writer put it, that coal companies were about to "strip mine the Northern Great Plains into a desert." The farmers had their guns, but the environmentalists had a more potent weapon: their rhetoric.

"The great coal and energy companies now descending upon the land I love are hardly defenseless," wrote one such defender, the University of Montana professor K. Ross Toole, in a book titled, *The Rape of the Great Plains.* "It is infuriating to hear them cry so piteously that they are misunderstood while their monstrous machines are eating at our vitals." Vainly did the coal companies argue that coal lands, like agricultural lands, are self-renewing—that mines (even strip mines) revert to their natural state after the mining process is over. As industrialists descending upon a pristine environment, the coal men were ill-equipped to wage any battle of public-relations imagery.

The homesteaders' unwillingness to sell or lease their surface lands forced Westmoreland Resources to relocate its planned mine site and abandon the rail spur route it had engineered. Ultimately, in June of 1972 the Crow tribe did grant Westmoreland Resources (as well as two smaller bidders, Shell Oil and AMAX) the right to mine specific tracts in exchange for a royalty of 17.5 cents per ton of coal. The next day Westmoreland Resources executed coal purchase contracts with four midwestern utilities to sell some 77 million tons over a thirty-year period beginning July 1, 1974. That meant that Westmoreland had two years to get its Montana mining operation up and running.

Westmoreland proceeded to invest $34 million to build a rail spur and a dragline, aimed at extracting and shipping 4 million tons of coal per year to power plants in Minnesota and Illinois. The Crow tribe's annual proceeds from that tonnage would have amounted to $700,000, or about $120 for every member of the tribe. But there was a catch: Westmoreland's sales contracts required the venture to obtain all necessary federal and state regulatory approvals by January 1 of 1974. Ted Leisenring and Pem Hutchinson hopefully presumed that the royalty agreement gave the Crows an economic incentive to support Westmoreland's applications to local, state, and federal regulators.

The venture, like Stonega's venture in Virginia in the 1890s, depended on access to railroads. But here, unlike in Big Stone Gap, there were no local civic boosters beating the drums in Westmoreland's behalf. The Burlington Northern Railroad agreed to build a spur line to the Westmoreland Mine, but first it had to acquire thirty-eight miles of right-of-way from landowners along the planned spur line. Ultimately the railroad entered into five condemnation suits and won all of them in U.S. District Court, but two landowners appealed the decision on the ground that the "spur" line was actually a main line, in which case an environmental impact statement would be required under the National Environmental Policy Act. Rather than suffer further delays as these cases wound through the federal bureaucracy, the railroad re-routed the spur line in a compromise with the landowners.

Meanwhile, one of the counties through which the spur would pass, citing safety concerns for cars and cattle alike, used a legal suit as leverage to force the railroad to build crossing signals which were unprecedented in Montana. As a result of these and other requirements, Hutchinson later noted, "This spur line has more environmental protections, culverts, safety signals and cattle underpasses built into it per mile than any rail line in history."

At the same time, in 1973 two environmental groups, the Sierra Club and Friends of the Earth, sued the U.S. government to restrain further coal mining in the Northern Plains states. Montana legislators, expressing similar concerns about damage to the environment, considered a bill placing a moratorium on further mining approvals. The bill narrowly failed in the Montana House, but the legislature did enact a new Strip Mining and Reclamation Act that voided all existing mining applications, including, of course, Westmoreland's. The company now found itself required to file new applications, the details of which would not be completed until the fall of 1973—scant months before the approval deadline specified in Westmoreland's contracts. The company had no choice but to prepare filings in advance that anticipated what the new regulations might contain.

But at that moment, a crisis in the Middle East intervened. On October 6, 1973, the state of Israel was attacked on two fronts by the armies and air forces of Egypt and Syria, joined soon after by other Arab nations. This "Yom Kippur War" was the fourth Arab invasion of Israel since Israel's creation in 1948—and this time, because the attack was timed for the Jewish holy day of Yom Kippur, Israel's forces were momentarily caught off guard. Within days the United States, Japan, and other democratic nations rushed funds and materials to Israel, and after heavy initial fighting Israeli troops were advancing on both fronts; in Egypt they crossed the Suez Canal and encircled most of the Egyptian army.

These tensions in turn created an opening for the largely Arab oil cartel known as the Organization of Oil Producing and Exporting Countries. OPEC, as it was called, had been founded at a conference in Baghdad in 1960 specifically to halt the steady erosion of world oil prices that had occurred since World War II. For its first ten years or so, OPEC had largely failed in its mission to boost oil prices. But the war in the Middle East—a region that supplied nearly 70 percent of the world's oil—provided the cartel with a golden opportunity. On October 19, less than two weeks after the Yom Kippur War broke out, OPEC imposed a total ban on oil exports to all nations—that is, virtually all Western industrialized nations—that had aided Israel. Three days later the United Nations Security Council ordered a cease-fire, but the Arab oil embargo remained.

The result was a severe oil shortage throughout the industrialized world, a dramatic upward spike in oil prices, long lines of automobiles at gas stations throughout Europe and North America, and a desperate search for alternative energy supplies. Industries and government policy makers who had long dismissed coal as dirty and dangerous suddenly remembered one big advantage in coal that they hadn't previously appreciated: its abundance. Where Americans depended heavily on oil from the rest of the world, with coal the opposite was true: The United States held more coal reserves than any other country in the world—three hundred years' worth by conservative estimates, up to five hundred years' worth by others. New methods of making coal cleaner and of extracting gas from coal were emerging. Once again coal was a cherished commodity—the key to America's independence from OPEC.

For the first time in a generation, coal men serendipitously found themselves acclaimed as economic saviors. And with this reversal of fortune, the success of Westmoreland's Montana gamble seemed assured: The Absaloka Mine in Montana shipped its first trainload of coal on July 1, 1974.

Now Ted Leisenring mapped plans to expand his companies' western operations. Pem Hutchinson was dispatched to Colorado to buy up Orchard Valley properties near Paonia, 250 west miles west of Denver, where Westmoreland this time launched not a joint partnership but a wholly owned mining subsidiary called Colorado Westmoreland Inc. In Washington Ted Leisenring seized the moment to lobby for tax breaks and the lifting of price controls to spur production. "Our oil supply is limited and our known natural gas reserves are running out," he testified to Congress in apocalyptic tones. "The promise of the atom is still years away." Conversely, he noted, "We have approximately 3 trillion tons of coal reserves in the United States—sufficient to last several hundreds of years." But these reserves, he insisted, remained largely untapped because "the financial community is reluctant to invest in new mines. The dollar incentive just does not exist when one considers the inherent risk. Much of this risk could be reduced through tax incentives."

After five months, the Arab oil embargo was lifted on March 18, 1974. But by then the embargo was no longer necessary for OPEC's purposes, because the OPEC nations, in tandem, had resolved to sharply reduce their oil production—an action that raised the price of crude oil from less than $3 a barrel to almost $9 by the end of 1974 (by the summer of 1980 it was nearly $43 a barrel). Since America's manufacturing economy was largely premised on a steady and reliable supply of low-cost energy, the OPEC cutbacks and price hikes threatened the entire nation's productivity. Now it was not only inflation but energy supplies that preoccupied Americans. To stimulate that supply, on March 27, 1974, President Nixon gave the coal companies what they had sought for two and a half years: removal of all controls on the price of coal.

In theory, in a free market the price of any commodity should rise and fall naturally through the aggregated decisions of millions of consumers responding to supply and demand. In practice, in the 1970s prices were being dictated by governments. Nixon's U.S. government had imposed price controls in 1971 (to fight inflation); the Arab nations had curtailed oil production in 1973 (to reduce supply and increase prices); and now Nixon lifted price controls (to stimulate supply). For the coal industry, this sudden policy shift at a moment of intense demand was the equivalent of feeding a banquet to a starving man.

Now coal companies everywhere—except, of course, those that were already locked into long-term supply contracts that prevented them from raising their rates—scrambled to capitalize on the boom, knowing full well that it wouldn't last forever. Producers like Westmoreland, which had averaged an anemic 6 percent return on capital over the previous three years, suddenly began to generate returns of 15 percent and higher.

Around this time John Shober, the president of Westmoreland's Penn Virginia Company affiliate, was having drinks one Sunday in a Philadelphia suburb with his father, a veteran of nearly fifty years with Westmoreland and its associated firms. Shober happened to mention that the price of metallurgical coal at Westmoreland's Hampton, West Virginia, division had jumped from $18 a ton to $44 within a week of the lifting of price controls.

"Oh my God," Pemberton Shober replied. "I'm going to sell my stock."

The son remonstrated that the way the market was taking off, coal stocks might rise even higher.

"John, the coal industry is an industry of cycles," his father replied. "You know, in the '30s if you got a five-cent increase per ton, you'd husband it and get out of debt because the bad times always come. I know exactly what's going to happen. These people in the industry are going to feel they've done something to earn this money, and they haven't. They're going to invest it in a lot of projects that aren't going to be justified, and I think it may kill them." He didn't specifically mention Westmoreland's Montana and Colorado ventures, but there was no missing his meaning.

Westmoreland at this time was, in the words of a Drexel Burnham analysts' report, "essentially an attractive organization, well positioned to participate in the growing long-term coal markets." Its sales, which had amounted to $173.5 million in 1973, leaped to $399 million in 1974 and $451.5 million in 1975, after the price controls were removed. Its profits, which had languished around $4 million to $5 million in the early 1970s, now jumped to $36.2 million in 1974 and then to $60.2 million in 1975. That is, over a four-year period its sales tripled, and its earnings rose fifteen times.

But of course Westmoreland was not immune to Newton's law that every action provokes an equal counter-reaction. American commuters grappled with the energy crisis by organizing car pools, turning down their home thermostats, and caulking their windows. Oil companies, lacking an adequate supply of their basic product, began buying coal mines and competing with coal companies. Coal customers developed alternatives to high-priced coal. In a Tokyo nightclub one night, a customarily polite Nippon Steel Company executive startled Ted Leisenring by pointedly telling him, "You American coal producers are able to extract very high prices from Japanese customers, including ourselves. But mark my words, we have the strongest possible incentive to find ways of making steel without using your coal." By developing new technologies to make use of lower-grade coking coal from Australia and Canada, the executive warned, Japanese steelmakers would drive down the price of coal dramatically.

The boom in coal prices produced one other unanticipated consequence: The United Mine Workers and the Crow Indians alike began to rethink the adequacy of their existing arrangements.

Shortly after the OPEC oil crisis began in 1973, the Crows enlisted the services of Charles J. Lipton, a lawyer who had advised twenty-two governments of emerging countries, from Botswana to Indonesia, on the drafting of natural resource agreements with energy operators. Lipton had no trouble convincing the Crows that their contracted royalty of 17.5 cents per ton—regardless of the quality of the coal or its selling price—was tantamount to a giveaway. "Thirty-five cents could buy an ice cream cone in New York and two tons of coal in Indian country," he told the *New York Times*.

Westmoreland, desperate to secure its commitments as well as the tribe's support, readily agreed to renegotiate the royalty provision. When formal renegotiations began early in February 1974, the Crow Mineral Committee had retained a coal consultant as well as Lipton's services as special counsel—what Pem Hutchinson later called "the most formidable negotiating team I have ever encountered." On March 12 the two sides reached a settlement: Westmoreland Resources agreed to pay royalties ranging from 25 cents per ton or 6 percent of the sales price—whichever was higher—up to 40 cents per ton or 8 percent of the sales price. This was the highest royalty ever arranged in North America for sub-bituminous coal, and it was approved unanimously by the Crow Mineral Committee. But it still required approval by the full tribal council.

The hearing took place on a bitterly cold March night nine days later in the high school gymnasium at Crow Agency. At the appointed hour of 7:30 P.M. Hutchinson and a younger colleague, Charles Brinley, both attired in business suits, took their seats at a small table in the front of the hall. Gradually, members of the community drifted in—not merely men and women, but also children, babies, dogs, and goats. A few of the women took seats in the front row, but most of the men hung out in the rear, smoking and talking. Others drank beer and ate fried chicken. Children and animals chased each other around the gym.

By 10 P.M., when the meeting finally got under way, a blizzard had gathered force outside, blowing frigid winds through the unheated gym. Under the format arranged by the tribe and the federal Bureau of Indian Affairs, Hutchinson was to make his pitch and explain the royalties to be paid to the tribe. Then the tribe members were to vote, each indicating approval or disapproval of the lease by dropping a coin in either of two boxes. Hutchinson, who was already perceived as the alien symbol of a faceless eastern corporation, was introduced to the audience not by his name but as "West Moreland." As he stood up to speak he was informed that each sentence he spoke would be translated into the tribal language, even though, as Hutchinson surmised, virtually everyone in the room understood English already. Nevertheless, Hutchinson gamely plunged ahead with his presentation, only to be interrupted again moments later when the electric power failed and the entire town, including the gym, went pitch dark.

In the darkness, members of the audience began calling to each other in animal sounds—first the calls of a coyote, then those of a wolf. If the whole scene had been deliberately designed to intimidate a visitor, it could not have succeeded better.

"Hey Westy," one disembodied voice shouted out of the darkness, "are you sure you want to do this?" Years later Hutchinson would recall backing up against the gym wall at that moment, shrinking down on his haunches and thinking to himself, "Wow, it's a long way to Chestnut Hill."

Since the power couldn't be restored, the meeting was adjourned, and before another meeting could be called, the Crow Mineral Committee members told Hutchinson that they could no longer stand behind the renegotiated terms. A new set of consultants, they said, had advised them that still better terms could be reached. This jockeying took place while the tribe's elections were taking place, and through April and May the Westmoreland people were hard put to determine whom they should deal with. The Crow Mineral Committee members, for their part, were reluctant to make a decision one way or the other. The next meeting wasn't arranged until May 21, and by the time the Crow Mineral Committee agreed to revised amended terms it was June 12. But twelve days later, at the required meeting of the whole tribe, the newly revised terms were resoundingly voted down. "Westmoreland's efforts since then to determine what aspect of the revised amended terms was the cause of the rejection have not been fruitful," the company dryly noted in a subsequent court filing.

Given the Tribal Council's vote, the Crow attorneys had no choice but to petition the Secretary of the Interior to declare the original leases null and void. The pretext for this request was their contention that Westmoreland had violated federal environmental regulations in the stripping process. Hutchinson, for his part, announced that Westmoreland's operations "have been developed over the past three years with the Crow tribe's full knowledge and cooperation," and that therefore Westmoreland would proceed under the original royalty agreement.

The case now shifted to Washington, where officials at the Department of the Interior astutely perceived that Westmoreland and the Crows weren't all that far apart. Both sides agreed to the department's offer to host a final negotiating session in Washington, subject to ground rules worked out in advance. Although the president now was not Franklin D. Roosevelt but Richard Nixon, one aspect remained unchanged: Coal operators and their adversaries still needed a higher authority to settle their disputes.

When the two parties arrived in Washington, they found that the department had scheduled meetings in a series of conference rooms over the next three days. As the negotiations proceeded, both sides noticed that each of the succeeding conference rooms was smaller and warmer than the preceding one. Finally, Hutchinson later recalled, they were assigned to a room so small and hot that they felt they had no choice but to reach an agreement quickly "to avoid ending up in a closet."

The agreement they reached—and which the Tribal Council now overwhelmingly ratified—provided for an 8 percent royalty, sweetened by a $500,000 bonus and an advance royalty payment of $628,000. Westmoreland began mining the tract in 1975 and by 1978 was extracting five million tons of coal a year, or nearly half of Westmoreland's annual total. For this consideration, the Crow tribe annually received about $200 per capita. The company's struggles with the Crows were over—perhaps because, by that time, the Crows were preoccupied with two other suits they filed in 1975 to invalidate the Shell Oil and AMAX leases altogether.

Back in the unionized East, still the source of most of Westmoreland's coal production, the last dictator of the United Mine Workers was giving a deposition in Washington in a civil suit against the union on September 6,

1973, when three FBI agents entered the hearing room. "There has been a warrant issued in Pittsburgh for your arrest," one of them announced to Tony Boyle, "on charges that you conspired with others to cause the murder of Joseph Yablonski." Boyle's lips twitched at the news, and his hands began shaking. "I don't know what this is all about," he mumbled as he was led through a crowd of reporters. Eighteen days later, Boyle swallowed a massive amount of a sedative called sodium amobarbital and went into a coma for three days.

His appeal of his earlier conviction for embezzlement was rejected by the U.S. Supreme Court in December. Later that month a U.S. District Court judge reduced his embezzlement sentence from five years to three but refused to grant Boyle a waiver of imprisonment due to his ill health. He was flown to a federal prison hospital in Springfield, Missouri, to begin serving his term.

Because of the widespread pre-trial publicity in western Pennsylvania, Boyle's trial for the murder of Yablonski was moved to the Philadelphia suburb of Media, where it began on March 25, 1974. Boyle, transferred to a nearby hospital under armed guard, offered a pathetic sight when he resurfaced in the courtroom. He appeared to have lost weight in prison: His suit was baggy, the jacket hung down over his shoulders, and his complexion was even more pallid than before. His defiant façade was gone, replaced by a wan smile, a limp wave of the hand, and a scarcely audible "Mornin' " to relatives and friends he spotted. In this feeble state it was difficult to imagine that he had ever posed a threat to anyone. Certainly Boyle was no match for Richard Sprague's high-powered prosecution team. When Boyle's lawyers presented a polygrapher's conclusion that Boyle was telling the truth when he denied involvement in the plot to murder Yablonski, the judge, in accordance with standard legal practice, refused to allow the report as evidence.

Like many other coal operators who had dealt with Boyle, Ted Leisenring couldn't quite believe that Boyle was capable of ordering a murder. In his own mind Ted presumed that Boyle in his terror of Yablonski had muttered something along the lines of Henry II's rhetorical question about Thomas à Becket— "Who will rid me of this meddlesome priest?"—and that Boyle's more violent henchmen (like Henry II's) had provided an answer beyond what Boyle had intended.

One day during Boyle's trial, as Ted was driving from Wilmington, Delaware, to his home on Philadelphia's Main Line, he found himself passing through Media. The Riddle Memorial Hospital, Boyle's temporary domicile, stood in plain view just off the Baltimore Pike. Now that Boyle was powerless and disabled, Ted found himself strangely drawn to his old antagonist. Boyle had often been difficult to deal with, but dealing with the

Crow tribe, the Powder River farmers, the Montana regulators, and Arnold Miller's chaotic new union could make an operator nostalgic for an ornery coal man like Boyle, whatever his failings.

Following an instinct, Ted pulled into the hospital parking lot and walked into the building. At the front desk, he asked for Boyle's room.

"Oh, you can't see Mr. Boyle," the receptionist replied. "He's not allowed to have any visitors except his immediate family."

Ted turned from the front desk and was walking through the lobby when he spied a man who appeared to be a hospital orderly.

"Do you know where Tony Boyle's room is?" Ted asked. For whatever reason—Ted's lack of guile, his conservative business suit, his soft-spoken demeanor, his vaguely authoritative air—the orderly's defenses were momentarily disarmed.

"He's up on the fourth floor," the orderly replied. "But I don't think they're letting anybody in there."

Ted thanked him, turned away, and took the elevator to the fourth floor. In the fourth-floor hallway he found a nurse who, similarly disarmed by his straightforward manner, directed him to Boyle's room. As he approached, a federal marshal emerged from the room to block his entrance.

"Excuse me," Ted said politely, "but I'd like to see Mr. Boyle. Is that possible?"

"No," the marshal replied. "Are you immediate family?"

"No, I'm not."

"What is your name?"

"Ted Leisenring. I've known Mr. Boyle for a long time down in Washington."

"I'm sorry," the marshal said, "but you can't come in here."

Ted began to walk away when he heard Boyle's voice, remonstrating with the marshal inside his room. "That was young Ted Leisenring!" he heard Boyle snarling. "That was Ted Leisenring out there. He's a friend of mine. What the hell's the matter—I can't even see my friends? What is this, Nazi Germany? A man isn't able to even have visitors in his hospital room?" It was a typical Boyle union hall performance, his voice rising as he whipped himself into righteous indignation, so that his voice could be heard as Ted walked away down the hall.

At the end of the hall by the elevators Ted found two women, one perhaps twenty years older than the other.

"I heard my husband, Tony, down there saying your name," the older woman said. "Are you Ted Leisenring?"

"Yes."

"This is my daughter," the woman said, indicating her red-haired companion.

"You must be Antoinette Boyle," Ted said, remembering the name of Boyle's lawyer daughter, whom he had heard of but never met.

"They've just got Tony in a terrible position," Ethel Boyle continued, and at her urging Ted sat down and listened to her for the next fifteen minutes. "He's not guilty," Ethel Boyle kept insisting. "He even took a lie detector test out in Chicago and it proved that he was telling the truth, that all the things they were saying about him weren't true. He didn't know anything about the murder. He had nothing to do with it. The lie detector test showed that. I'll even give you a copy of the lie detector test." In effect, a wife desperate to save her husband was attempting to press an old adversary into his service.

"Well, you know, that would be interesting, Mrs. Boyle," Ted said uncomfortably. On a scrap of paper he scribbled his name and address, and after the trial she mailed him a copy of Boyle's lie detector test. But he never followed up on it. On April 10 the jury convicted Boyle of first-degree murder. That conviction was subsequently overturned by the state Supreme Court on the ground that the judge had improperly disallowed defense evidence (including the results of Boyle's lie-detector test) at the trial, but in 1978 Boyle was tried and convicted again in the same court before the same judge. Ted never got to see Boyle again. Shortly after his conviction, Boyle was dead at the age of seventy-seven. Now Ted would confront the radically different but equally frustrating experience of dealing with Boyle's successor.

CHAPTER 15

Riding the Roller Coaster

The 1973 OPEC oil crisis found the United Mine Workers as bitter and disorganized as the Crow tribe. Its new president, Arnold Miller, couldn't control the rank and file. Its international executive board was divided between Miller loyalists and holdovers from the Boyle era. John L. Lewis's great virtue had been his ability to make a deal and stick to it. But now there was no one who could reliably speak for the union.

In the UMW's new reformed leadership, Ted Leisenring believed he saw an opening for a new era in union-operator relations. Why not, he reasoned, make a reciprocal gesture? Thus in his capacity as chairman of the Bituminous Coal Operators Association, Ted offered the BCOA presidency to Joseph Brennan, the union loyalist who had worked for John L. Lewis and his successors since the late 1950s.

Brennan was the son of Mart Brennan, a UMW district president in the Pennsylvania anthracite region and an International board member. His family roots in the UMW extended back almost to the union's origins in 1890. Joe Brennan himself held a bachelor's degree in economics from Notre Dame and an MBA in marketing from American University. He had served on the UMW's bargaining team during the 1971 national contract negotiations. He understood the union's inner workings and was completely trusted by union men and operators alike. In effect Brennan would occupy a unique position as a protégé of both John L. Lewis and Ted Leisenring. But Ted believed Brennan's presence in the BCOA president's chair would signal to labor that the operators had the interests of both parties at heart.

Unfortunately for this strategy, Arnold Miller was too preoccupied with the ideal of democratizing the UMW to reciprocate such signals. In the process of cleaning out Boyle's cronies, Miller replaced many veteran district vice presidents who, whatever their other failings, at least understood the nuances of give-and-take negotiations with mine operators. The vice presidents Miller installed in their place, conversely, were impractical idealists who were largely incapable of delivering tangible results either to miners or operators. The result was that the union lacked a united negotiating front, and the operators felt little confidence in Miller's negotiating team.

The UMW's three-year contract with the Bituminous Coal Operators Association expired on November 12, 1974, in the midst of the OPEC oil shortage, when the union's bargaining power was probably as great as at any time over the previous past quarter-century. Coal prices were up, and coal companies were making so much money that they were eager to pay almost any price for labor peace. For the moment their greatest concern was not pay scales but wildcat strikes, which had increased dramatically amid the leadership vacuum perceived by many miners after Arnold Miller's election in 1972. The following year the coal industry lost two million man-days to wildcat strikes; by 1975 the number of lost days had risen to 2.8 million. These strikes hadn't troubled coal operators too much when coal prices were low and coal stockpiles were high. But now the reverse was true. For coal miners with any sort of grievance, wildcat strikes became a popular form of extortion against operators who were suddenly desperate to keep their mines functioning.

Only a coal man could understand an irrational phenomenon like a wildcat strike—a point that became painfully clear during the energy crisis as coal companies were acquired by larger energy companies and other conglomerates whose leaders had no experience with coal. Shortly after Joe Brennan arrived at the Bituminous Coal Operators Association he received an angry call from the chief executive of one such parent company, who demanded to know why workers were striking at his coal subsidiary in West Virginia.

"Well, there's a picket line in front of your coal mines," Brennan explained.

"Wait a minute," the CEO broke in. "Do I have an issue with the union?"

"No, the union likes you," Brennan said.

"Is it a contract matter?"

"No, the contract's fine."

"Well, tell me again why we're on strike."

Brennan tried to explain that the UMW's International lacked control over its locals.

"That's exactly what my president told me," the CEO replied, "and you don't make any more sense than he does." Brennan would later cite the conversation to illustrate the futility of explaining coal strikes to "a CEO who lived in the real world."

The UMW's initial contract demands, presented to the BCOA on September 3, 1974, contained more than two hundred proposals, including demands for improved fringe benefits, wage and cost-of-living increases, improved health and safety provisions, and a sizable royalty increase for the UMW's Welfare and Retirement Funds. In effect the union sought to scrap the 1971 contract altogether and negotiate a new contract from scratch. The BCOA, for its part, agreed in theory to wage adjustments in the 1971 agreement but resisted the UMW's proposed smorgasbord of non-economic issues.

Amid the wrangling that followed, a rift developed between Miller and the middle-level officers who made up the UMW's Bargaining Council. Consequently, negotiators didn't even get around to discussing wages until late October, barely two weeks before the contract expired—this at a time when the Miller regime's new democratic bargaining process would require weeks for final approval. Under this process, any tentative agreement needed to be approved first by the Bargaining Council and then by the whole UMW membership.

When the old contract expired on November 12, more than 120,000 UMW members—following the long union tradition of "No contract, no work"—walked off the job. In a round-the-clock bargaining session, the parties reached a tentative agreement the next day. But the UMW's Bargaining Council rejected the new agreement on November 20. Not until November 26 did the council approve yet another agreement, by a vote of 22 to 15.

This agreement was a tribute to Miller's "start from scratch" approach. Among the significant advances won by the UMW were a week of paid sick leave each year, a sickness and accident benefit plan, pension increases of as much as $250 a month for currently retired miners, a wage package with a cost-of-living allowance to protect workers against inflation, the guaranteed right of an individual to withdraw from a hazardous area, a streamlined grievance procedure, and an increase in the number of total paid days off from twenty to thirty. These were new and imaginative provisions that recalled the foresight of John L. Lewis in his prime.

But under Miller's new procedure, the contract still required ratification by the union's general membership. Local unions arranged meetings between December 1 and 4 to discuss and vote on the new contract. But between the members' confusion over the terms (not to mention the process itself), a snowstorm in the East, and the opening of deer-hunting season in Pennsylvania, only 65 percent of eligible union members actually cast ballots. The final tally listed 44,754 votes for ratification and 34,741 opposed. This was sufficient to end the UMW's nationwide soft-coal strike after twenty-four days, but the low turnout was a harbinger of trouble ahead: Only 36 percent of the 121,536 miners covered by the contract had voted to approve it; nearly two-thirds of those covered had voted against it or hadn't voted at all.

Having overthrown a dictatorship, Miller needed time to allow a democratic culture to take hold. But the coal operators couldn't afford the luxury of time. Their economic survival had always depended above all on their ability to dig out as much coal as possible whenever the price was right. Now the coal industry was riding a roller coaster, and the coal operators were finding, as many a U.S. president had found, that in the short run it's often more expedient to deal with a dictator than a democracy.

At the peak of their roller-coaster ride the coal operators were happy to give lucrative new benefits to the union, for the industry's prices and profits were soaring as never before. Even incompetent operators made money in this climate; and operators who had their acts together, such as Westmoreland, registered profit margins that were almost embarrassing. In 1975, Westmoreland reported the highest return on equity of any publicly traded industrial company in the United States. "Westmoreland is demonstrating that it is not one of those companies interested in owning coal merely for the thrill of it," an admiring report by Drexel Burnham Lambert suggested the following year. Ted Leisenring himself was widely lionized as the long-awaited coal executive who had finally brought modern management principles to his industry. "I don't believe in pigeonholing people and I seldom refer to organization charts," he was quoted approvingly in a coal trade magazine. "This is why many of our executives are able to do each other's jobs. When someone is away for any extended period of time, someone else can step in and do almost as well."

As the price of oil climbed through the 1970s, previously unthinkable concepts began to receive serious consideration. Coal, for example, could be converted to gas. This process was so capital-intensive—a coal gasification facility might cost anywhere from $250 million to $500 million—that most coal companies had never explored the idea. But with oil climbing to $30 or even $40 per barrel, the concept began to seem economically feasible. And if coal companies couldn't afford a layout of that magnitude, the major oil companies could. Thus the coal companies became takeover targets—either because they were sound investments or because they owned desperately needed fuel reserves.

In the ensuing scramble, many coal operators sold out to oil and energy companies for hefty prices. Representatives of oil and gas firms began showing up at Westmoreland's Philadelphia headquarters to make their pitches. At the peak of the boom Ted Leisenring received an offer for Westmoreland Coal from Standard Oil of Indiana. The proposal would have made him the largest single shareholder in what was then one of America's dozen largest oil companies—a company that traced its lineage back to John D. Rockefeller himself.

Many coal men, including some of Ted's colleagues and relatives, believed that such a boom would never occur again and, therefore, the time had come to cash in their gains, get out of the coal business and move on to new endeavors. But Ted Leisenring refused to sell Westmoreland to Standard Oil of Indiana or anyone else. The fact was that he was constitutionally incapable of selling his birthright, whatever the price. "When you have five generations of coal in your blood," he later rationalized, "it's very hard to shift gears." He engaged in little analysis of the pros and cons of a sale, and as Westmoreland's controlling stockholder he didn't have to. Against all logic he resolved to keep Westmoreland independent. "It's human nature, I guess," he explained. "You fight for what you inherit."

The UMW's hard-won 1974 contract did not bring an end to wildcat strikes, as the operators (and Arnold Miller) had hoped. Instead, these outlaw strikes, which had once been limited to individual mines or companies, now became epidemic, sometimes spreading from a single mine to an entire state. Bands of roving pickets traveled from one state to another, closing down mines as they went. Such was the leverage of coal at this time—and, consequently, of the men who mined it—that practically any grievance could become a pretext for a wildcat strike, whether or not it was related to coal. When the federal government imposed alternate-day restrictions for filling automobile gas tanks, miners in southern West Virginia struck in protest. In another case, again in southern West Virginia, mine workers struck to protest the content of textbooks in public schools, and soon miners across the country were striking in sympathy. Another wildcat strike targeted a new prohibition against studded tires during the winter.

Such issues defied the control of any coal company, not to mention the UMW's international officials. In effect the UMW's locals and districts had become vehicles for political protest rather than conventional labor negotiations. In other cases the strikes revolved around attempts to legitimize the right to hold wildcat strikes. In March of 1977 Westmoreland's Bullitt mine in Wise County, Virginia, was struck when the company refused two miners' requests to designate "alternative" birthdays for themselves—a paid holiday under the UMW contract—because they had been engaged in wildcat strikes during their actual birthdays. The "birthday grievance dispute," as it came to be called, snowballed into a strike that lasted seven weeks.

The net result was that during 1976 the industry lost 19 million tons of production due to wildcat strikes, and the UMW's Pension and Benefit Funds lost $36.8 million in royalty income. In the first four months of 1977 the losses were 69 percent higher than for the same period twelve months earlier.

The union's 1974 national contract expired in December 1977, and Ted Leisenring, as chairman of the Bituminous Coal Operators Association, shouldered a leading responsibility for negotiating a new contract. As negotiations approached, he publicly staked out his position: "The operators sought one thing in return" for the benefits of the 1974 contract, he said, "and we seek it more than ever in 1977: stability of production along with a return to higher productivity." The anarchy he perceived in the union was no source of comfort to him. "It is far better to reach agreement with a strong union than to suffer the mutual losses of dealing with a weak or divided union," he insisted in a speech in Bluefield, West Virginia, in September. But the coal operators were at least partly to blame for the weakness of the International UMW: As companies like Westmoreland moved their operations to the non-unionized West, only 48 percent of American coal was being mined by UMW members, a sharp drop from 76 percent a decade earlier.

The contract talks formally opened on October 6 at the Capitol Hilton Hotel in Washington. Three years earlier, at the height of the OPEC oil embargo, Miller and his team had been eager, and the operators had been on the defensive. Now the opposite was the case. Arnold Miller, who had survived a hotly contested election four months earlier, came to the table unprepared, virtually alone, and committed by his campaign pledges to demands—most notably a right to strike in mid-contract—that the operators were sure to reject. Three years earlier Miller had opened the talks by presenting the operators with an inch-thick set of demands and supporting background data; now he distributed only a three-page list of bargaining goals to the press, and no list at all to the operators. Most of the first meeting consisted of a slide presentation on industry economics presented by the BCOA's president, Joe Brennan. Two months later, with both sides still far apart, the existing contract expired, and 180,000 mine workers across the country walked off their jobs.

Such a walkout in the absence of a contract was standard procedure, of course. But in the weeks that followed it became clear that settling this walkout would be different. In the UMW, as in all organizations and governments, a divide had long existed between the international officers (who had a vested interest in the health of the union as a whole) and the rank-and-file members (who were preoccupied with personal and local issues). John L. Lewis had solved the problem by squelching the rank and file, and Boyle had tried to do the same. Arnold Miller, in contrast, was determined to return power to his constituents in the mines. It was a noble principle in the long run, but in the short run it meant empowering miners who felt little vested interest in the health of either the international union or the coal industry.

In Miller's zeal to democratize the bargaining process, he had set up a two-tier bargaining system in which agreements reached by Miller and his negotiating committee required approval by a committee representing each of the roughly three dozen UMW districts, prior to ratification by the full UMW membership. In a further effort to broaden participation, Miller insisted on rotating the bargaining committee members, so that each week Ted Leisenring and his fellow coal negotiators found themselves bargaining with a new set of adversaries who hadn't been privy to the previous week's conversations. In effect Miller's process meant that the BCOA was negotiating with UMW international officers, with district officials, and with rank-and-file miners—each with a different set of priorities—all at the same time. "A nightmare," Ted Leisenring called it.

John L. Lewis had understood the cyclical nature of the coal industry and the need for operators to harvest their profits in boom times against the privations of the busts. But even Lewis in his forty-year regime had never encountered profits like these. Ted Leisenring found himself poorly equipped to argue against large wage and benefit increases at a time when his own company's earnings were at an all-time high. "Don't tell us what you can afford," he heard from across the bargaining table. "We get your earnings statements. We know how much money you're making."

Ted would limply reply that the industry hadn't made much money in the '60s and might not make much in the '80s. "If you want a company that's strong and can afford you the security of jobs over a long period, that's one thing," went his mantra. "But if you want to take it out of our hides so that we don't have the capital to put in new mines or equip our own mines over the long period, then your membership is going to suffer with us." With Lewis, such an argument had been unnecessary; but with these new adversaries, it mostly went over their heads.

The operators, for their part, were similarly splintered between huge and diversified companies such as Consolidation Coal (which exercised de facto veto power by virtue of its tonnage) and small and midsize operators. The smaller independent firms, mostly family-owned, were eager to settle because coal was their only source of income, and they threatened to bolt and settle separately if their needs weren't addressed. Ted Leisenring's Westmoreland Coal, which ranked thirteenth in tonnage out of thirty-nine hundred U.S. coal companies, found itself charged with the thankless task of placating both factions.

But another new problem plagued the BCOA's negotiators. Since the OPEC oil embargo and the subsequent boom in coal, many leading coal operators had been acquired by multinational energy conglomerates such as Exxon, Gulf, and Conoco, whose executives had little use or respect for coal men. Peabody Coal, one of the industry's largest companies, had recently installed as CEO Elmer Hill, who blamed the strike on the coal industry's backward leadership and resolved to negotiate his own separate contract behind the BCOA's back. Naively assuming that Miller alone could reach an agreement on behalf of the UMW, Hill met privately with Miller and asked him, "What do you really want?" Of course he failed to reach an agreement and succeeded only in undermining Ted's committee's credibility with Miller.

As the negotiations wore on into 1978, the strike took on a life of its own. Sympathetic independent farmers and truckers organized convoys from the Midwest to bring food to strikers in mine areas. The United Auto Workers contributed more than $2 million to the UMW's strike fund. Yet Miller, who had always shrunk from the spotlight, now seemed genuinely terrified by the nationwide attention. In his first term as UMW president he had often retreated from Washington to his home base in Charleston, West Virginia. One Charleston-based operator, Herbert Jones of Amherst Fuel Company, later recalled that during his presidency Miller would "drop in at our office unannounced, for lunch, and sit there for three hours, subjecting us to monologues about what he wanted to do."

Like Boyle, Miller now "began seeking an enemy under every bed," as one former aide put it. But where Boyle, in paranoid mode, withdrew to a small inner circle of trusted confederates, Miller shut himself off from virtually everyone. For all his theoretical belief in the openness of democratic structures, he confided in no one, causing rumors and misinformation to spread among his own UMW officials. His attendance at bargaining sessions became sporadic. Despite his own call for round-the-clock bargaining, during critical periods in the talks he was often nowhere to be found. For days on end his own aides sometimes had no idea where he was. At one point he was said to have escaped the pressure of the talks by driving around the District of Columbia Beltway while negotiations were in progress. Increasingly he escaped the pressures of Washington altogether for the friendlier confines of the Heart O' Town Motel in Charleston, West Virginia.

As the weeks stretched into months, the chaos only intensified. On January 12, 1978—in the sixth week of the strike—the union's negotiators infuriated the operators by reintroducing the "right to strike" provision, an issue that the operators thought had been taken off the table. Twelve days later the industry's negotiators walked out, largely because of secret negotiations being conducted by a senior UMW aide with a group of small coal operators. At another point Joe Brennan, the BCOA president, took Miller aside and asked, "Arnold, OK, what is it that you want?" Notwithstanding the months of talks before and during the strike, Miller replied, "We haven't prepared that yet. We're not ready to talk about that yet. What will you give us?"

In the vacuum, random acts of violence broke out in mining communities throughout Appalachia and the Midwest. At one point during the strike, Penn Virginia's president John Shober visited Big Stone Gap, against a local aide's advice, to inspect a land reclamation project aimed at planting trees and restocking fish ponds on mine properties abandoned by Westmoreland Coal. Penn Virginia was technically not a coal operator—it merely owned the land on which Westmoreland Coal operated its mines—and Shober's small truck bore a large "Penn Virginia" logo to emphasize that point. But such distinctions were lost on the locals (as, indeed, they were lost on anyone not involved in the arcane financial structures of the coal business).

Shober and a local aide were sitting in the truck examining a fish crop in a pond when suddenly the rear window shattered and a bullet lodged between the two men under the dashboard.

"Is he a good shot?" Shober asked after catching his breath.

"Yeah," his aide replied. "They know what they're doing. We'd better get out of here." The implication was clear: The strikers were shooting not to kill but as a warning.

But such a shooting delivered another implicit message as well: In some perverse way, even in the heat and confusion of the strike, the miners and operators knew and respected each other. Shortly after that shooting incident, Shober returned to Big Stone Gap for a quick inspection trip. Dressed in a business suit, he was being driven through a mine area when he and his Wise County aide found their road blocked by a group of men unloading pipe from a truck. The workers, seeing only the car and the business suit, deliberately slowed their work to prolong the delay of what they presumed were alien outsiders. But when the car inched closer and one of the men came over, the mood changed immediately.

"Hello, Sammy, how are you?" Shober's local aide said to the man.

"Why, Bill," the other replied. "Goddamn, if I'd have knowed it was you, we would have let you through."

"Listen," Bill said, "here's John Shober, he's one of us, and I want you to treat him right." Nothing more needed to be said. As Shober later observed, if you were one of them, there was a real camaraderie whether you were management or union. In that basic decency lay the mystical appeal of coal to miners and managers alike.

In the thick of the contract negotiations, Ted and Julie Leisenring went to see *Harlan County USA,* a remarkably involving documentary about the plight of Kentucky miners, focusing on a 1973 UMW attempt to organize a non-union mine. The film, directed by Barbara Kopple, included actual footage of a company foreman shooting at pickets from his truck, as well as a bizarre conversation in which a miner and a New York City cop compared their salaries and fringe benefits. The film tended to portray the mine operators and their allies as cardboard villains, but Ted and Julie found it so fascinating—Ted believed he recognized some of the people in it—that they later returned to see it a second time.

Shortly afterward the Leisenrings went to Vail, Colorado, for a four-day ski outing with Pem Hutchinson and his wife, Pamela. The Hutchinsons in turn brought along William States Lee, the president of Duke Power Company, which was the largest single consumer of steam coal from Westmoreland's Virginia properties. Lee was a widely respected second-generation utilities executive (his father had been chief engineer of Duke Power), and his company had been honored by the Edison Electric Institute, the industry's trade association, for the safety of its power plants.

One night in Vail when their group assembled a small party for cocktails and dinner in their apartment, the conversation turned to *Harlan County USA*, which had subsequently won the Academy Award for best documentary film of 1976. In her customary outspoken manner, Julie Leisenring pitched eagerly into the conversation.

"Yes, it's one of the most exciting movies I've ever seen," she said. "And I was really amazed to find that it was made by a woman." In her enthusiasm for the quality of the film she had overlooked the subject and its relation to the assembled guests.

"Before you go any further," Bill Lee interjected, "I want you to know that that movie is the damnedest diatribe against my company that's ever been made."

Suddenly the room went silent as the guests contemplated what was happening: The wife of the chairman of Westmoreland Coal was being confronted by the president of Westmoreland's largest customer. Not just personal relationships but millions of dollars' worth of coal business could hang in the balance.

"Do you know," Lee continued, "that that coal mine in the film was owned by one of our subsidiaries?"

"Yes, I do remember hearing that," Julie answered.

"Those people that they were sticking the needle into in that motion picture," Lee said, his face reddening, "that was *my* company. My company has never been so maligned as by that picture."

Julie refused to back down. "Well, I'm sorry if that doesn't suit you," she said, "but that's the way I feel about it. I thought it was a damn good movie." Mortified over the embarrassment she had caused her husband and his company, Julie retreated to the kitchen while everyone else remained frozen in their places in silence.

Finally, Bill Lee broke the silence. "Oh, Ted," he said, "I'm sorry I upset your wife so badly." Impulsively, Lee followed Julie into the kitchen, put his arms around her and kissed her on the cheek. What was true in Big Stone

Gap applied in Vail as well: Among coal people, one moment's crisis could evaporate in the next.

At President Jimmy Carter's request, on February 15—more than two months into the strike—the negotiations moved into the White House, in the hope that the change of venue and Carter's moral suasion would help move things forward. Miller and three UMW vice presidents sat on one side of the table facing Ted Leisenring and three other coal operators. For forty-five minutes they waited for the president. When Carter arrived, he spoke for about fifteen minutes about the importance of joining ranks and ending the strike for the good of the country. To Ted Leisenring, Carter's homilies evoked a preacher lecturing a Sunday school class. There was no dialogue. When Carter was finished, he simply announced, "Gentlemen, it's nice to meet you," shook hands all around and left.

Carter turned the meeting over to one of his aides, whom the negotiators found no more effective. The next day Carter revealed the only apparent arrows in his quiver: He could seek an injunction under the Taft-Hartley Act to force the miners back to work for eighty days while negotiations continued; he could seek legislation empowering him to seize the mines, as Harry Truman had done in 1946; or he could call for binding arbitration. But Carter never used these options, for good reason. "Each option is only as good as the miners' willingness to comply with ultimatums," *Newsweek* magazine noted at the time. "Historically, neither fines nor jail sentences nor bayonets have moved the miners to dig a single lump of coal"—or, at least, not since Truman had nationalized the mines in 1946.

Ultimately Miller and the BCOA endorsed three different contract settlements during the strike, only to see all three of them rejected—two by the UMW's Bargaining Council, and one by the UMW's general membership. The endless jockeying and posturing on both sides made it impossible to discover the real reasons for anyone's opposition. On March 17 the *New York Times* quoted Ted Leisenring's friend Herbert Jones, president of Amherst Fuel Co., the largest independent coal concern in West Virginia, to the effect that "we are not going broke, but it will be very difficult to keep from showing a loss at the end of the year." When Ted Leisenring read that story in Philadelphia, he circulated it to his fellow Westmoreland executives with a note: "I spent Mon.-Tues. 3/20-21 with Herbert and he wasn't so pessimistic as this. I suspect some of his remarks are for rank-and-file consumption." The truth of what happened during the nearly four months of the strike, Joe Brennan later observed, "will never be known—ever—because there were too many people involved who now, for a lot of other reasons, either don't remember or don't want to talk about it."

Not until March 27, 1978, did the miners return to work with a new contract, after 111 days on strike—the UMW's longest strike since the

1920s. To reach a final agreement, the UMW dropped its high-priority demand for the right to strike, and the BCOA dropped its demands for a contractual right to hire, fire, or discipline striking miners. The previous arbitration precedent—allowing operators to discharge leaders of wildcat strikes—remained intact. In effect the critical issue of wildcat strikes that had so vexed both sides remained unresolved.

At Westmoreland, Ted Leisenring sought to seize this moment by developing what he called "an ongoing and intensive communications program" with workers "to identify emerging problems and to resolve them before they fester and erupt into wildcat strikes." This notion sounded eerily similar to the feel-good programs that Stonega Coke & Coal had developed in the 1920s to teach foremen about "giving the individual more attention" and "making the workman feel that his superiors have a real interest in his welfare," and it might reasonably have been expected to be equally ineffective.

Yet to everyone's surprise, the whole issue of wildcat strikes now ran out of steam on its own. Only in retrospect do the reasons make sense: Miners who typically made $12,000 to $14,000 a year were eager to recoup the $3,000 to $4,000 they had lost during the strike. And the new contract gave them a 37 percent wage increase over the next three years, a powerful incentive to stay on the job.

The 1978 contract also coincided with the easing of the OPEC oil embargo. As the demand for coal and the price of coal subsided, so did the operators' pressing need for miners. A wildcat strike that idled a mine for a few days could now be shrugged off, so it was no longer a potent weapon. For whatever reason, in the three years following the 1978 contract agreement the number of wildcat strikes in the coal industry declined by 90 percent, to their lowest level in ten years.

By 1978, UMW workers, who had once dug 70 percent of America's coal, were producing barely 50 percent. The union was unraveling, yet coal operators continued to cling to the hope that they could find another John L. Lewis who could effectively speak on behalf of all mine workers. Westmoreland Coal's officers professed to be encouraged by the rise to the UMW's international vice presidency of Sam Church, himself a former Westmoreland miner who had no qualms about demonstrating his loyalty to his old employer. Shortly after the 1978 national strike was settled, the UMW called a strike to organize a coal mine just across the Virginia line in Kentucky, on land owned by Penn Virginia Corporation. Although Penn Virginia and Westmoreland were separate corporations, they shared many stockholders and officers (including Ted Leisenring himself), and Penn Virginia owned the land on which Westmoreland operated its mines (as well as the unmined coal). But the striking miners, seeing no distinction between the two companies, threw up picket lines at Westmoreland's mines as well. Church's response was immediate and telling: He flew to southwestern Virginia and led a contingent of Westmoreland employees through his own union's picket line. It was, as Joe Brennan would later suggest, Church's way of saying that Westmoreland had always kept its word to the union, and the union wouldn't forget that.

Arnold Miller was only fifty-five years old at the time of the 1978 settlement, but the combination of his black lung disease and the pressures of the strike had aged him. Four days after signing the 1978 agreement, he suffered a slight stroke. Two weeks later, while recuperating in a hospital, he suffered a mild heart attack. Several more heart attacks followed in 1979. These attacks slowed his gait and slurred his speech. But his leadership was equally impaired by the very democratic processes he had introduced, which now empowered his rivals to challenge him. In August of 1979 his own local in Cabin Creek, West Virginia, declined to elect Miller a delegate to the UMW International convention, causing Miller to declare, belatedly, "The president of this union has got to have more control." In November, not yet halfway through his second term, he resigned the International presidency.

Miller's successor was Sam Church, a development that Ted Leisenring seems to have grasped as a cause for optimism. "We can sense a growing respect and trust between our management and local, district and international union officials," Ted said hopefully in a speech in 1979.

Yet neither the UMW contract nor Sam Church's accession eased the coal industry's problems. Japanese steel companies had followed through on their warning to Ted: They had developed a cheaper way to make steel, using a lower grade of bituminous coal from Australia and Canada. American steel companies, desperate to compete with the Japanese, refused to pay the higher costs of the new UMW contract; instead they shopped elsewhere—especially in West Germany—for the bituminous coking coal they needed. Electric utilities were canceling or postponing their plans to expand coal-generated steam plants while they waited for the government to formulate regulations for air pollution controls. In the summer immediately following the four-month UMW national strike, Westmoreland and other operators in Virginia and West Virginia were hampered again by a three-month strike of clerks against the Norfolk & Western Railway, which brought many coal shipments to a standstill. For the calendar year 1978, just three years after Westmoreland had reported the highest return on equity of any publicly traded U.S. company, the company's mining operations actually lost $9.8 million. Between 1977 and 1980 Westmoreland laid off fifteen hundred employees, or about 30 percent of its total work force, at its unionized mines in Virginia, West Virginia, and Pennsylvania.

Ted Leisenring presumed he had seen it all before. "One of the main advantages of being in business for several generations," he remarked to a reporter in December 1978, "is that we are thoroughly familiar with the cyclical nature of the coal industry and we are able to react rationally and calmly to the ups and downs of the business."

Six months later he sounded anything but calm. Utilities were spending fortunes to equip power plants with scrubbers to limit sulfur oxide emissions into their air, as required by the Environmental Protection Agency. The waste sludge produced by these scrubbers was now being converted into stabilized landfill and other by-products. Yet coal continued to be rejected by government policy makers and environmentalists, and demand for coal remained so low that most U.S. mines were running at barely half their capacities. In his frustration, Ted lashed out at one of the perceived culprits: Jimmy Carter.

"We have a president who is an environmental idealist at heart," Ted said petulantly at a college commencement address. "He and Admiral [Hyman] Rickover understand how a submarine works, but he has never had anything to do with coal, and philosophically he has little use for it. . . . The Environmental Protection Agency, with the needling of the environmental extremists, has declared as a policy that 'coal is dirty,' and therefore unacceptable."

But the genuinely new and unprecedented threat to the industry was barely noticed by Ted and his fellow operators. As one concession in the 1977–78 UMW strike settlement, the coal companies had accepted an obligation to provide lifetime medical care for all union members. At the time, that was just another blip on the horizon—a humane if expensive gesture agreed upon during one of those up-cycles when cost seemed irrelevant. No one paused to calculate how the cost might grow as coal production declined and a shrinking number of coal companies were required to provide medical care for an expanding population of former miners. Thousands of small and medium-sized coal companies, Westmoreland among them, were about to confront the law of unintended consequences.

CHAPTER 16

Nowhere to Hide

John L. Lewis had once imagined a brighter day when his miners would no longer work in mines at all, but could spend their days in the sunshine instead. In Lewis's vision, mining machines would replace most human mine workers, so that relatively few miners would be needed. Those who were needed would earn a comfortable living in a less strenuous environment, while the great mass of ex-miners would find new careers in other fields. Something akin to that process was now taking place in southwestern Virginia among the descendants of Bob Givens.

By the late 1970s Bob's son Saylor Givens had watched most of his six sons depart from Westmoreland Coal. Lanis, the oldest, was newly retired after working forty-two years in the mines. Charles had moved to Chicago, and the family had lost touch with him. Ray had moved to Indiana to work in a glass factory. After the Imboden mine closed in 1956, James had moved to New Jersey to run an apple orchard for a large fruit exporting company, only to die there of bone cancer from the insecticide spray used in the orchard. Of the six Givens brothers, only Don and his next-oldest brother, Robert, known as "Rome," remained at Westmoreland. Of Saylor's grandchildren, only Lanis's son Terry had gone into the mines. (Don's daughters, by virtue of their gender, had been spared that decision.)

In the years since Don had first encountered Ted Leisenring in the Imboden mine in 1949, Don's accumulated experience and seniority had brought him job stability and promotions at a time when many younger men were being laid off. After becoming a foreman at Westmoreland's Wentz mine in 1970, when he was forty, Don was sent that year to Westmoreland's Bullitt mine as its general mine foreman, and over the next four years he was promoted to superintendent—the mine's top manager. In the mid-'70s, as mines that had been closed were reopened to accommodate the OPEC boom, Don spent one-year stints as superintendent at two other nearby Westmoreland mines. When surgery on his back—a consequence of twenty-five years spent in low-ceilinged underground spaces—rendered Don incapable of bending over, the company returned him to Wentz as assistant superintendent of the evening shift—not in the mine, but aboveground.

Saylor Givens himself, retired since 1955, continued to contemplate the world from the porch of his little frame house next door to Don's wife and their two daughters in Powell Valley. In seeming defiance of every law of medicine and nature, Saylor had survived forty-six years in the mines and sixty-six years of smoking unfiltered Camel cigarettes, without accident or contracting black lung disease or lung cancer.

One cool Sunday morning in the '70s, when Saylor was about eighty, Don came over to find his father sitting on his sofa with tears in his eyes. When Don asked what was wrong, Saylor replied, "I got strangled on a blamed cigarette." A moment's choking had startled him like nothing in his long life in the mines. Saylor resolved never to smoke again from that moment, a resolution he kept for the fourteen remaining years of his life.

Also gone from the mines were the three sons of Saylor's brother Albert, who had left Wise County to mine coal in Amonate, West Virginia, in the 1940s. The accidental death there of Worley Givens in 1943 had not deterred his brothers Robert and Ed. They remained at Amonate and eventually became supervisors, until a mine explosion there in the mid-1950s finally convinced them to give up mining for good. Robert moved to Dayton, Ohio, and took a job working in the Delco battery plant. Ed also moved to Ohio to work in a factory; eventually he went into the roofing business for himself and became a successful contractor.

After Worley's violent and premature death in West Virginia, his father-in-law made good on his promise to move Worley's family as far from the mines as possible. From 1946 on, Worley's widow, Clara Linkous Givens, raised her four children on her parents' newly acquired farm near Sparta, Ohio, a village about thirty-five miles northeast of Columbus. Her third daughter, Gladys, who was two years old when her father died, subsequently married a local boy named Don Belcher, who became a carpenter. They remained in Sparta and raised four children there. Their third son, Timothy Wayne Belcher, born in 1961, grew like his Givens relatives into a strapping and athletic young man who stood six foot, three inches tall and weighed 220 pounds. At Sparta-Highland High School he played shortstop on the baseball team, but when he wrote letters to some fifteen small colleges in Ohio seeking some type of baseball scholarship he received no reply, not even a letter of rejection.

Eventually Tim Belcher enrolled at Mount Vernon Nazarene College, a school of barely a thousand students not far from his home, where he evolved into a pitcher blessed with a wicked curve ball, a hard slider, and a fastball that was consistently clocked at ninety-four miles per hour. In the final seven starts of his junior season in 1983 he threw a no-hitter, two one-hitters, two two-hitters and two three-hitters. By the end of that season as many as forty professional scouts could be found among the typical crowds of 150 to 200 that attended Mount Vernon Nazarene's home games. In the June 1983 Major League draft of amateur players, the Minnesota Twins—the very first team in the very first round—opened the proceedings by calling out Tim Belcher's name. This twenty-one-year-old young man whose grandfather had lived and died on his knees in the mines was about to embark on a career in Major League Baseball.

Westmoreland Coal Company's transfer of Don Givens to an outdoor job at the Wentz mine meant that by the late 1970s, among the Givens men, only Don's brother Rome and their nephew Terry were still working underground. But underground mining itself had changed dramatically. A big-city visitor to Big Stone Gap in 1980 was amazed to find that a Westmoreland mine there resembled less the inside of a mountain than a dark office building.

"None of the miners use picks and shovels," the visitor, a high school senior from a Philadelphia suburb, recounted afterward. "Everything was done by machinery. The whole effect of being down in the mine was like being in the Broad Street Subway, only it [the mine] was longer and narrower." The crude nineteenth-century "man-drops" that formerly transported workers down the shaft into their rooms—the stuff of Hollywood films like *How Green Was My Valley* and *The Molly Maguires*—had long before been replaced by small railroad cars descending a gently inclined path. The air underground, said the young visitor, "was very cool and very clean, probably because so many safety precautions have been taken and so much modern technology has been introduced into mining procedures during the past fifty years." He was especially surprised by the miners themselves—most of whom, he said, appeared to be in their mid-twenties and "really enjoy what they're doing."

The mines were also much safer places since the grim days of 1943, when Don's cousin Worley Givens was killed in West Virginia by a falling chunk of slate. Throughout the 1970s, fatalities in American mines declined sharply, both in absolute numbers and in proportion to the work force. Ted Leisenring attributed these improvements to a new, conscious emphasis on preventing mine disasters and strengthening mine supports and warning systems against roof collapses. "I submit that while coal mining is hazardous," he declared to a conference on coal safety, "it does not have to kill people."

Yet the statistical improvements brought small comfort to the families of those random mine workers who continued to die there. Rome Givens, like his brother Don, had worked for Westmoreland since the late 1940s. In 1979 he was a strapping fellow just past fifty, a widower who had lost his wife to kidney failure some years earlier. He worked at Pine Branch, the mine that Don in the '60s had relished for the ease with which its soft coal could be cut and loaded. The morning of February 1 found fresh snow on the ground, and Rome was assigned to join an extra crew that was scattering gravel on the slippery track descent into the mine. At the bottom of the dark incline, Rome was sitting in a cutout attached to the front of an empty motorized coal car when a young miner, who was operating the motor in the rear, accidentally threw the wrong gear switch, sending the motor car hurtling forward at full speed. In the darkness, the heavy motor car crashed into a loaded three-car coal train standing idle on the track just ahead. Rome's chest was crushed, and he died instantly. His oldest daughter, Vickie, was left to raise her two sisters as well as their teenage brother.

Rome's death was ascribed to human error, but the following year, a federal commission appointed by President Carter identified Westmoreland Coal as having the worst safety record in the coal industry. In November 1980, five miners were killed when an explosion ripped through a Westmoreland mine in Robinson, West Virginia. A federal court later convicted five mine supervisors of safety violations related to the accident, and Westmoreland itself was fined $300,000 by the West Virginia Department of Mines. To Ted Leisenring, who was widely perceived as the conscience of the industry, this devastating blot on his record constituted one more reason to withdraw from Westmoreland's eastern mines and stake the company's future on the West. From a business perspective, the new UMW contract of 1978, with its 37 percent wage increase, no longer made sense for a coal operator struggling to begin competing with less expensive foreign coal. Thanks to the industry's overeager expansions during the boom of the mid-1970s, coal operators now possessed far more production capacity than they needed. Westmoreland's Montana and Colorado mines— low in sulfur and devoid of unions—were up and running. Why, in that case, should another man have to die below ground in any Eastern mine when clean coal could be mined plentifully and safely on the surfaces of Montana and other western states?

The professional baseball world that Tim Belcher entered in 1983 was in many respects the antithesis of the coal industry in which his grandfather Worley Givens had labored and died. Coal mining was a vital but disorganized and aggressive industry whose managers and labor leaders alike seized every opportunity to undercut and outbid their rivals. Individual mine workers may have lacked mobility or highly marketable skills, but at least in theory they were free to drift from one mine to the next, offering their services to the highest bidder (which is how Worley Givens and his brothers moved from the Stonega mines to the West Virginia mine where Worley was killed).

Baseball, conversely, was merely a game, and its primary performers possessed no useful talent other than a strange ability to run around a field throwing, hitting, and catching baseballs. A lone miner digging a single chunk of coal from the earth performed an intrinsically valuable service, but a great baseball player's talents were worthless outside the context of an organized league. Baseball's value as a commercial enterprise depended entirely on its ability to deliver a controlled and meaningful environment consistently from one season to the next and from one generation to the next.

For this reason, since 1922 Congress had specifically exempted Major League Baseball from the provisions of federal anti-trust laws. Unlike coal operators and other business owners who were specifically forbidden to form the sort of price-fixing syndicates that Henry Clay Frick and Ned Leisenring had formed in the 1880s, this exemption granted Major League franchise owners a collective monopoly that spared them from competing with each other anywhere but on a baseball field. Their players were bound by baseball's "reserve clause" to a single team throughout their careers, unless a team owner sold a player's contract to another team. For most of modern baseball history even the most celebrated and highly paid baseball stars—Babe Ruth, say, or Ted Williams or Willie Mays—lacked the right enjoyed by the lowliest coal miner: the right to shop among prospective employers for the best deal. If a player was unhappy with his team or his employer, his only options were to suffer in silence or find another line of work.

Nevertheless, most baseball players readily accepted this exploitation for the same reason that most miners accepted theirs: because it was a better deal than the available alternatives. When Congress exempted baseball from anti-trust laws in 1922, baseball was America's only organized professional sport, and barely 10 percent of all Americans held high school diplomas. Thus uneducated young athletes were more than willing to give up their bargaining power in exchange for the unique opportunities a baseball career offered them (just as, say, many rural men in the nineteenth century preferred work in dark and dangerous coal mines over the undependable life of sharecropping and drifting). Without a credible league system, the ballplayers' talents would be worthless. It was presumed that baseball would lose its credibility (and consequently its value) if it couldn't enforce competitive parity among its teams—that otherwise the richest teams in the biggest cities could simply lure the best players away from their competitors.

But two generations later, when baseball franchises were valued at tens of millions of dollars and owners routinely moved teams from one city to another to exploit market opportunities, the original rationale for baseball's antitrust exemption no longer seemed defensible. Baseball had evolved from a struggling sport into a lucrative entertainment business. Free-market competition for players, it now seemed, could not undermine the game's integrity any more than the team owners had already undermined it.

The system experienced its first serious challenge in 1969, when the St. Louis Cardinals traded a highly regarded outfielder named Curt Flood to the Philadelphia Phillies. Instead of acquiescing to the trade, as baseball players (and even Babe Ruth) had helplessly done for generations, Flood did the unthinkable: He refused to report to his new club. Instead he wrote to the commissioner of baseball, Bowie Kuhn, asking Kuhn to declare him a free agent.

"I am not a piece of property . . . to be bought and sold," Flood declared. "Any system that produces that result violates my basic rights as a citizen." When Commissioner Kuhn refused his request, Flood filed a lawsuit against Major League Baseball, contending that the reserve clause, by prohibiting a player from choosing his team, violated federal antitrust laws.

Flood's suit eventually reached the U.S. Supreme Court, where it was rejected in 1972. But the sharp division of the court—the vote was five to three—suggested to all concerned that baseball's claim to be a sport, as opposed to a business like any other, would not hold up much longer. Three years later an arbitrator granted free agent status to two pitchers, Andy Messersmith and Dave McNally. In the wake of that decision, owners negotiated an agreement with the players' union under which players with six seasons of Major League experience became "free agents" who could freely negotiate with teams of their choice.

The arcane kinks of this new system were still being worked out through trial and error when Tim Belcher was first exposed to it in 1983. As a consequence, Belcher inadvertently became a center of controversy even before he pitched his first professional baseball game; and before he turned twenty-three he had learned more about baseball's bizarre legal and contractual maneuverings than many veteran players acquired in their entire careers.

The night before the Major League Baseball draft of June 1983, Belcher received a phone call from George Brophy, a Minnesota Twins vice president, about the bonus the Twins were prepared to offer Belcher as an inducement to sign their contract. Brophy suggested a bonus around $90,000. Belcher found that figure insulting for high-level draft picks, who customarily received anywhere from $100,000 to $150,000. The conversation degenerated into a shouting match, and Belcher refused to commit himself to a figure.

"They thought they would intimidate a little country boy," Belcher told a reporter. The conversation poisoned Belcher's feelings about working for the Twins, and although the Twins did indeed draft him the next day and ultimately offered him $125,000 to sign, he rejected their offer, becoming the only first-round draft choice who did not sign a contract. Instead Belcher spent the summer of 1983 touring Europe, Japan, and South America with a United States amateur team. In the process, his team won a bronze (that is, third place) medal at the Pan American games, and Belcher himself became the first member of the Givens family to cross an ocean since his ancestors had been transported from the British Isles two centuries earlier.

The New York Yankees drafted Belcher in a "supplemental phase" draft in January 1984, and he promptly signed their contract, only to find himself seized from the Yankees roster a month later by the Oakland Athletics as compensation for the Athletics' loss of another free-agent player to the Yankees. Belcher's professional minor league debut with the Muskegon Muskies that spring was rained out for three consecutive nights, and when it finally arrived he gave up three runs before being removed in the fifth inning. "I'm twenty-two years old and I have a lot to learn," he acknowledged at the time. "But I think I can make it [to the Major Leagues] within three years." Three years later, in September 1987, Belcher was traded to the Los Angeles Dodgers, and just three days later he made his Major League debut, recording a victory as a relief pitcher against the New York Mets.

In the early 1980s Westmoreland Coal seemed to be bouncing back again. Overseas companies that had previously imported only metallurgical coal for steelmaking suddenly began demanding steam coal to fuel their utilities. The United States, with its hundred million tons of excess capacity left over from the overanticipated boom of the mid-'70s, was the logical place for foreign companies to shop, for two reasons: America's excess steam coal was available at bargain-basement prices, and its shipping rates were cheaper (at least from the eastern United States) than from coal producers in Australia or South Africa.

For these reasons, in the first five months of 1980, America's steam coal exports doubled, and Westmoreland, thanks to its long cultivation of foreign markets stretching back to the days of Daniel Wentz in the 1920s, again led the pack. Ted Leisenring's second in command, Pem Hutchinson was traveling the globe to warm welcomes. The *Wall Street Transcript* proclaimed Ted Leisenring and Pem Hutchinson America's leading coal executives. The company was again making annual profits of $30 million to $40 million—barely half the level of the '70s peak boom years, but healthy nevertheless.

In such a heady environment, even hardened coal men like Ted and Pem couldn't help having their heads turned by their press notices and their financial statements. Once again, friends, competitors, and energy conglomerates alike approached medium-sized coal companies like Westmoreland with offers to buy. This time one of the coal men who welcomed them was Ted's savvy older cousin, John Leisenring Kemmerer Jr. Since his father's death in 1944 John Kemmerer had expanded his family's Kemmerer Coal Company into one of the most successful coal firms in the western states, with the largest open-pit coal mine in the world. Yet at the peak of the 1981 market, when he was seventy, John Kemmerer sold the company to Gulf Oil for $325 million. Then he began agitating to sell his stake in Westmoreland's western operations (which he owned through his inherited stock in Westmoreland's Penn Virginia affiliate).

Penn Virginia's president, John Shober, found himself thinking: *John's smarter than I am. Why is he selling out?* But Kemmerer—reluctant to criticize his cousin Ted's judgment—simply claimed that he needed liquidity for his family commitments. Only when Shober pressed him did Kemmerer finally come clean. "The stakes are getting too big for me," he told Shober. "I can't compete in what's going to happen, and I don't want to take the risks."

Yet Ted remained opposed to selling Westmoreland. He and his lieutenants were older now—Ted was fifty-five, and Pem Hutchinson was fifty—but they were not ready to retire or to uproot themselves and their families to the Midwest or the South.

In the midst of a two-month UMW strike in the spring of 1981, a local UMW financial secretary in West Virginia wrote an encomium to the *Charleston Gazette* praising Westmoreland for continuing to pay employees' health insurance premiums—$200 per month or more—throughout the strike. "I've been with Westmoreland thirty years," wrote Arthur H. Hill, whose local represented Westmoreland's Hampton No. 3 mine, "and they've been good to me and my family, and I just thought people should know what kind of company they are. . . . No other coal company has been doing anything like that for its employees."

This kind of intangible goodwill, accumulated over ninety years in the Virginias, was priceless, Ted knew. Yet he could no longer ignore more concrete calculations. The boom in steam coal exported to Europe and Asia, like all coal booms, ended as soon as the U.S. coal industry's excess coal inventory declined, forcing coal prices to rise yet again. Foreign customers lost interest in American coal as soon as it was no longer cheap. American customers, increasingly, had no interest unless it was clean. To Ted, his company's future survival seemed to boil down to two essentials: low-sulfur coal, mined in non-union collieries. To that end, within a few years in the 1970s his company had poured more money into Westmoreland mining operations in Montana and Colorado than four generations of Ted's ancestors had invested in Pennsylvania and the Virginias over nearly a century and a half. Now Westmoreland's mines were operating at barely half their capacity, and something had to give.

Ted had been loyal to Westmoreland's eastern mines to a fault, but in the face of financial necessity he began phasing out these mines altogether. In 1984 Westmoreland wrote off a $60 million loss connected to the closing of its West Virginia mines, whose coal was no longer needed by America's shrunken steel industry.

Ted's western strategy had been predicated on the hope that the world's oil supply was finite and its price was bound to rise, bringing sharply higher coal prices in their wake, which in turn would justify the hundreds of millions in borrowed funds that Westmoreland and its partners had invested out West.

But the much-awaited upward spike in coal prices never arrived. For one thing, there was too much coal in the world and too much competition among coal producers. For another, Montana and Colorado coal was much lower in heating value and wasn't as energy efficient as Virginia or Pennsylvania coal, so Westmoreland couldn't charge as much for it.

But the rising price of oil also triggered another by-product that Ted hadn't predicted. As oil prices rose, oil men felt justified in developing expensive new technologies for finding and recovering oil. Ted's Philadelphia Main Line neighbors, the Pews of Sun Oil, for example, had gambled half a billion dollars in the 1970s on the painstaking process of scraping oil from tar sands in northern Canada; thanks to rising oil prices, that gamble was paying off. Other oil companies were developing advanced drilling and detection techniques, which revealed that the world's accessible oil reserves were far larger than anyone had previously estimated. Britain's North Sea oil field, for example, had once been presumed to hold no more than twenty-five years' worth of production; now the estimate was adjusted to many times that amount. The so-called global oil shortage had turned out to be no shortage after all. As a consequence, Westmoreland's western mines failed to generate the huge cash returns necessary to pay off its loans and attract new investors. Instead, most years they barely turned a profit.

Things soon got worse. One of Westmoreland's largest customers of western steam coal, the Northern Indiana Public Service Company, filed suit in 1985 for relief from its long-term contractual obligation to buy more coal from Westmoreland than it wanted. Northern Indiana Public Service needed less coal now because it was producing less electricity, a response to the decline of steel companies around Chicago. The shipments were halted in 1987, and that same year Westmoreland shut down its Colorado operations just eleven years after they opened, taking a write-off of $48.9 million.

For all his usually intuitive grasp of coal matters, Ted had failed to anticipate developments in other fields, such as the decline of steel and the rebirth of oil. He was a coal man in a multidisciplinary energy age. Oil companies that had jumped into the coal business in the '70s felt no qualms about unloading their coal subsidiaries in the '80s and '90s once they soured on the industry, but Ted was wedded to coal. Ted's own son and namesake, young Ted (formally Edward Wickham Leisenring, as opposed to Edward Barnes Leisenring), born in 1953, had joined Westmoreland's sales company, sold export coal to Westmoreland's overseas customers, earned an MBA at a Wharton School weekend program and helped to develop a Westmoreland subsidiary for cogeneration (an innovative energy system that enables businesses and institutions to generate their own heat and electricity without depending on a utility company). But as Westmoreland's prospects declined, young Ted opted for a career in a new field—computers—and Ted had to agree that it was the only wise thing for his son to do.

The last straw was the company's obligation, specified in the 1978 UMW contract and its successors, to provide lifetime medical benefits to active as well as retired miners and their families. At the time of the contract, Ted and most other coal operators had assumed, not unreasonably, that they would be responsible only for the care of their own present and former employees. But as mines shut down and companies closed, the question arose as to who was responsible for the "orphans"—the former miners whose companies were no longer in business. In 1989 West Virginia's U.S. Senator John D. Rockefeller IV, with support from senators of three other nearby mining states, proposed legislation that would require the surviving coal companies that had signed the UMW contract to contribute to a common fund for "orphaned" miners in addition to providing health benefits for their own employees. In effect Rockefeller's law would obligate a shrinking number of coal companies to share the burden of an expanding population of orphans.

Over the next three years, as Jay Rockefeller lobbied for his proposal, some media observers perceived in his efforts an attempt to atone for the sins of Jay Rockefeller's great-grandfather, the original John D. Rockefeller, in provoking the Ludlow Massacre of 1914. Jay Rockefeller never specifically alluded to the Ludlow massacre, but he pursued his rescue plan with the fervor of a redemptive crusade.

"This has been a very emotional day," he announced in 1992, when he finally succeeded in attaching his plan to a comprehensive federal energy bill. "I know a lot of people have bad thoughts about what goes on in politics and public life. But if you dig in your heels long enough, you can do good for people." But of course the cost of the mandated health benefits would be borne neither by Rockefeller's family nor by American taxpayers. In effect a Rockefeller had simply shifted the burden of his Ludlow guilt to a long list of other coal operators like the Leisenrings, as well as their shareholders.

The cost of the rescue plan was staggering, especially for smaller companies. As the size of the obligations mounted, the pace of coal company failures accelerated. As coal operators shut down unprofitable subsidiaries, their medical obligations were apportioned among their remaining profitable subsidiaries, which consequently soon became unprofitable subsidiaries. Now coal operators closed down not for lack of business, but to avoid payments to the UMW's welfare and pension funds. By 1986, when Ted Leisenring turned sixty, Westmoreland was spending $12.8 million on medical and dental benefits alone, with the prospect of exponential rises in the years to come.

As Ted began thinking about his retirement, he could foresee nothing but declining prospects for his company. He could sell Westmoreland to a competitor or expand it into another field, but at his age the thought of shifting gears did not excite him. In 1988, after twenty-seven years at the head of the company, he turned his CEO job over to Pem Hutchinson, who proved equally helpless to reverse Westmoreland's fortunes. In 1993 Westmoreland suffered a staggering loss of $93 million; the following year an auditor issued a report expressing doubt that the company could survive. By this time Ted himself had lost faith in the company's survival and sold most of his Westmoreland stock.

"I saw the handwriting on the wall," he later explained. "I couldn't believe the company would come back."

On November 9, 1994, Westmoreland Coal sought protection from its creditors under Chapter XI of the Federal Bankruptcy Code, listing liabilities of $136 million, most of them owed to the UMW's medical fund. America's oldest and most illustrious coal company had reached the end of its long journey. Or so it seemed.

Epilogue
A Hyacinth Blooms at Imboden

In the nineteenth century the relentlessly entrepreneurial Judge John Leisenring had insisted on creating a separate partnership or company for each mine he operated, so as to assure that the failure of any single mine wouldn't drag down the others. For the sake of more efficient management, his great-grandson Ted Leisenring had subsequently consolidated many of those operations under the Westmoreland Coal umbrella. But now Westmoreland's bankruptcy inadvertently vindicated Judge John Leisenring's judgment. Throughout the twentieth century the Leisenring family's Penn Virginia Corporation had continued to function as a separate company—specifically, as the landlord for Westmoreland's coal and other operations. So Westmoreland's Chapter XI filing of 1994 left the assets of Penn Virginia intact and unaffected.

In theory, a Chapter XI bankruptcy filing offers a company temporary refuge from its creditors while the company reorganizes itself in the hope of regaining its viability. In practice, Chapter XI places the company in the hands of court-appointed trustees who usually severely hamper the actions of the company's officers.

Upon Pem Hutchinson's retirement in 1994, control of the unappetizing remains of Westmoreland Coal Company passed to Christopher Seglem, a bright Philadelphia lawyer in his late forties with a Princeton B.A., a Yale law degree, and a Wharton M.B.A. He had impressed Ted Leisenring and Pem Hutchinson during a brief tour in the 1970s as regional counsel to the Environmental Protection Agency's Mid-Atlantic Region, based in Philadelphia, and had come aboard in 1980 as a vice president to operate Westmoreland's Colorado operation. Seglem's strongest credential as a coal executive at this moment, as well as his greatest weakness, was his lack of direct experience in the mining of coal. Precisely because he was neither a coal man nor a businessman, he seemed salubriously unfazed by the magnitude of Westmoreland's problems. To Seglem, Westmoreland represented an opportunity to try his hand at running a company, to attach himself to coal's mystical heritage, and to otherwise inject new adventure into his formerly desk-bound eastern white-collar routine. Like John Leisenring in the 1820s, Seglem didn't fully understand what he was getting into, but he perceived a potential opportunity where few others did.

In short order Seglem closed most of Westmoreland's remaining eastern operations and moved its corporate headquarters from Philadelphia to Colorado Springs, Colorado (not incidentally the hometown of Seglem's wife). His strategy was to buy time with the Chapter XI reorganization and hunker down behind the company's Montana operations until the coal industry's next boom cycle returned. If Westmoreland could survive that long, Seglem theorized, the company's accumulated losses could eventually be offset against future profits for income tax purposes, possibly sparing Westmoreland years of corporate taxes. If such a scenario came to pass, Westmoreland would enjoy a significant advantage over its more diversified competitors.

But this rosy scenario tended to minimize the huge potential liability that had provoked Westmoreland's bankruptcy filing in the first place: the millions, and perhaps hundreds of millions, potentially owed by the company to the retired miners under the federal Coal Act. Only a man with no coal experience, most observers agreed, would have jumped so eagerly into such a hopeless situation.

In his retirement with Julie in their restored former nineteenth-century stable and nearby cattle farm on the western end of Philadelphia's Main Line, Ted Leisenring caught up on his traveling and his reading and remained active in some of his pet projects, including his work with the Eisenhower Fellowships, a program that brings potential foreign leaders to the United States for two months of exposure to American leaders and institutions. When he looked back on his career he sometimes reflected that everything had come to pass pretty much as he had foreseen. The collapse of the U.S. steel industry had destroyed the major market for bituminous coking coal from Pennsylvania and Appalachia. And as mounting evidence linked coal to acid rain and global warming, public officials and private citizens alike presumed that coal of all sorts was a relic of the past. Yet it remained indispensable. The modern world needed electricity every bit as much as it needed clean air; and at the dawn of the twenty-first century coal remained the required fuel for more than half of America's electric power, just as it had been since the 1930s.

Congress tried to grapple with this conundrum in 1990 by amending the Clean Air Act to require power plants to reduce their emissions of sulfur dioxide and nitrogen dioxide, the major causes of acid rain. That way, existing power plants could continue using coal, just as long as they cleaned it up. Many power plants did indeed respond by installing scrubbers and adopting other new technologies to make coal-burning cleaner. But most of them responded by switching their coal supply from Appalachia to the low-sulfur coal basins of the West, just as Ted had predicted in the early 1970s. In due course many underground eastern mines shut down, and by 1999 overall coal production from surface mines in the West had eclipsed that from the East. By 2002 the Powder River Basin alone was generating one-tenth of America's electricity.

By then no member of the Leisenring family remained to reap the possible benefits of Ted's foresight. "The sixth generation of the family"—that is, Ted's son—"got out of the coal industry because we were not of a sufficient size and capital to compete with the giants," Ted reflected in a 1998 interview.

Why, then, had Ted resisted selling Westmoreland for a handsome profit back in the mid-'70s, or even the early '80s, when he had the chance?

"I suppose I'm a Don Quixote," he replied, without rancor or apparent regret. "I can say I nobly fought for a cause that I believe in." That cause was not coal per se—which would continue to survive, if only by virtue of its availability and its vital role in generating electricity—but a particular coal mining culture: "Coal as I knew it—in the East, in underground mines—is rapidly going. Everybody who was connected with it can see that that industry will be gone. It's a different industry."

On a sentimental journey to the Connellsville coal district in western Pennsylvania, Ted and Julie Leisenring visited the village of Leisenring, one of the company towns founded by Ted's grandfather in the 1880s. At the Leisenring company store—no longer company owned, of course, but still functioning as a community general store—they bought a few souvenir ceramic mugs bearing the Leisenring name, then sat on the front steps of the porch with two retired coal miners, a white man and a black man. Like Ted, they betrayed no regret over the choices they had made and the course their lives had followed.

"What's your name, fella?" one of them asked affably. Ted told him.

"Well, I'll be damned," the miner replied. "Was it your granddaddy and your great-granddaddy who put these mines in?"

They wound up chatting for hours about work in the local mines, which had passed from the Leisenrings to Henry Clay Frick and Andrew Carnegie and eventually to a subsidiary of United States Steel Corporation. Then they talked about Ted's mining experiences in Virginia. That Ted belonged to the ownership class that had profited from their labors did not concern the two old miners. They simply accepted, Ted later recalled, that coal companies did business the way they had always done business, and that was simply a fact of life. That Ted was descended from the mines' former owners was not a source of bitterness to them; it was a source of amusement.

Some 149,000 people were still making a living in coal mines across the United States in 1990, when the Clean Air Act amendments were passed. But a decade later that number had been cut almost in half, to 75,000. The closing of the high-sulfur unionized mines of the East dealt yet another crippling blow to what was left of the United Mine Workers of America. Under John L. Lewis the UMW had been the nation's most powerful union, with more than 700,000 members in the 1940s. But by the end of the twentieth century its membership had declined to 30,000.

Yet the remaining miners tended to cherish their jobs, and they rarely quit voluntarily. Although the work underground necessarily remained dark, damp, and dirty, the basic wage was $18.91 per hour, plus benefits worth an additional $6 an hour. Whatever else could be said about the risks of mining, miners were no longer poor: The typical miner made $50,000 a year, and more than half of all miners had at least some college education. John L. Lewis's UMW and the other industrial unions that had led organized labor's struggles in the mid-twentieth century were no longer a significant factor in American economic life. But Lewis's ultimate hope—for a world in which miners were few in number but made a comfortable living, rather than one in which multitudes of miners barely subsisted—had indeed come to pass.

Coal people no longer unload their carts in Mauch Chunk, the Lehigh Valley company town where Josiah White and the first John Leisenring launched the anthracite industry. But Mauch Chunk survived the death of anthracite by reinventing itself as a tourist attraction, even to the extent of changing its name. After the world-famous Native American athlete Jim Thorpe died in 1953 and his native state of Oklahoma failed to erect a monument suitable to his widow, she cast about elsewhere for a final resting place for her husband. Thorpe had never set foot in Mauch Chunk (he played college football at the Carlisle Indian School, some one hundred miles to the west). But the town, struggling for its economic life in the wake of anthracite's decline, was eager to accommodate Mrs. Thorpe. Consequently, in 1954 Thorpe was laid to rest in a large granite mausoleum there, and Mauch Chunk and its suburb of East Mauch Chunk were merged under a new name: "Jim Thorpe."

Yet the town's primary tourist attraction today remains its restored Victorian downtown district, which retains a vivid sense of Mauch Chunk in the days when anthracite was king. Long after Josiah White's gravity-based switchback railroad was judged obsolete for hauling coal, it continued to operate as a tourist attraction into the 1930s, and in 2001 it was revived again, so that visitors can take a forty-minute ride to and from the vicinity of White's original mines. Tourists may also visit the ornate mansion of the Lehigh Valley Railroad founder Asa Packer, or inspect the old city jail where, in 1877, four men were hanged for their part in the Molly Maguire murders.

Neither Josiah White's mansion (later inhabited by Judge John Leisenring) nor Mary Leisenring Kemmerer's mansion still stands. The grand Mansion House hotel, which the first John Leisenring ran in the 1820s, is gone as well, although its remains housed a beer distributorship as late as 1976. But the Kemmerer mansion's carriage house survives, along with its playground, its park (now called Kemmerer Park) and many of its original Victorian carriage paths, retaining walls, iron fencing, and vegetation. And the two John Leisenrings and their extensive families all lie buried together in a fenced-off, half-acre section of the hilltop cemetery overlooking the town they sustained for most of the nineteenth century.

Eckley, Pennsylvania, another anthracite mining town created by Judge John Leisenring and his partners in 1854, survives today as Eckley Miners' Village, a tourist attraction complete with a state-operated coal museum. In a town that housed fifteen hundred miners and their families at its peak in the 1880s, some forty homes and two churches still stand, a few of them available for inspection (and perhaps half a dozen still inhabited). Paramount Pictures used Eckley as the setting for its 1970 film *The Molly Maguires,* starring Sean Connery and Richard Harris. In the process, the film crew returned much of the town to its nineteenth-century appearance, concealing all electric and telephone lines and constructing a huge breaker that hovers over the town to this day.

In Wise County, Virginia, Saylor Givens died of prostate cancer in 1989 at the age of ninety-four, eighty years after he first went to work for Stonega Coke & Coal Co. and thirty-four years after he retired. Saylor's house, which his son Don had built next to his own home in Powell Valley, Virginia, was dismantled after his death. "I didn't want anyone else renting it near me," Don explained. "I like space." At this writing, Don Givens and his wife, Shelby, have lived on their fifteen-acre spread in Powell Valley for forty-six years.

Pocahontas Fuel Company's mine in Amonate, West Virginia, where Worley Givens was killed in 1943, was shut down in 1994. The Pocahontas Company itself was sold shortly afterward to the coal industry's giant, Consolidation Coal Company. The houses in the company town of Amonate were sold long before, in the mid-1950s. Today the West

Virginia part of the town has been largely dismantled, with the exception of two existing company houses, but on the Virginia side most of the homes are still occupied.

After the mines closed in southwestern Virginia, some of Westmoreland's company towns decayed into ghost towns, but others survived—in part because of the low cost of their small wooden homes, and in part because of the staying power of their planners' original vision. Unlike modern suburban communities, these towns predate the automobile, and consequently their houses are clustered closely together, so neighbors can casually drop in on each other on foot without any need to climb in and out of their cars. The abandoned mines themselves and their industrial effluvia reverted with surprising speed to their natural state, so that today these once-grim towns enjoy a backdrop of the same mountain vistas draped in the same vivid spring greens and fall oranges that greeted the first members of the Givens family upon their arrival perhaps two centuries ago. The mines, the beehive coke ovens, the railroad spurs that carried coal from the mines to the main lines, the spot at the defunct Stonega colliery where Stonega's superintendent John Taggart was killed in an explosion in 1896—all have been covered over by such an extent of growth and brush that a casual passerby would hardly suspect that anything noteworthy had transpired there.

The company town of Imboden, where Saylor Givens raised his family and Don and Shelby Givens grew up together, remains very much alive. One of its inhabitants is Hagy Barnett, a cousin of Don Givens' who bought his home at Imboden for $900 after that mine closed in 1956. "We thought it was a good place to raise our children," his wife, Hyson, says. The Barnetts, now in their mid-seventies, continue to live there.

Over four generations, more than twenty members and relatives of the Givens family worked in the mines for Stonega and its successor, Westmoreland. Perhaps not coincidentally, the branches of the Givens family that roamed farthest afield from the mines were those of the two Givens men who were killed in the mines. Steve Givens, a high school student when his father, Rome, was crushed to death by an errant coal car at Pine Branch in 1979, subsequently entered the University of Virginia's nearby Wise campus and became a chemist, first in upstate New York and later in Delaware.

Tim Belcher, grandson of the ill-fated Worley Givens, began the 1988 baseball season—his first full season in the major leagues—as a relief pitcher for the Los Angeles Dodgers, but he soon moved into the team's starting rotation. By the end of the season he had won twelve games while losing just six; the Dodgers had finished first in the National League's Western Division; and Belcher himself finished third in the voting for the league's Rookie of the Year. He won two more games in the League Championship Series that year and one more in the World Series, which the Dodgers won by upsetting the heavily favored Oakland Athletics.

After the 1991 season Belcher was traded to the Cincinnati Reds, where sportswriters found him a prickly combination of outspokenness and puckish good humor. When he complained to Hal McCoy of the *Dayton Daily News* about the droppings left on the field by team owner Marge Schott's dog, McCoy wrote about it and promptly found himself barred from the press dining room by the thin-skinned owner; Belcher responded by ordering a pizza to be delivered to McCoy in the press box. When his forearm swelled up after being hit by a line drive in 1993, he quipped, "I look like Popeye. All I need is Olive Oyl and some spinach." A month later, Belcher was removed from a game after being hit by a line drive (while pitching) and by a pitch (while batting). After the game, when reporters flocked to Belcher's locker for his comments, they found that he had already departed, leaving instead a thoughtful "cardboard interview" taped to his cubicle:

> Media members,
>   To assist you in your quest for quotes, let us eliminate the needless and often time-consuming volley of intellect while crowding around a cramped cubicle. Please except [*sic*] the following as my "serious and truthful" account of the game just ended. . . .

Having qualified for free-agent status following the 1993 season, Belcher spent the rest of his career successfully peddling his services in succession to the Detroit Tigers, the Seattle Mariners, the Kansas City Royals, and the Anaheim Angels. Like his nineteenth-century ancestors, Belcher became something of a drifter; the difference was that he was now making well over a million dollars a year, and sometimes double and triple that amount. Throughout most of the 1990s, in fact, Belcher earned more money as a pitcher than the Westmoreland Coal Company earned from its mining operations. By 1999 and 2000, when he finished his career with the Anaheim Angels, Belcher was making $4.6 million per season.

Belcher ultimately won 146 regular-season games for seven different major league teams before announcing his retirement, after fourteen seasons, during spring training in 2001, when he was thirty-nine. "I have not lost my desire to compete," he said at the time, "only the ability to keep up." At this writing Belcher works for the Cleveland Indians as an assistant to the general manager, and he lives two miles from his parents on a farm near Sparta, Ohio, where he grew up. His grandmother, Clara Givens, never remarried after her husband was killed in the Amonate mine in 1943; today, at eighty-seven, she lives with Tim Belcher's parents—her daughter and son-in-law. Of her famous grandson she says simply, "We're very proud of Tim."

In the fall of 2001, when he was seventy-five, Ted Leisenring returned to Big Stone Gap for a reunion with three of the miners he first met underground at Imboden in 1949. At the Harry W. Meador Jr. Coal Museum—named for the son of the Stonega executive who first arranged for Ted to work in the mines—Ted sat around a stove swapping stories with Don Givens and his wife Jean, Don's cousin Hagy Barnett and his wife Hyson, and Beecher Powers, a Westmoreland foreman who in retirement became the museum's curator. They reminisced about the days in the 1930s and early '40s when Stonega employed more than five thousand men in Virginia, when the bars in the town of Appalachia were so crowded on a Saturday night that "you couldn't walk through." They shared stories about Jack Deaton, the union boss, and how he intimidated miners and whipped supervisors in the bathhouse. Don Givens recalled Ted Leisenring as "a tall hunky boy" when he first came into the mines. They remembered Rome Givens and other men they had known who had died in mine accidents, but they also joked about their own flights of carelessness underground and the dressings-down they had received from their supervisors. They talked about how they passed their idle time when the mines operated only two or three days a week—fishing, playing baseball, or working around the house. They joked self-deprecatingly about their lack of education: Hyson Barnett's first name, she explained with a gentle smile, stemmed from her parents' desire to name her after their favorite flower; and since they didn't know how to spell *hyacinth,* they simply spelled her name as they had always pronounced the word.

With the passage of time, bitter and sweet memories alike seemed to coalesce into a single warm nostalgic brew. But of course these men had never been complainers to begin with. When they spoke of Hagy Barnett's one-armed miner uncle Sam Warden, they dwelled not on his tragic loss of a limb but on the company's generosity in giving him his house rent-free for life: "You couldn't beat Westmoreland," Don Givens said. Hagy Barnett remembered how, when the Imboden mine closed in 1956, he had moved to Dayton, Ohio, to work at the Delco battery plant, only to return eighteen months later because "I couldn't wait to get back here." They mentioned that Westmoreland, in its last years in Virginia, had created a program to hire miners' sons to work the mines over their summer vacations in order to earn money for college.

"It was a company that took care of their people," Beecher Powers said. "When the union would strike, the company would give the men enough money to live on while they were striking. That's right. They might be off two months, but the company gave you enough money to live on while you were striking. Then you'd pay it back when you went back to work. Not many people do that."

After a series of blackouts and power disruptions caused general panic in California in 2000 and 2001, the newly inaugurated President George W. Bush—an oil man who had never been much of an environmentalist to begin with—declared that the nation's energy security mattered more than protecting the environment. The Environmental Protection Agency, responding to a new administration and new imperatives, pointed out that, thanks to technological improvements, coal-fired electricity plants generated 33 percent less pollution than they had thirty years earlier, and more electricity could be squeezed from the same amount of coal, so that coal-fired plants had actually improved their environmental efficiency by 70 percent over those three decades.

In response to these encouraging signals, utilities, desperate to satisfy Americans' unquenchable appetite for electric power, began reconsidering their plans and switching from gas-powered plants to those powered by the ever-plentiful and relatively inexpensive coal. Within the space of a few months in 2001, twenty-two new coal-powered electric plants were proposed. In May 2001 the Bush administration unveiled an energy plan that championed coal as the favored source of electric generation for decades to come. A year later the Environmental Protection Agency issued new rules that absolved many older, dirtier plants from any need to upgrade their equipment. In January 2002 federal regulators approved construction of the largest railroad project since Abraham Lincoln's presidency: a nine-hundred-mile, $1.4 billion line to help move western coal to power plants in the East and Midwest. Automated "unit trains"—as many as 135 aluminum coal cars, each loaded by radio control in about seventy-five minutes with minimal human labor—now ship Powder River coal as far away as Georgia for less than it once cost Stonega to ship coal from Virginia to the Midwest. This technological advance all but removed any economic justification for further coal mining in Appalachia. Nevertheless, in the new rush for coal some companies raced to reopen operations in places like West Virginia and Kentucky, permanently marring landscapes by cutting off mountaintops for the coal, dumping the refuse into the valleys and waters below, destroying hundreds of miles of streams, killing fish and wildlife, and precipitating flash floods. "It's boom time for coal," the *Wall Street Journal* declared in May of 2001. And any boom would have its share of charlatans and greedy opportunists.

Westmoreland Coal Company emerged from more than four years of Chapter XI bankruptcy protection in March of 1999, having paid off all of its creditors in full, with interest. Bankruptcy had put an automatic hold on Westmoreland's staggering obligation to pay lifetime health benefits to retired miners—its own as well as those who worked for defunct coal companies. Two months earlier, using the bankruptcy protection as a negotiating tool, Westmoreland's chief executive, Christopher Seglem, had succeeded in settling the company's health benefit obligations, thus setting the stage for Westmoreland's discharge from bankruptcy just as coal came back into favor again.

Now Westmoreland's huge prior operating losses became an accounting asset, to be carried forward and applied to shelter the company's future earnings from income taxes. After reporting a loss of $1.5 million in 2000, Westmoreland earned $3.5 million on sales of $245 million in 2001, ranking it tenth among the nation's coal producers. Westmoreland's stock price, once down in the $1 to $2 range, rose above $16 per share in 2002. "We now do expect to be profitable going forward," Seglem announced in the fall of 2001.

A year later Westmoreland won the 2002 Platts/*Business Week* Global Energy Award for its acquisition and successful integration of two smaller coal companies, and Seglem was named a finalist in that contest's competition for "CEO of the Year." After nearly two hundred years, the "rock that burns" had at last released its inexorable grip on the descendants of John Leisenring and Bob Givens, only to seduce a new acolyte in the person of Christopher Seglem. Westmoreland had survived once again. And coal itself still endured.

Principal Characters

Ayers, General Rufus A. (1849–1926). Pioneer developer of Wise County, Va. Involved in formation of Virginia Coal & Iron Co., 1879–82. Represented Olinger heirs in Virginia who sold land to General Imboden, subsequently acquired by Leisenrings. Died May 14, 1926, Radford, Va.

Belcher, Timothy W. (1961–   ). Major League Baseball pitcher; grandson of Worley Givens, great-great-grandson of Bob Givens. Born Sparta, Ohio. In 14-year career, 1987–2000, won 146 games for Los Angeles Dodgers, Cincinnati Reds, Chicago White Sox, Detroit Tigers, Seattle Mariners, Kansas City Royals, Anaheim Angels.

Bertsch, Daniel, Sr. (1801–1877). Father-in-law of Judge John Leisenring. Born Lockport, Pa.; to Mauch Chunk as blacksmith, 1827. Construction contractor on Lehigh Canal and Lehigh & Susquehanna RR, 1828–45; contract operator for LC&N mines, 1845–65. His daughter Caroline married Judge John Leisenring, May 12, 1844.

Bertsch, Daniel Jr. (1827–1891). Brother-in-law of Judge John Leisenring. President of Upper Lehigh Coal Co. With Leisenring group, director of Virginia Coal & Iron Co. from its founding in 1882 to his death on Dec. 17, 1891.

Boyle, W. A. (Tony) (1901–1985). John L. Lewis's assistant at UMW, 1948–60; president of UMW 1963–72; convicted of ordering 1969 murder of union rival Joseph Yablonski, 1974.

Brennan, Joseph P. (1935–   ). Labor and coal association official. Joined UMW research department 1957, director 1968–73; National Coal Assn., VP/economics and planning, 1973–75; president, Bituminous Coal Operators Assn., 1975–99.

Carnegie, Andrew (1836–1919). Steel magnate. Born Dunfermline, Scotland, Nov. 25, 1835; to United States, 1848, settled in Allegheny, Pa. Entered iron business 1865; steel 1873. Bought controlling interest in H. C. Frick & Co. 1882; consolidated holdings into Carnegie Steel Co., 1889, and put Frick in charge. Merged company with U.S. Steel, 1901, and retired. Died Aug. 11, 1919, Lenox, Mass.

Frick, Henry Clay (1849–1919). Coke operator in western Pennsylvania. Born West Overton, Pa., Dec. 19, 1849. Organized Frick & Co., 1871; sold controlling share to Andrew Carnegie, 1882; bought out Leisenring group holdings around Connellsville, 1889; chairman, Carnegie Steel Co., 1889–1900. Died Dec. 2, 1919, New York.

Givens, (Charles) Saylor (1894–1989). Virginia coal miner; one of five sons of Bob Givens. Worked mines at Arno and Imboden for Stonega Coke & Coal Co., 1909–56. Father of six miners–Lanis, Ray, James, Charles, Robert, and Don–whose cumulative service with Stonega and its successor, Westmoreland, totaled 164 years.

Givens, Don (1930–). Virginia coal miner and superintendent; youngest son of Saylor Givens. With Stonega Coke & Coal Co. and successor, Westmoreland Coal, 1949–85.

Givens, Robert (**"Rome"**) (1928–1979). Virginia coal miner, fifth son of Saylor Givens. With Stonega Coke & Coal Co. and successor, Westmoreland Coal, 1949–79. Killed in mine accident at Pine Branch, Va., Feb. 1, 1979.

Givens, Worley (1909–1943). Virginia coal miner; grandson of Bob Givens, nephew of Saylor Givens. Worked for Stonega Coke & Coal Co., then Pocahontas Fuel Co. Killed in mine accident at Amonate, W.Va., July 26, 1943.

Gowen, Franklin B. (1836–1889). Born Philadelphia. District attorney of Schuylkill County, Pa., 1861–63; Reading Railroad general counsel, 1863–69; president of railroad, 1869–86. Expanded railroad's coal holdings. Special prosecutor in Molly Maguires murder trials, Pottsville, Pa., 1876.

Hutchinson, S. Pemberton III (1931–). Joined Westmoreland Coal Co.'s General Coal sales arm in 1961. VP of General Coal, 1968; president, 1975. VP of Penn Virginia Corp., 1968–71; president of Westmoreland Resources, 1972–1974; president of Westmoreland Coal Co., 1981–88; CEO, Westmoreland Coal, 1988–94.

Imboden, General John D. (1823–1895). Confederate general, Virginia lawyer, politician, land promoter. Bought 47,000 coal acres in Wise County, Va., 1880, induced Connellsville coal operators Abraham and Christian Tinstman to form Tinsalia Coal & Iron Co. Sold to Leisenring group's newly formed Virginia Coal & Iron Co., 1882. The Imboden seam and the coal town of Imboden, Va., are named for him.

Kemmerer, John Leisenring Sr. (1869–1944). Grandson of Judge John Leisenring. Born Upper Lehigh, Pa. Took charge of father Mahlon Kemmerer's mining interests, especially in Wise County, Va. Dissolved family's anthracite operations, 1928, but retained Kemmerer Coal Co.

Kemmerer, John Leisenring Jr. (1911–2002). Born New York City. Chief executive of Kemmerer Coal Co.; sold it to Gulf Oil, 1981.

Kemmerer, Mahlon S. (1843–1925). Judge John Leisenring's son-in-law. Worked under him as engineer for LC&N, 1862–70, married Leisenring's oldest daughter Annie, 1868. Father-in-law set him up in M. S. Kemmerer & Co., mining concern, 1876; led to diverse mining interests. Formed Kemmerer Coal Co. in Wyoming, 1895.

Kennedy, Thomas (1887–1963). Longtime UMW vice president under John L. Lewis. President of UMW District 7, 1910–25; secretary of UMW International, 1925–34; UMW International vice president 1947–60; succeeded Lewis as president, 1960; died Jan. 20, 1963, Hazleton, Pa.

Knode, Ralph H. (1893–1963). Stonega and Westmoreland executive and coal industry leader. Deputy administrator for fuel for American Expeditionary Force in France in 1918. From 1920s spent entire career with Leisenring group, eventually chairman of executive committee of Stonega Coke & Coal, Virginia Coal & Iron, Wentz Corp., Westmoreland Coal. Director of all from 1920s. Close friend of John L. Lewis and of George Love, chairman of Consolidation Coal Co. and Chrysler Corp. Served two terms as president of National Coal Assn., 1949–53. Died Dec. 7, 1963, Bryn Mawr, Pa.

Leisenring, Edward B. (Ned) (1845–1894). Oldest surviving son of Judge John Leisenring. Born Ashton, Pa. Assistant engineer, LC&N, 1862–69. Superintendent of LC&N's Newport mines (previously owned by his father), 1869–71, Honey Brook Coal Co. and successors, 1871–84; ran most of family's coal companies after father's death in 1884, most notably Connellsville Coke & Iron and Virginia Coal & Iron Co. President also of LC&N, 1893–94. Died Sept. 20, 1894, Hamburg, Germany, four months before birth of his only son, known as E. B. Leisenring Sr.

Leisenring, Edward B., Sr. (Ted) (1895–1952). Born in Nice, France, Jan. 12, 1895. Educated at Hotchkiss and Yale (grad. 1917). Joined family's Hazle Brook Coal Co. 1919, in engineering corps. Mine superintendent, vice president, 1923; president, 1926. With Stonega Coke & Coal Co., director, 1924; VP, 1926; chairman, 1929–51. Also chief executive of General Coal Co., Penn Virginia Corp, Wentz Corp., Westmoreland Coal Co., and Westmoreland Inc. Died June 16, 1952, Bryn Mawr, Pa.

Leisenring, Edward B., Jr. (Ted) (1926–   ). Great-grandson of Judge John Leisenring; third of the "E. B. Leisenrings." Born Bryn Mawr, Pa., Jan. 25, 1926. Served U.S. Navy, World War II; B.A. from Yale, 1949; married Julia Bissell, 1950. Worked in mines of Stonega Coke & Coal, Big Stone Gap, Va., Sept. 1949–Dec. 1951. Also worked for Westmoreland Coal Co. at Irwin, Pa., in 1952. Went to its Philadelphia office, June 1952. Director, 1954; assistant VP, 1955; executive VP, 1959; president, 1961; CEO until 1988; chairman until 1992. Also president and CEO of Penn Virginia Corp. and chairman of General Coal Co.

Leisenring, John, Sr. (1793–1853). Patriarch of Leisenring family. Moved to Mauch Chunk, Pa., about 1828; ran Mansion House hotel for Lehigh Coal & Navigation Co., proprietor, 1832–35; postmaster, 1832–47. Hardware merchant in Mauch Chunk, 1835–51. Died July 9, 1853, Mauch Chunk, Pa.

Leisenring, John, Jr. (Judge) (1819–1884). "Boy Wonder of the Anthracite." Born Lehighton, Pa., Feb. 5, 1819; moved to Mauch Chunk with parents, age nine. With Lehigh Coal & Navigation Co. engineer corps, 1836; supervised construction of Lehigh & Susquehanna RR, 1837–40, extension of LC&N railroads, 1843.

Chief engineer, Lehigh Valley RR, 1852–54. Formed partnership of Sharpe, Leisenring & Co. to lease and operate coal mines, 1854–60; supervised mine and town at Eckley, Pa. Named chief engineer of LC&N, 1860, and moved back to Mauch Chunk; also expanded private mining operations, organizing Lehigh Luzerne Coal Co. (1868–69) and Lehigh & Wilkes-Barre Coal Co. (1872–73). Judge of Pennsylvania District Court, 1871–76. Bought coking coal properties in western Pennsylvania and southwestern Virginia, 1880. Died August 22, 1884, Mauch Chunk, Pa.

Lewis, John L. (1880–1969). Longtime UMW president. Born Lucas Co., Iowa. President of UMW local, Panama, Ill., 1908–09; agent for UMW District 12 (Illinois), 1909–11; AFL representative in Ohio, Pennsylvania, and West Virginia, 1915–17; UMW statistician, 1917, business manager, 1917, vice president, 1917–19; president, 1919–60. Died June 11, 1969, Alexandria, Va.

Miller, Arnold (1922–1985). UMW reform leader. Born West Virginia, worked in mines 1938–70. Succeeded Tony Boyle as UMW president, 1972; served 1972–79, resigning because of ill health. Died July 13, 1985, Charleston, W.Va.

Packer, Asa (1805–1879). Coal entrepreneur. Born Groton, Conn. Arrived in anthracite region 1822, acquired coal lands and canal boats; founded Lehigh Valley Railroad; endowed Lehigh University, 1866; built mansion at Mauch Chunk, Pa.

Pass, Albert E. (1921–). UMW District 19 (Tennessee) secretary treasurer from 1952; implicated in 1969 murder of Joseph Yablonski; convicted 1973 and sentenced in 1974 to three consecutive life terms. Born Rockwood, Tenn.

Seglem, Christopher (1946–). Westmoreland Coal Co. executive. Practiced law in Philadelphia; regional counsel of Environmental Protection Agency Mid-Atlantic region; executive deputy attorney general of Pennsylvania. Joined Westmoreland as VP/general counsel, 1980; president, 1993; chief executive, 1994; chairman and CEO, 1996–present.

Shober, John (1933–). Coal executive, distant cousin of Pemberton Hutchinson; father worked in coal for Leisenring group nearly 50 years. Graduated from Yale, 1955; early computer programmer with Univac 1. Joined Leisenring group's Penn Virginia Corp., 1970, as VP/operations; later became president, CEO.

Sprague, Richard (1925–). Special prosecutor in Yablonski murder case. Born Baltimore, Dec. 22, 1925. Assistant district attorney, Philadelphia, 1956–74. Won convictions of Tony Boyle and associates for 1969 Yablonski murders, 1973–74.

Taggart, John K. 1851?–1896. Connellsville and Stonega superintendent. Born Northumberland, Pa. Ran Leisenring group's Connellsville Coke & Coal operation, 1880–89; Virginia Coal & Iron, 1890–96. Killed in explosion at Stonega, Va., May 23, 1896.

Taggart, Ralph (1887–1951). Stonega and Westmoreland executive, son of mine superintendent John K. Taggart, born at Leisenring, Pa. Married Virginia Bullitt, daughter of Joshua Fry Bullitt, one of founders of Big Stone Gap. Joined engineering corps of Stonega Coke & Coal in Virginia, 1909; general superintendent, 1913; general manager, 1917; chief of Virginia operations, 1923; executive vice president (in Philadelphia), 1929; also president of Westmoreland Coal Co., 1929.

Wentz, Colonel Daniel Bertsch (1872–1926). Son of Dr. John Wentz and grandson of Judge John Leisenring. Born Jeddo, Pa., Sept. 4, 1872. Managed Virginia Coal & Iron Co. after death of superintendent John Taggart, 1896–1904; made president of Leisenring group's bituminous coal companies, 1904; president of all companies after father's death, 1918. As lieutenant colonel, directed Fuel & Forage Division of American Expeditionary Force in France, World War I. A founder of National Coal Association and its president, 1920–21. Died Feb. 8, 1926, Wyncote, Pa.

Wentz, Dr. John S. (1838–1918). Son-in-law of Judge John Leisenring. Born Whitemarsh Township, Pa., March 25, 1838. U.S. Army surgeon in Civil War, followed older brother George, also a doctor, to Leisenring group's Eckley, Pa., about 1867. Married Judge John Leisenring's second daughter, Mary, 1871; became partner in John Leisenring & Co. and superintendent of its Eckley mines, 1875. Father-in-law set him up in J. S. Wentz & Co., 1884, with brother-in-law Edward B. Leisenring as silent partner. On death of E. B. Leisenring, Wentz took charge of most other Leisenring family coal companies, 1894. From Philadelphia, supervised active development of Stonega Coke & Coal properties in Virginia. Died July 1, 1918, Philadelphia.

White, Josiah (1781–1850). Anthracite coal pioneer. Born Mount Holly, N.J., March 4, 1781. To Philadelphia as apprentice, 1796. Successfully experimented with anthracite coal, 1813; revived moribund Lehigh Coal & Navigation Co., 1818; built mines, first railroad, canal on Lehigh River and mining town of Mauch Chunk, Pa. Returned to Philadelphia 1832 but remained active in the LC&N. Died Nov. 15, 1850, Philadelphia.

Yablonski, Joseph A. (1910–1969). UMW official in Pennsylvania. Born Pittsburgh. Opposed W. A. (Tony) Boyle for presidency, 1969; murdered on UMW's orders, Dec. 30, 1969, Clarksville, Pa.

Notes

Intro

Page vi

the source of all their blessings: For refreshing my memories of Pottsville circa 1950, I am grateful to the Pottsville writer Ione Geier, author of "Energy Pulsed through Downtown Pottsville."

dirty business in every sense of the word: Burt, *The Perennial Philadelphians*, p. 199.

Prologue

Page 1

punctuated by long breaks for meals and trips to the tavern: Franklin, *The Autobiography*, p. 74, also 78–79, discusses the reputation he established for working through the entire day and implies that his pace was the exception in a time when most merchants and craftsmen took long breaks for meals and trips to the tavern. Also see Warner, *The Private City*, pp. 7–8, on the slow pace of Philadelphia life in the late eighteenth century.

Page 3

as soon as he was old enough to go off on his own: For Josiah White's background and early years, see Morton, *Josiah White*, pp. 18–31.

"but at home—" that is, in New Jersey—"by farming": Morton, *Josiah White*, p. 39.

a plow, sheaves of wheat and a ship under full sail—reflected the city's priorities: Weigley, *Philadelphia*, p. 208.

unloaded on Philadelphia's Front Street wharfs: Weigley, *Philadelphia*, p. 212, mentions that a total of 3,564 ships came up the Delaware River to Philadelphia in 1805, an average of better than ten per day, since the port was closed by ice for several weeks in the winter.

Page 4

the agricultural odors of wood, charcoal, manure, and beer being brewed: Descriptions of Philadelphia in 1796–1802 are taken from, among other sources, Richardson, "The Athens of America," pp. 208–257; Rottenberg, *The Man Who Made Wall Street*, p. 20; William S. Stevens, "Philadelphia 1802," pp. 46ff.; Morton, *Josiah White*, p. 37.

"Railroads may sometimes supersede canals": *Philadelphia Aurora*, May 1, 1801, cited in Morton, *Josiah White*, p. 50.

"and supping in Baltimore, all the same day.": Morton, *Josiah White*, p. 58. This prediction was apparently made in 1805.

"machines are unknown.": Morton, *Josiah White*, p. 69.

Page 5

"Then I'll quit in toto.": Morton, *Josiah White*, p. 52.

had been tamed by human ingenuity: Morton, *Josiah White*, pp. 77–78.

Page 6

few existing furnace grates could withstand it: Hagner, *Falls of Schuylkill*, p. 43.

his arrest as an impostor and a swindler: Frank Hall, *Mines and Mining*, p. 8.

but would not burn: Groner, *American Business and Industry*, p. 70.

Page 7

two were sent to debtors' prison: *World Book Encyclopedia*.

from the German duchy of Saxony in 1748: Johann Conrad Leisenring came from the town of Hildberghauer, Saxony.

Josiah White and his wire mill at the Falls of Schuylkill: Mimeographed biography of John Leisenring in Westmoreland Coal Co. files at HLA.

Page 8

"before it required renewing.": Hagner, *Falls of Schuylkill*, p. 43; Richardson, "The Athens of America," pp. 235–237; DiCiccio, *Coal and Coke in Pennsylvania*, p. 11; Burt, *The Perennial Philadelphians*, pp. 200–201; Frank Hall, *Mines and Mining*, p. 8. Hall has the millhands experimenting in the morning and discovering the blaze after returning from their midday dinner; the other accounts place the discovery at night. White's partner Erskine Hazard, in an account published by the Pennsylvania Historical Society sometime prior to 1873, says the millhands spent "a whole night" trying to make a fire; see Richardson, *Memoir of Josiah White*.

to revive the moribund Lehigh Coal Mining Company: On White: See Simpson, *The Lives of Eminent Philadelphians.*

Chapter 1

Page 13

"A rock that burns.": The phrase originated with DiCiccio, *Coal and Coke*, p. 4.
"the Moon is visited.": Eisely, "Human life," in *Encyclopaedia Britannica*, 1984.
just as rivers and earthquakes had shaped its past: "Forget nature. Even Eden is engineered." *New York Times*, Aug. 20, 2002, p. F1.

Page 14

"a chunk of coal made valuable under pressure.": Anonymous quote cited in DiCiccio, *Coal and Coke*, p. 1.

Page 15

an entity of almost infinite variety: For origins of coal, see Goldstein and Melnick, "Coal Report."
more effective than wood or charcoal: DiCiccio, *Coal and Coke*, pp. 4–6.
how they were formed by geological forces: DiCiccio, *Coal and Coke*, p. *5.*
covering the fire with earth: Schreiner, *Henry Clay Frick*, p. 11.

Page 17

or 13 percent of America's land area: DiCiccio, *Coal and Coke*, p. 7.

Page 18

one of the great bargains of history: DiCiccio, *Coal and Coke*, p. 15
equal to the best walnut wood: *The Bituminous Coal Industry with a Survey of Competing Fuels,* p. 35, cited in DiCiccio, *Coal and Coke*, p. 21.

Page 19

from America to Europe: Groner, *American Business and Industry*, p. 86.

Chapter 2

Page 21

as well as a reputation as an authority on coal: Morton, *Josiah White*, p. 98.
no more strenuous than helping their parents on the farm: Morton, *Josiah White*, p. 119. For a good description of early methods of extracting coal, see DiCiccio, *Coal and* Coke, pp.19–21.

Page 22

"and how much deeper it is, we do not know": Morton, *Josiah White*, p. 124.

presumed to have been influenced by the devil.": The superstition of tossing spilled salt over one's left shoulder originated because the devil was presumed to be lurking there. See Montagu and Darling, *Ignorance of Certainty*, p. xii.

established its headquarters community of Mauch Chunk.: Mauch Chunk is now known as Jim Thorpe.

Page 23

too shallow for barges to use in either direction: Zagofsky, "Lehigh River and Canal."

support to reconstruct their canal: "History of Josiah White," White Haven Chamber of Commerce (from Iinternet).

Page 24

"he no longer had time to bid us goodnight.": Burt, *Perennial Philadelphians*, p. 201.

the first savings bank in the United States: Burt, *Perennial Philadelphians,* pp. 201–202; Morton, *Josiah White*, p. 98.

to form the Lehigh Coal & Navigation Co.: The merged company received its charter February 22, 1822. See Morton, *Josiah White*, p. 134.

after paying off their debts: Morton, *Josiah White*, p. 122.

"You have our permission—to ruin yourselves!": Morton, *Josiah White*, p. 110.

understood what the term meant: Morton, *Josiah White*, p. 111.

Page 25

any river could be navigated at any level of water: "History of Josiah White," White Haven Chamber of Commerce (from Internet).

"the craft floated safely and smoothly over the rapids.": "Some Switch Back Railroad History," White Haven Chamber of Commerce (from Internet).

Page 26

named for White himself: The Lehigh Canal was completed in 1829. See Morton, *Josiah White*, p. 168.

"exceeds any other fuel known.": Philadelphia *Democratic Press,* March 21, 1821, cited in Morton, *Josiah White*, p. 134.

Page 27

by May it was completed: Encyclopaedia Britannica.com says April.

Page 28

"swift as the wind!": Morton, *Josiah White*, p. 171.

"The meals are wonderful.": Morton, *Josiah White*, pp. 170–171.

Chapter 3

Page 29

from the Lackawanna River to city markets: Morton, *Josiah White,* pp. 149–150.

Page 30

and returned to Philadelphia: Morton, *Josiah White,* p. 180. White died Nov. 15, 1850.

"The whole work is now done.": Richardson, *Memoir of Josiah White,* p. 13. See physical description of White on page 125.

Page 31

"I have been afraid of banks.": Jackson letter to Nicholas Biddle, 1832, cited in Rottenberg, *Man Who Made Wall Street,* p. 28.

"plenty of iron to cool her enemies!": Morton, *Josiah White,* pp. 226–227. Also DiCiccio, *Coal and Coke,* p. 25.

Page 32

more than two hundred railroad charters had been granted in eleven states: DiCiccio, *Coal and Coke,* p. 34.

to the Lehigh and Delaware Rivers and their various canals: DiCiccio, *Coal and Coke,* p. 34.

who manufactured hand engines for volunteer fire companies: Burt, *Perennial Philadelphians,* p. 189.

was superior to anything New York could provide: DiCiccio, *Coal and Coke,* p. 35.

Page 33

94,000 in 1880: Rottenberg, *Man Who Made Wall Street,* p. 40; also DiCiccio, *Coal and Coke,* p. 38.

had all converted to coal-burning locomotives: DiCiccio, *Coal and Coke,* p. 34.

nearly one-fourth of all mined coal was used by railroads: DiCiccio, *Coal and Coke,* p. 37.

Page 34

by 1880, 48.6 percent: DiCiccio, *Coal and Coke,* p. 38.

Page 35

the world's largest corporation twice over: DiCiccio, *Coal and Coke,* p. 35.

"I think I shall derive much benefit from being here a while.": John Leisenring, Cape May, to Anna Maria Leisenring, Mauch Chunk, Aug. 9, 1840, at HLA.

Chapter 4

Page 38

who had learned from him on the job: Biographical notes concerning Canvass White (1790–1834) are taken from *Whitford's History of New York Canals,* p. 1170.

a relatively comfortable $2.50 per day: The pay rate is mentioned in a letter from the first John Leisenring to his son John, Nov. 5, 1837, at HLA.

"if you choose to be careful enough.": John Leisenring to his son John, Nov. 5, 1837, at HLA.

Page 39

and a few Mauch Chunk businessmen: Baer, *Guide,* p. 12, at HLA.

Page 40

"and looked down on all Philadelphians as rather middle class and stuffy.": Burt, *Perennial Philadelphians,* pp. 203–205; Wainwright, "The Age of Nicholas Biddle," in Weigley, *Philadelphia,* 272–273. The quote is from the autobiography of Robinson's grandson, George Biddle, cited in Burt, *Perennial Philadelphians,* p. 204. Moncure Robinson was born in 1801 and died at the age of 90.

Page 41

by blood or marriage: *Brief History of the Associated Companies,* p. 1, at HLA.

later succeeded by John's son Edward B. Leisenring: See Edward B. Leisenring obituary in unidentified Mauch Chunk newspaper, Sept. 28, 1894, at HLA.

Page 42

just south of Upper Lehigh: Baer, *Guide,* pp. 13–14, at HLA.

and them to take a course at Philadelphia Polytechnical College: For biographical details about Edward B. ("Ned") Leisenring (1845–1894), see his obituary in an unidentified Mauch Chunk newspaper, Sept. 28, 1894, on file at HLA.

plus housing and fuel: Typescript biography of Judge John Leisenring, at HLA.

Page 43

of the nation's 36,500 miners: DiCiccio, *Coal and Coke,* p. 31.

Chapter 5

Page 47

what his contemporaries described as an almost hypnotic charm: Lewis, *Lament for the Molly Maguires,* pp. 24–25.

a public official's meager salary: Lewis, *Lament for the Molly Maguires,* p. 36.

and then, in 1869, its president: Lewis, *Lament for the Molly Maguires,* pp. 24–25.

Page 48

more than 60 percent of the total in the anthracite region: See "History of the Reading System," Hagley Library summary, p. xiv, at HLA. Rottenberg, *Man Who Made Wall Street*, pp. 129–130.

Page 49

drowning miners in groundwater floods: Holt, "The Life and Labor of Coxe Miners," p. 9.

they determined their own working hours: DiCiccio, *Coal and Coke*, pp. 47–48.

Page 50

a leading cause of mine workers' deaths during that period: DiCiccio, *Coal and Coke*, pp. 50, 52.

"No damned foreman can look down my shirt collar.": DiCiccio, *Coal and Coke*, p. 48.

if he survived: Holt, "Life and Labor of Coxe Miners," p. 9.

absconded with the union's funds: Burt, *Perennial Philadelphians*, p. 205.

a lasting union presence at the company: Sean Adams, "Different Charters, Different Paths," pp. 78–90.

in northeastern Pennsylvania alone: DiCiccio, *Coal and Coke*, p. 57.

Page 51

was deducted from the workers' pay: DiCiccio, *Coal and Coke*, p. *85*.

"in fact he was everything.": DiCiccio, *Coal and Coke*, p. 127.

risked losing their homes: DiCiccio, *Coal and Coke*, p. 127.

less than 3 percent of all miners' dwellings nationwide had bathtubs or showers: DiCiccio, *Coal and Coke*, p. 91.

Page 52

"Of course, if you were killed in the mines, the company paid your family nothing.": Schreiner, *Henry Clay Frick*, pp. 34–35.

"for he can engage in mining or not.": Korson, *Coal Dust on the Fiddle*, p. 229, cited in DiCiccio, *Coal and Coke*, p. 121.

"unwilling to permit self-respecting men to work.": DiCiccio, *Coal and Coke*, p. 56.

Page 53

"he does not think so much of it.": Holt, "The Life and Labor of Coxe Miners," p. 12.

and wounding forty others: Finley, *Corrupt Kingdom*, p. 120.

a miner or operator could rarely be certain whom he had offended or why.": Lewis, *Lament for the Molly Maguires*, pp. 4–5.

but the numbers may have been much higher: DiCiccio, *Coal and Coke*, p. 58.

Schuylkill County alone experienced 142 unsolved homicides and 212 felonious assaults between 1862 and 1875: Lewis, *Lament for the Molly Maguires*, p. 4.

Page 54

could defeat the threat posed by the unions: Lewis, *Lament for the Molly Maguires*, pp. 24–27.

after discharging or quarreling with miners: Lewis, *Lament for the Molly Maguires*, p. 35.

had been a major victim of the Mollies' sabotage: Lewis, *Lament for the Molly Maguires*, p. 36.

Page 55

"might not be influenced by sympathy.": Lewis, *Lament for the Molly Maguires*, pp. 246–247.

"I should undertake the task.": Lewis, *Lament for the Molly Maguires*, pp. 257–260.

nineteen men had been hanged and others imprisoned as Molly Maguires: Burt, *Perennial Philadelphians*, pp. 205–206.

"have never received a dollar in cash from their labor.": *Philadelphia North American*, May 25, 1881, quoted in Lewis, *Lament for the Molly Maguires*, p. 7.

Page 56

a rare case of battling unions by killing with kindness: *Autobiography of Joseph Smith Harris (1835–1910)*, president of Lehigh Coal & Navigation Co., p. 242, at HLA.

the fund had accumulated a balance of nearly $47,000: *Autobiography of Joseph Smith Harris (1835–1910)*, president of Lehigh Coal & Navigation Co., p. 240, at HLA.

Chapter 6

Page 58

"electric call bells, speaking tubes and all modern conveniences abound.": *Mauch Chunk (Pa.) Democrat*, Feb. 22, 1879, at HLA.

were all founded in the two decades after 1890: Baltzell, *Protestant Establishment*, pp. 127–128.

Page 59

a task that had previously required three months: DiCiccio, *Coal and Coke*, p. 67.

compared to the three-year life of the iron rails it would replace: Groner, *American Business and Industry*, p. 167.

"Steel is king!": DiCiccio, *Coal and Coke*, p. 67.

Page 60

and in 1883 coal would take the lead over wood for the first time: DiCiccio, *Coal and Coke in Pennsylvania*, pp. 61–62.

"or so easily mined.": Rochester, *Labor and Coal*, p. 80, cited in DiCiccio, *Coal and Coke*, p. 41.

and organized the Connellsville Coke & Iron Co.: Baer, *Guide*, pp. 5–8, at HLA. Also see DiCiccio, *Coal and Coke*, pp. 22–25. The name "Tinstman" is spelled "Tintsman" in some accounts, but I use the former spelling because it shows up more frequently in that family's genealogies. The fact that their company was named "Tinsalia" (rather than "Tintsalia") also inclines me to believe that "Tinstman" is the proper spelling.

Page 61

"and nobody thanks me for my efforts in a common cause then, and no one ever will unless I get rich.": J. D. Imboden to Major G. Marshall McCue, Pittsburgh, Dec. 21, 1879. Imboden papers, Alderman Library, University of Virginia. Cited in Henson, "General Imboden." Also see *Hagley Library's Guide*, pp. 8 and 28, at HLA.

Page 62

"This will be a second Connellsville in five years in coke, and a Johnstown in iron and steel.": J. D. Imboden, Big Stone Gap, Va., to Major H.C. Wood, June 17, 1880. Imboden collection at University of Virginia.

"But if I do this we shall be dependent on charity for mere existence.": J. D. Imboden, Pittsburgh, to Annie H. Imboden, July 11, 1880, Imboden papers at University of Virginia.

"Oh! What a relief it will be when eternal money strain is ended, as it will be after my Company is fairly at work.": J. D. Imboden, Pittsburgh, to Annie H. Imboden, July 19, 1880, Imboden papers at University of Virginia.

"I only receive what funds I am obliged to have to live on and keep my family economically.": J. D. Imboden, Bristol, Tenn., to Major H. C. Wood, Dec. 25, 1880. Imboden collection at University of Virginia.

Page 63

"to supply the market with one million tons a year for a thousand years.": Contemporary biography of Judge John Leisenring, at HLA, p. 4. The main sources for my account of the Leisenrings' purchase of the Virginia coal lands are "Formation of Virginia Coal & Iron Co.—1879–82—Recital by General R.A. Ayers," at HLA; and Baer, *Guide*, pp. 8 and 28, at HLA.

"that ruined my father.": Virginia Coal & Iron Co. letter to stockholders, April 28, 1884, with note by Imboden's son scribbled in margin, at HLA.

"it is doubtful whether any income could be derived from the property for a long time.": E. B. Leisenring, Mauch Chunk, to J. D. Imboden, Richmond, Va., July 29, 1884, at HLA.

Page 64

Less than a month later, Judge John Leisenring was dead: Baer, *Guide*, p. 8, at HLA. John Leisenring died on Aug. 22, 1884.

"becoming weaker day by day, passing away without suffering pain.": E. B. Leisenring, Mauch Chunk, to J. D. Imboden, Bristol, Tenn., Aug. 26, 1884, at HLA.

"we want you to be interested.": E. B. Leisenring, Philadelphia, to J. D. Imboden, Jan. 8, 1885, at HLA.

Chapter 7

Page 67

"the most methodical thinking machine I have ever known.": The aging Schwab (1862–1939) made these remarks in the mid-1930s in a series of long interviews with Sidney B. Whipple of the *New York World-Telegram*, cited in Kenneth Warren, *Triumphant Capitalism*, p. 374. Warren erroneously identifies the newspaper as the *New York World-Telegraph*.

Page 68

whiskey had made him the wealthiest man in the area by the time he died in 1870: Warren, *Triumphant Capitalism*, p. 7.

prompting his grandfather to hire him, in 1869, as chief bookkeeper at the Old Overholt distillery: Kenneth Warren, *Triumphant Capitalism*, p. 8.

" I see no reason why I should not become a millionaire during my lifetime.": C.W. Wardley, "The Early Development of the H. C. Frick Coke Co.," address to Westmoreland-Fayette Historical Society, June 18, 1949, p. 79, in Westmoreland Coal file at HLA.

Page 69

"He had the soul of a bookkeeper.": Josephson, *Robber Barons*, p. 47.

123 acres of coal land for more than $50,000: Josephson, *Robber Barons*, pp. 260–264.

in which he held a one-fifth interest: Josephson, *Robber Barons*, pp. 260–264; also Warren, *Triumphant Capitalism*, p. 12.

"and all that was weakly human was to be stripped and flung aside.": Josephson, *Robber Barons*, 260–264.

Page 70

Frick bought out his uncle's interest: Schreiner, *Henry Clay Frick*, pp. 33–34. Also Josephson, *Robber Barons*, pp. 260–264. Josephson places the explosion in 1870, Schreiner a few years later.

and Mellon himself had known Frick's mother when she was a girl: Warren, *Triumphant Capitalism*, p. 12.

"knows his business down to the ground.": Warren, *Triumphant Capitalism*, p. 13.

and, in his opinion, a worthy risk: Wardley, "The Early Development of the H. C. Frick Coke Co.," pp. 80–81, at HLA; also Josephson, *Robber Barons*, pp. 260–264.

But Frick—who by this time owned four hundred acres of coal lands and two hundred coke ovens- refused to close down: The figures are taken from Warren, *Triumphant Capitalism*, p. 13.

for goods they purchased at Frick's company stores: Josephson, *Robber Barons*, pp. 260–264; also DiCiccio, *Coal and Coke*, p. 129.

he had a business plan, and he intended to follow it: Schreiner, *Henry Clay Frick,* p. 25.

Page 71

and Frick controlled four-fifths of their output: Warren, *Triumphant Capitalism,* pp. 10–11.

along the Allegheny and the Monongahela in and around Pittsburgh: Holbrook, *Age of the Moguls,* pp. 81–88.

Frick's largest customer, the steel baron Andrew Carnegie, wrote in his memoirs: Josephson, *Robber Barons,* pp. 260–264.

and learned that he was worth a little more than $1 million: Holbrook, *Age of the Moguls,* pp. 81–88.

and the two businessmen began to think of themselves as partners rather than rivals: Warren, *Triumphant Capitalism,* p. 19.

Page 72

"for any crumbs you may in your own good nature see fit to shower upon us.": Warren, *Triumphant Capitalism,* p. 23.

Page 73

"therefore should I be careful to choose that life which will be the most elevating in its character.": Thorndike, *The Very Rich,* p. 330.

Carnegie resolved to buy him out as well: Thorndike, *The Very Rich,* p. 330; also Holbrook, *Age of the Moguls,* pp. 81–88.

" a man with a positive genius for management.": Schreiner, *Henry Clay Frick,* p. 38.

about one-third of them from the Thaw family of Pittsburgh: Holbrook, *Age of the Moguls,* pp. 81–88.

Chapter 8

Page 75

less than that of the neighboring regions: *History of Fayette County, Pa., 1882,* p. 411.

"the purer and better it is found.": *History of Fayette County, Pa., 1882,* p. 411.

Page 76

"limited only by the means of transportation.": *History of Fayette County, Pa., 1882,* p. 411.

up to the company's ovens at the company town of Leisenring: Baer, *Guide,* p. 26, at HLA.

"Your property has the elements for one of the best future paying enterprises in the country.": Annual report, Feb. 10, 1881, in *History of Fayette County, Pa., 1882,* p. 411.

operated a coal mine nearby at Tamaqua in the early 1850s: Baer, *Guide,* p. 19, at HLA.

"and was very reserved in his business matters.": *Big Stone Gap (Va.) Post*, May 28, 1896.

with a potential capacity of five thousand tons: *History of Fayette County, Pa., 1882*, pp. 412, 520.

Page 77

"and its adaptability for sustaining heavy burdens.": Connellsville Coke & Iron Co. circular, Leisenring, Pa., Oct. 1, 1882, at HLA.

"uniform quality and preparation can always be depended upon.": Connellsville Coke & Iron Co. circular, Leisenring, Pa., Oct. 1, 1882, at HLA.

"or more than two-thirds the capacity of the Connellsville district.": The four syndicate participants were: H. C. Frick & Co. (2,784 ovens), McClure Coke Co. (1,146 ovens); Colonel James Schoonmaker (780 ovens), and Connellsville Coke & Iron Co. (764 ovens). See article in *Connellsville (Pa.) Daily Courier*, May 1914, on file at HLA.

Page 78

in the process expanding his control to one-third of the region's total productive capacity: Baer, *Guide*, p. 26, at HLA.

his wealth was estimated at more than $1 million: Baer, *Guide*, p. 26, at HLA. Judge John Leisenring died on Aug. 22, 1884. See Baer, *Guide*, p. 32, at HLA.

Page 79

who succeeded to most of the judge's offices as head of the Leisenring group: Baer, *Guide*, pp. 8–9, at HLA.

at a salary of $2,500 a year: Excerpt from board of managers' meeting, Dec. 7, 1869, at HLA.

when the LC&N acquired the property: E. B. Leisenring, "Report of Superintendent Newport Mines," Jan. 1, 1870, at HLA.

from complications of the second childbirth: Baer, *Guide*, p. 9, at HLA.

"he had become entirely familiar with all its details.": Edward N. Leisenring obituary in Mauch Chunk newspaper, Sept. 28, 1894, at HLA.

Page 80

had failed to achieve any of their goals for long: Aurand, *Molly Maguires to United Mine Workers*, p. 96–114, cited in DiCiccio, *Coal and Coke*, p. 56.

"and our voices must be heard in the legislative halls of our land.": DiCiccio, *Coal and Coke*, p. 56.

in the bituminous coal fields of western Pennsylvania: DiCiccio, *Coal and Coke*, pp. 57–58.

Page 81

and the inflated prices charged at company stores: Warren, *Triumphant Capitalism*, p. 44. Also Schreiner, *Henry Clay Frick*, p. 44.

three-quarters of the coke ovens around Connellsville were idle: Warren, *Triumphant Capitalism*, p. 44.

offered to settle with his workers: *Hagley Library's Guide*, p. 26–27, at HLA.

"We do want to go along there regularly now.": Warren, *Triumphant Capitalism* p. 44.

the workers' complaints about company stores and short-weights would be addressed later: Schreiner, *Henry Clay Frick*, pp. 44–46; also Warren, *Triumphant Capitalism*, p. 44.

whenever a grievance arose: Baer, *Guide*, p. 26, at HLA.

Page 82

"and get in a set which you can depend on.": Edward B. Leisenring, Mauch Chunk, to J. K. Taggart, May 3, 1886, at HLA.

"their new demands are unfair and unjust and cannot be granted.": Edward B. Leisenring, Mauch Chunk, to J. K. Taggart, June 7, 1886, at HLA.

"I would not offer objections upon the above conditions.": Edward B. Leisenring, in Leisenring, Fayette County, Pa., to J. K. Taggart, Aug. 3, 1886, at HLA.

"and never heard anything further about it.": Edward B. Leisenring, Pittsburgh, to J. K. Taggart, Oct. 15, 1886, at HLA.

Page 83

"if you want to keep him.": Edward B. Leisenring, Pittsburgh, to J. K. Taggart, Oct. 15, 1886, at HLA.

"you better have your eye on someone to replace him at any rate.": Edward B. Leisenring, Mauch Chunk, to J. K. Taggart, Oct. 21, 1886, at HLA.

"the sooner the fight must come.": Edward B. Leisenring, Mauch Chunk, to J. K. Taggart, Oct. 29, 1886, at HLA.

"What do you think about it?": Edward B. Leisenring, Mauch Chunk, to J. K. Taggart, April 12, 1887, at HLA.

Page 84

"there is no use dealing with them.": Edward B. Leisenring, Birmingham, Ala., to J. K. Taggart, May 1, 1887, at HLA.

"it might create an impression, as you say, that our Company are very anxious to have the men go to work and want to compromise.": Edward B. Leisenring, Mauch Chunk, to J. K. Taggart, May 23, 1887, at HLA.

"I do not think I will have anything more to do with it.": Edward B. Leisenring, in Leisenring, Fayette County, Pa., to J. K. Taggart, June 2, 1887, at HLA.

at an estimated loss of $250,000 per hour: James Howard Bridge (Carnegie's secretary), *The Inside History of the Carnegie Steel Company*, quoted in Schreiner, *Henry Clay Frick*, p. 51.

Page 85

"I object to so manifest a prostitution of the Coke Company's interests in order to promote your steel interests.": Frick to Carnegie, May 13, 1886, and June 7, 1886, cited in Schreiner, *Henry Clay Frick*, p. 52.

the miners had lost $689,000 in wages: Warren, *Triumphant Capitalism*, pp. 44–45.

most notably at the Leisenring shaft: *Pittsburgh Times*, June 27, 1887, cited in Warren, *Triumphant Capitalism*, pp. 44–45.

"However, we don't want to do it unless compelled to.": Edward B. Leisenring, Pittsburgh, to J. K. Taggart, June 27, 1887, at HLA.

"I hope the men themselves in the whole region will conclude to give up the fight next week.": Edward B. Leisenring, Pittsburgh, to J. K. Taggart, June 27, 1887, at HLA.

"and careful in using the cipher so it will not confuse us.": Edward B. Leisenring, Mauch Chunk, to J. K. Taggart, July 7, 1887, at HLA.

Page 86

and in case of trouble you can depend on them to stick: Edward B. Leisenring, Mauch Chunk, to J. K. Taggart, July 7, 1887, at HLA.

"to the leases signed by the parties who are to be turned out.": Edward B. Leisenring, Mauch Chunk, to J. K. Taggart, July 13, 1887, at HLA.

"unless they are necessary.": Edward B. Leisenring, Mauch Chunk, to J. K. Taggart, July 21, 1887, at HLA..

"to decide whether to resume or not.": Edward B. Leisenring, New York, to J. K. Taggart, July 11, 1887, at HLA.

" wire me promptly.": Edward B. Leisenring, New York, to J. K. Taggart, July 11, 1887 at HLA.

Page 87

"as it is expensive to have them sent so far.": Edward B. Leisenring, Mauch Chunk, to J. K. Taggart, July 13, 1887, at HLA.

"I told him to supply him with the coal for his own use.": Edward Leisenring, Mauch Chunk, to J. K. Taggart, Oct. 16, 1888, at HLA.

"and keep them together.": Edward Leisenring, Mauch Chunk, to J. K. Taggart, July 15, 1887, at HLA.

"you must make preparations to feed them until they get to work.": Edward Leisenring, Mauch Chunk, to J. K. Taggart, July 21, 1887, at HLA.

Page 88

"you can send them $100 for me personally.": Edward Leisenring, Mauch Chunk, to J. K. Taggart, Feb. 28, 1889, at HLA.

"I have myself paid several thousand dollars sending men out, policemen, &c.": Edward Leisenring, Mauch Chunk, to J. K. Taggart, July 21, 1887, at HLA.

"so that I know exactly how things are going on at all the places.": Edward Leisenring, Mauch Chunk, to J. K. Taggart, July 21, 1887, at HLA.

"but I suppose I forgot to do it.": Edward Leisenring, Mauch Chunk, to J. K. Taggart, July 29, 1887, at HLA.

"but for a few days longer.": Edward Leisenring, Mauch Chunk, to J. K. Taggart, July 15, 1887, at HLA.

Page 89

"on the same basis that the new men went in on.": Edward Leisenring, Mauch Chunk, to J. K. Taggart, July 21, 1887, at HLA.

for a 12.5 percent pay increase: Schreiner, *Henry Clay Frick*, p. 53.

"and have the others removed their goods from the side of the road?": Edward Leisenring, Mauch Chunk, to J. K. Taggart, Aug. 13, 1887, at HLA.

"notwithstanding there may be one or two kickers.": Edward Leisenring, Mauch Chunk, to J. K. Taggart, Oct. 3, 1887, at HLA.

"I do not know what has been done by the other operators on the subject.": Edward Leisenring, Mauch Chunk, to J. K. Taggart, Aug. 2, 1887, at HLA.

"I would prefer to make our own wage arrangements with our own men.": Edward Leisenring, Mauch Chunk, to J. K. Taggart, Aug. 2, 1887, at HLA.

"I am tired and disgusted with this meeting committees and arriving at no conclusions.": Edward Leisenring, Mauch Chunk, to J. K. Taggart, Aug. 15, 1887, at HLA.

Page 91

"and sent their mules off to pasture.": Edward Leisenring, Mauch Chunk, to J. K. Taggart, Oct. 3, 1887, at HLA.

"At least it is so claimed by our Agents.": Edward Leisenring, Mauch Chunk, to J. K. Taggart, Jan. 20, 1888, at HLA.

"so that there is a decent profit in the business.": Edward Leisenring, Mauch Chunk, to J. K. Taggart, Jan. 20, 1888, at HLA.

Page 92

"when we get ready to do so.": Edward Leisenring, Mauch Chunk, to J. K. Taggart, Feb. 20, 1888, at HLA.

"and shows what little men some people can be.": Edward Leisenring, Mauch Chunk, to J. K. Taggart, Feb. 20, 1888, at HLA.

resumed his efforts to buy out his competitors: For the collapse of the pool, see Baer, *Guide*, p. 26, at HLA; also *Connellsville (Pa.) Daily Courier*, May 1914, clip in Westmoreland file at HLA.

"in order to keep up the organization and give them employment.": Edward Leisenring, Mauch Chunk, to J. K. Taggart, May 4, 1888, at HLA.

Page 93

"please advise me.": Edward Leisenring, Mauch Chunk, to J. K. Taggart, Sept. 4, 1888, at HLA.

"but at the same time I want them to have the impression that we are always honorable and just to our employees.": Edward Leisenring, Mauch Chunk, to J. K. Taggart, Sept. 11, 1888, at HLA.

Page 94

"The store is only a secondary consideration.": Edward Leisenring, Mauch Chunk, to J. K. Taggart, Oct. 8, 1888, at HLA.

"as in other places in the region where similar wages are paid.": Edward Leisenring, Mauch Chunk, to J. K. Taggart, Oct. 8, 1888, at HLA.

"and he and I understand one another perfectly.": Edward Leisenring, Mauch Chunk, to J. K. Taggart, Oct. 8, 1888, at HLA.

"would gladly have him in our Co.": Edward Leisenring, Philadelphia, to J. K. Taggart, Oct. 30, 1888, at HLA.

Page 95

"However, I hope all this will be remedied.": Edward Leisenring, Philadelphia, to J. K. Taggart, Oct. 30, 1888, at HLA.

"Yes at least that much if not more.": Edward Leisenring, Mauch Chunk, to J. K. Taggart, Nov. 12, 1888, at HLA.

"and get him interested in that way.": Edward Leisenring, Mauch Chunk, to J. K. Taggart, Nov. 24, 1888, at HLA.

"we would do so rather than develop [it].": Edward Leisenring, Mauch Chunk, to J. K. Taggart, April 12, 1889, at HLA.

"I will try to come out sometime next week.": Edward Leisenring, Mauch Chunk, to J. K. Taggart, March 26, 1889, at HLA.

"wherever you go.": Edward Leisenring, Philadelphia, to J. K. Taggart, May 18, 1889, at HLA.

but in steel as well: Warren, *Triumphant Capitalism*, p. 54

from raw materials to finished product: Josephson, *Robber Barons*, pp. 260–264.

Page 96

"nor do I know what he proposes to do about his wages.": Edward Leisenring, Mauch Chunk, to J. K. Taggart, March 26, 1889, at HLA.

"We are certainly willing to reimburse your Company for the coal removed.": Edward Leisenring, Mauch Chunk, to H.C. Frick, Esq., Pittsburgh, May 10, 1889, at HLA.

"it would be very satisfactory to me.": Edward Leisenring, Mauch Chunk, to J. K. Taggart, June 18, 1889, at HLA.

"on the first day of August.": Edward Leisenring, Mauch Chunk, to J. K. Taggart, July 12, 1889, at HLA.

which by that time had expanded to nine thousand acres of coal and 1,500 coke ovens: Connellsville (Pa.) *Daily Courier,* May 1914, clip on file at HLA.

"That's what we will pay you.": Author's interview with E. B. Leisenring Jr., April 22, 1998; also see *Forbes,* Sept. 1, 1971, p. 30.

The price, according to some published accounts, was more than $3 million: *Big Stone Gap* (Va.) *Post,* Nov. 19, 1896. Since Big Stone Gap was the Leisenring group's Virginia headquarters, it seems likely that someone in or close to the group was the source of this figure.

Page 97

"no prospect of any dividends.": *Connellsville (Pa.) Daily Courier,* May 1914, clip on file at HLA.

he now dominated a coke industry that had grown in gross revenues from less than $4 million in 1880 to nearly $8 million in 1889: *Connellsville (Pa.) Daily Courier,* May 1914, clip on file at HLA. The exact figures are $3,948,000 in 1880 and $7,974,000 in 1889. The number of coke ovens in the region similarly rose from 7,211 in 1880 to 14,458 in 1889. C. W. Wardley, in "The Early Development of the H. C. Frick Coke Co.," June 18, 1949, p. 81 (HLA), mentions that all but two of Frick's forty coke plants were acquired by purchase. The exceptions were the Adelaide plant (named for Frick's wife) and Standard No. 2. Clay F. Lynch, in talk to Westmoreland-Fayette Historical Society, June 18, 1949, says Frick & Co. undertook no other construction until about 1904. Talks are on file at HLA.

"and it will win.": Schreiner, *Henry Clay Frick.* Holbrook, in *The Age of the Moguls,* pp. 85–86, places the assassination attempt at the Hussey Building in downtown Pittsburgh.

Page 98

he "displaced his feelings for Martha.": Judith Dobrzynski, "Mourning Became Frick as an Art Collector," *New York Times,* Oct. 9, 1998, page E1. The book cited is Martha Frick Symington Sanger, *Henry Clay Frick. An Intimate Portrait.*

would become the backbone of the newly created United States Steel Corporation: Lynch, comments to Westmoreland-Fayette Historical Society, June 18, 1949, p. 84, at HLA.

Chapter 9

Page 101

but few of them actually settled there: Herrin, *From Cabin to Camp,* p. 10.

or even Virginia's original lost colony at Roanoke: For a discussion of Melungeon origins, see Melungeon Heritage Association website, Wise, Va., rootsweb.com/~mtnties/melungeon.html.

Page 102

the last surviving remnant of America's original rugged individualists: William Goodell Frost, "Our Contemporary Ancestors in the Southern Mountains," *Atlantic Monthly,* March 1899, pp. 311–319, cited in Herrin, *From Cabin to Camp,* p. 1.

sometime after the Civil War: Author's interview with Don Givens, Oct. 27, 2001.

Page 103

leaving him stranded there: Edward ("Ned") Leisenring describes this exchange in a letter to J. K. Taggart, January 9, 1890 (possibly 1891— the date is indistinct), at HLA. Also see Baer, *Guide*, p. 34, at HLA.

to Eastern investors invading this part of the world: Prescott, *Virginia Coal & Iron Co.*, chapter 8.

Page 104

"when they can get good regular dividends": Edward B. Leisenring, Mauch Chunk, to J. K. Taggart, Jan. 9, 1890, at HLA.

"except make address Big Stone Gap, Virginia.": Edward B. Leisenring, Mauch Chunk, to John O. Tombler, Philadelphia, June 25, 1890, at HLA.

"unless an earthquake swallows it.": William Canty, "Derby Goes Union," p. 1, at HLA.

Page 105

be working for him.: Baer, *Guide*, pp. 34–35, at HLA.

since the case now involved an out-of-state firm: Report of J. K. Taggart, 1894, at HLA, cited in Herrin, "From Cabin to Camp," p. 22.

"The sucker who buys the land without investigation of titles.": Prescott, *Virginia Coal & Iron Co.*, chapter 8.

set the standard for the industry: Canty, "Derby Goes Union," p. 2, at HLA.

Page 106

to proceed with mine shafts and ovens: Canty, "Derby Goes Union," p. 4, at HLA.

"which cannot now be estimated.": *Big Stone Gap (Va.)Post*, Sept. 26, 1890, at HLA.

but by April the Louisville & Nashville and the Norfolk & Western were both operating in Wise County as well: Prescott, *Virginia Coal & Iron Co.*, p. 6.

and produce its first coke: Canty, "Derby Goes Union," p. 3, at HLA.

Page 107

he would be ready: Prescott, *Virginia Coal & Iron Co.*, p. 3.

and contained a similar high grade of coking coal: Prescott, *Virginia Coal & Iron Co.*, p. 5.

later author of the coal-country novel *The Trail of the Lonesome Pine*: Canty, "Derby Goes Union," p. 4, at HLA.

"so that we can go on and keep our accounts regularly?": Edward B. Leisenring, Mauch Chunk, to John O. Tombler, Philadelphia, Oct. 4, 1890, at HLA.

Page 108

for the prep schools favored by the Leisenrings, Wentzes, and Kemmerers: Author's interview with S. Pemberton Hutchinson, Sept. 16, 1998. Also see Shiflett, *Coal Towns*, p. 34.

even if they gobbled up what little level farmland existed in the region: Herrin, "From Cabin to Camp," p. 23.

with the latest comforts and conveniences: Canty, "Derby Goes Union," p. 1, at HLA.
"we are masters of the situation.": J. K. Taggart, Big Stone Gap, to Edward B. Leisenring, Nov. 22, 1893, at HLA.

Page 109

By the time they arrived, Ned had already died, at the age of forty-nine: Edward B. Leisenring obituary in Mauch Chunk paper, Sept. 28, 1894, at HLA.
But to him would fall the burden of developing the Virginia Coal & Iron Company: Baer, *Guide*, p. 15, at HLA.

Page 110

was contracted to build five hundred coke ovens at Pioneer: Herrin, "From Cabin to Camp," p. 36.
the first coal was mined there: Prescott, *Virginia Coal & Iron Co.*, chapter 10, p. 5.
a contraction of the words "Stone Gap.": Prescott, *Virginia Coal & Iron Co.*, chapter 10, p. 1.
The growing town at this point consisted of some one hundred miners' houses, offices, a company store, and a stable: *Big Stone Gap (Va.)Post*, Nov. 19, 1896, at HLA.

Page 111

"What a blow to his devoted wife, his daughter Helen and two sons, Jack and Ralph.": Prescott, *Virginia Coal & Iron Co.*, chapter 10, p. 3. Also see *Big Stone Gap (Va.) Post*, May 28, 1896.
as the company's land agent: Prescott, *Virginia Coal & Iron Co.*, chapter 10, at HLA.
"than was Mr. Taggart himself.": *Big Stone Gap (Va.)Post*, Nov. 19, 1896, at HLA.
as far away as Wisconsin, Illinois, and Alabama: Prescott, *Virginia Coal & Iron Co.*, chapter 10, p. 5.
more than doubling that of the runner-up, Great Britain: DiCiccio, *Coal and Coke*, p. 66.

Page 112

adopted by Edison's rival George Westinghouse: Groner, *American Business and Industry*, p. 179.

Page 113

but to auto factories, offices, schools and hospitals: Zieger, *John L. Lewis, Labor Leader*, p. 7.
had left in 1884: Baer, *Guide* p. 16, at HLA. John Wentz died July 1, 1918.
in a bewildering array of mining conditions and markets: Zieger, *John L. Lewis, Labor Leader*, p. 13.
coal or men: Zieger, *John L. Lewis, Labor Leader*, p. 10.

Chapter 10

Page 115
for a family of eight: Zieger, *John L. Lewis,* pp. 1–2.

Page 116
booking traveling companies and exhibits: The figure on worker idleness is taken from DiCiccio, *Coal and Coke,* p. 119.
"No great man can have friends.": Zieger, *John L. Lewis,* p. xviii.

Page 117
"that strikes may become unnecessary.": Finley, *Corrupt Kingdom,* p. 23.
"as makes present-day civilization.": DiCiccio, *Coal and Coke,* p. 133.
and a treasury of $1 million: DiCiccio, *Coal and Coke,* p. 136.
killed in a disaster in Hanna, Wyoming, in 1903: Zieger, *John L. Lewis,* p. 5.
and, ultimately, ruthlessness and guile: This incident is cited by Murray Kempton in *Part of Our Time: Some Ruins and Monuments of the Thirties* (New York: Modern Library, 1998), quoted in *New York Review of Books,* Feb. 18, 1999, p. 4.

Page 118
but Lewis's biographers suggest that her true achievement lay in providing Lewis wih a stable and nurturing home: Zieger, *John L. Lewis,* p. 6.
became a city police magistrate: Zieger, *John L. Lewis,* pp. 6–7.

Page 119
for the rest of his life: Author's interviews with Don Givens, Oct. 27, 2001, and Nov. 7, 2002.

Page 120
and more reliable than cutting trees: Author's interviews with Don Givens, Oct. 27, 2001, Sept. 12, 2002, and Nov. 5, 2002.

Page 121
and the nearby town of Appalachia: *A Brief History of the Associated Companies,* p. 4, at HLA.
as well as the company's own miners: *A Brief History of the Associated Companies,* p. 4, at HLA.

Page 122
going to work for Stonega C&C: Herrin, "From Cabin to Camp," p. 56.
to prevent the grime of coal from intruding into their homes: Herrin, "From Cabin to Camp," pp. 58–59.
carved out of the wilderness: Shiflett, *Coal Towns,* p. 37.

for class feeling and behavior: Corbin, *Life, Work, and Rebellion*, p. 61.
for nearly a generation: Baer, *Guide*, p. 128, at HLA.

Page 123

"of which so much depends.": George F. Baer (1842–1914), open letter to press, August 1902, quoted in *New York Times*, March 5, 1978.
It is time that vigorous measures were taken.: Collier and Horowitz, *The Rockefellers*, p. 108.

Page 124

Worthy of the support of every man who loves his country.: Collier and Horowitz, *The Rockefellers*, p. 108.
The violence didn't end until President Wilson ordered federal troops into the area: This account of the Ludlow massacre is taken largely from Collier and Horowitz, *The Rockefellers*, pp. 106–114.

Page 125

and the number of mining deaths was not decreasing but rising: DiCiccio, *Coal and Coke*, p. 121, says nearly 50,000 miners died in U.S. mining accidents from 1870 to 1914. Zieger, *John L. Lewis: Labor Leader*, p. 9, says 26,434 miners were killed between 1890 and 1917; he also says that accidental mine deaths averaged 1,600 annually "well into the 20th Century."
An additional 12,500 miners were maimed each year: Zieger, *John L. Lewis*, p. 9.
"That's the way we handle these Slovaks.": Author's interview with John Shober, son of Pemberton Shober Jr., Oct. 1, 1998.
"or squat on a pile of slate.": Zieger, *John L. Lewis*, p. 8.

Page 126

almost 19 percent of all "transportation men" brought into the area left without working: Herrin, "From Cabin to Camp," pp. 78–79.
freshly baked coke had to be pulled from the ovens with long-handled ladles by hand: Baer, *Guide*, p. 126, at HLA.
"only a cottage is a room in the mines.": Herrin, "From Cabin to Camp," p. 85.
spent all of the town's rent receipts on house repair and sanitation: Herrin, "From Cabin to Camp," p. 62.

Page 127

"or in conversation with their friends.": Stonega annual report, 1917, p. 3, cited in Canty, "Derby Goes Union," p. 6, at HLA.
as far away as Cincinnati: Canty, "Derby Goes Union," p. 6, at HLA.
"and other disaffection.": *Coal Age* magazine, 1915, cited in Herrin, p. 84.
might seep into the houses: Herrin, "From Cabin to Camp," p. 62.
but also a strain on community order: Herrin, "From Cabin to Camp," p. 64.

"and trash of all kinds were recklessly strewn about.": Herrin, "From Cabin to Camp," p. 62.

first-aid teams were segregated by race: *Hagley Library's Guide,* pp. 126–127, at HLA.

Page 128

and hired additional teachers: 1917 Stonega annual report, cited in Canty, "Derby Goes Union," pp. 4–5, HLA.

to figure out what he would say next: Author's interview with E. B. Leisenring Jr., Aug. 5, 2002. He identified the close friend as J. B. Morrow.

Page 129

had increased by more than half, to 8,921: DiCiccio, *Coal and Coke,* p. 137.

"if recognition is granted.": Stonega annual report, 1917, p. 5, cited in Canty, "Derby Goes Union," p. 18, at HLA.

"as given in April and May.": Stonega annual report, 1917, p. 4, cited in Canty, "Derby Goes Union," p. 6, at HLA.

hoped it would take care of their labor shortage: Stonega annual report, 1918, p. 5, cited in Canty, "Derby Goes Union," p. 6, at HLA.

as fierce as ever: Stonega annual report, 1918, p. 7, cited in Canty, "Derby Goes Union," p. 6, at HLA.

Page 130

allowed the company: Canty, "Derby Goes Union," p. 7, at HLA.

became the leader of the national association: Canty, "Derby Goes Union," p. 7, at HLA.

Page 131

to attain that rank: Associated Companies press release announcing death of Knode, December 1963.

sought an armistice in November 1918: Associated Companies press release announcing death of Knode, December 1963. The reference to Rickenbacker was provided by E.B. Leisenring Jr. in interviews with the author.

and made a chevalier of the French Legion of Honor: Baer, *Guide,* p. 17, at HLA.

Page 132

after the war ended: *Encyclopedia of Biography,* Vol. 16, p. 238. Also author's interview with E. B. Leisenring Jr., April 24, 1998.

became a mine superintendent: Author's interview with E. B. Leisenring Jr., April 22, 1998, pp. 6–7.

Page 133

"Our ship made port today.": Zieger, *John L. Lewis,* p. 19.

in the give-and-take of the capitalist system: DiCiccio, *Coal and Coke,* p. 140.

"I will not fight my government, the greatest government on earth.": DiCiccio, *Coal and Coke,* p. 140.

Page 134

and they could hardly expect to escape Lewis's notice: Stonega annual report, 1941, cited in Canty, "Derby Goes Union," p. 2, at HLA. Proportion of miners is from Shiflett, *Coal Towns*, p. xii.

he had helped to be found: Hagley Summary, pp. 16–17, at HLA.

Chapter 11

Page 135

in order to ship coal directly to eastern markets: See *A Brief History of the Associated Companies*, p. 7, at HLA.

they no longer needed quite as much: DiCiccio, *Coal and Coke*, p. 151.

with only two-thirds of the amount of coal required in 1913: Canty, "Derby Goes Union," p. 4, at HLA.

Page 136

as they had in 1900: DiCiccio, *Coal and Coke in Pennsylvania*, p. 151.

in Rio de Janeiro, São Paulo, and Santos: Author's interview with E. B. Leisenring Jr., Dec. 24, 2002.

"to any degree of industrial activity.": Shober, "John L. Lewis: Power Baron," pp. 67–68.

"would decline to just 6 percent.": DiCiccio, *Coal and Coke*, p. 152.

Page 137

on a full-time basis: Finley, *The Corrupt Kingdom*, pp. 59–60; DiCiccio, *Coal and Coke*, p. 150.

"at the nearest railroad switch.": Irvine, "Wise County in War Time."

compared to 15 percent in most other industries: Shober, "John L. Lewis: Power Baron," p. 69.

that is, less than one railway car per working day: Canty, "Derby Goes Union," p. 3, at HLA.

a local historian suggested: Irvine, "Wise County in War Time."

accounted for more than 5 percent of the market: DiCiccio, *Coal and Coke*, pp. 156–57, says Stonega C&C was listed as the tenth-largest coal company in 1929. See *A Brief History of the Associated Companies*, p. 4, at HLA, for Stonega's 1929 output.

"Twice too many mines and twice too many miners.": Lewis to the U.S. House, 1922, quoted in Canty, "Derby Goes Union," p. 4, at HLA.

Page 139

was still being loaded by hand: DiCiccio, *Coal and Coke*, pp. 158–59.

he subsequently explained: DiCiccio, *Coal and Coke*, p. 194.

"Companies can no longer countenance superficial observance of processes by foremen.": *Coal Age*, February 1931, p. 62, cited in Dix, *What's a Coal Miner to Do?*, p. 81.

Page 140

"everybody knew where he went next and such as this.": Dix, *What's a Coal Miner to Do?*, pp. 90–91.

just before he was gunned down: Zieger, *John L. Lewis*, pp. 26–27.

all but vanished from West Virginia: Zieger, *John L. Lewis*, pp. 27–28.

Page 141

"[and] that is what the country was founded for.": *Appalachia (Va.) Independent*, Oct. 4, 1922, cited in Canty, "Derby Goes Union," pp. 8–9, at HLA.

and by December 1 every remaining Stonega C&C employee had signed: Canty, "Derby Goes Union," pp. 8–9, at HLA.

"our field is the logical source of supply.": Canty, "Derby Goes Union," p. 4, at HLA.

"Next time we'll make it five.": Zieger, *John L. Lewis*, pp. 32–33.

Page 142

"in the decade after World War I.": DiCiccio, *Coal and Coke*, pp. 171–173.

Soon other northern operators were copying Pittsburgh Coal's tactics: Zieger, *John L. Lewis*, p. 33.

"more unsanitary than a modern swine pen.": Zieger, *John L. Lewis*, p. 33.

to negotiate their own separate contracts with the Central Competitive Field: Zieger, *John L. Lewis*, p. 33.

in order to compete with the South: Baer, *Guide*, p. 120, at HLA.

assumed the presidency in 1919: DiCiccio, *Coal and Coke*, p. 176.

Page 143

in which Howat's supporters shouted "Mussolini!" at Lewis: Zieger, *John L. Lewis*, pp. 38–39.

had only one dues-paying member: Finley, *Corrupt Kingdom*, p. 64.

Page 144

"to the dogs below.": Zieger, *John L. Lewis*, p. 42.

"lazy and disunited.": DiCiccio, *Coal and Coke*, p. 171.

"union man hunted union man.": Zieger, *John L. Lewis*, p. 51.

Page 145

"to mark the graveyards of almost twenty thousand jobs.": Zieger, *John L. Lewis*, p. 52.

in the nearby *Crawford's Weekly*: *Crawford's Weekly*, Aug. 5, 1921, p. 5, cited in Canty, "Derby Goes Union," pp. 7–8, at HLA.

Page 146

with the local mine management: Canty, "Derby Goes Union," pp. 9–10, at HLA.

imposing the yellow-dog contract that summer: Canty, "Derby Goes Union," p. 8, at HLA.

"his superiors have a real interest in his welfare,": Stonega annual report, 1922, p. 2, cited in Canty, "Derby Goes Union," p. 9, at HLA.

"We kinda insisted on 'em coming.": Derby superintendent B. E. "Brownie" Polly, interview, 1974, cited in Canty, "Derby Goes Union," p. 9, at HLA.

to prevent them from spreading: Canty, "Derby Goes Union," p. 4, at HLA.

and certain lower-level straw bosses: Stonega annual report, 1922, p. 37, cited in Canty, "Derby Goes Union," p. 10, at HLA.

Page 147

to septic tanks: Stonega annual report, 1922, p. 37, cited in Canty, "Derby Goes Union," p. 11, at HLA.

along the valley's only road: Stonega annual report, 1925, p. 53, cited in Canty, "Derby Goes Union," p. 11, HLA.

they could fester into violence or strikes: Canty, "Derby Goes Union," pp. 10–11, at HLA.

Page 148

where good coal was difficult to find: Canty interview 1973 with former Derby superintendent Oakley Powers, cited in Canty, "Derby Goes Union," pp. 11–12, at HLA.

never felt a pressing need for foreign workers: Canty, "Derby Goes Union," p. 12, at HLA.

he suddenly took ill and died: *Encyclopedia of Biography,* Vol. 16, p. 240.

Page 150

had launched the family's coal holdings a century earlier: Baer, *Guide,* p. 78, at HLA.

when Derby opened in 1922: Canty, "Derby Goes Union," p. 13, at HLA.

"We regret to give you a reduction in wages.": Joe Blanton, interview, 1974, cited in Canty, "Derby Goes Union," p. 14, at HLA.

rather than owners who had a stake in the company's survival: *Brief History of the Associated Companies,* p. 7, at HLA.

Page 151

to retire from the coal business in 1932: DiCiccio, *Coal and Coke,* p. 182.

Page 152

in the three years after 1929: Bernstein, *Lean Years,* p. 360, cited in Canty, "Derby Goes Union," p. 14, at HLA.

that had last been reduced in 1927: Annual report, 1941; *Crawford's Weekly,* July 2, 1932, p. 1, cited in Canty, "Derby Goes Union," p. 14, at HLA.

assured Ted Leisenring in December 1930: J. D. Rogers to E. B. Leisenring, Dec. 19, 1930, cited in Canty, "Derby Goes Union," p. 15, at HLA.

Page 153

"by one miner only.": Stonega annual report 1933, p. 5, cited in Canty, "Derby Goes Union," p. 14, at HLA.

"We have gotten nothing to eat at this job for ten days.": DiCiccio, *Coal and Coke,* p. 179.

were drawing relief "wages" of 20 cents an hour: *Crawford's Weekly,* Nov. 19, 1932, p. 1, cited in Canty, "Derby Goes Union," p. 15, at HLA.

"you can see that seeds of revolt are being sown.": DiCiccio, *Coal and Coke,* p. 179.

"to furnish continuous employment.": Stonega , annual report, 1931, p. 3, cited in Canty, "Derby Goes Union," p. 15, at HLA.

Page 154

folded for lack of company support: Canty, "Derby Goes Union," p. 15, at HLA.

in contrast to rival operators: Canty, "Derby Goes Union," p. 15, at HLA.

or provide occasional jobs: Canty personal interviews in 1974, cited in Canty, "Derby Goes Union," pp. 16–17, at HLA.

had long been an active fund-raiser for that group: *Crawford's Weekly,* Dec. 6, 1933, p. 1, cited in Canty, "Derby Goes Union," p. 17, at HLA; also Dr. C. B. Bowyer, physician in charge at Stonega Hospital and administrator of the Tubercular Fund, letter to general manager J. D. Rogers, Sept. 18, 1931, cited in Canty, p. 17, at HLA.

"created for the Virginia Tuberculosis Association.": C. B. Bowyer to J. D. Rogers, Sept. 18, 1931, cited in Canty, "Derby Goes Union," p. 17, at HLA.

Page 155

"but we're really afraid of the publicity.": J. D. Rogers to R. E. Taggart, Sept. 19, 1931, cited in Canty, "Derby Goes Union," p. 18, at HLA.

"This ain't no time for action, now.": William Duncan, quoted in Canty, "Derby Goes Union," p. 19, at HLA.

Page 156

"only after thorough investigation of the economic effect of the decision.": *Appalachian Coals Inc. v. U.S.,* 288 U.S. 344, cited in Canty, "Derby Goes Union," p. 20, at HLA.

"to get all they can for their labor.": *Crawford's Weekly,* March 29, 1933, cited in Canty, "Derby Goes Union," p. 20, at HLA.

Page 157

realized otherwise: Canty, "Derby Goes Union," p. 21, at HLA; also see Zieger, *John L. Lewis,* pp. 61–62.

"it took away their breath.": Zieger, *John L. Lewis,* p. 65.

membership in the United Mine Workers had quadrupled: DiCiccio, *Coal and Coke,* p. 185.

"from among the ranks of Virginia miners.": *Coalfield Progress,* Norton, Va., July 13, 1933, cited in Canty, "Derby Goes Union," pp. 21–22, at HLA.

Joe Blanton recalled many years later: Joe Blanton, 1974 interview, cited in Canty, "Derby Goes Union," p. 22, at HLA.

Page 158

remarked the weekly *Coalfield Progress*: *Coalfield Progress,* Norton, Va., July 13, 1933, p.1, cited in Canty, "Derby Goes Union," p. 22 , at HLA.

"was down there.": Joe Blanton, 1974 interview, cited in Canty, "Derby Goes Union," p. 22, at HLA.

and also to rebut the organizer's claims: *Coalfield Progress,* Norton, Va., July 13, 1933, cited in Canty, "Derby Goes Union," p. 22, at HLA.

Despite these efforts, between 150 and 200 men took the UMW oath that day: *Big Stone Gap (Va.) Post,* July 13, 1933, p. 1, cited in Canty, "Derby Goes Union," p. 22, at HLA.

To be sure, the company took pains to find other excuses for these dismissals: District attorney's document of January 1934 in the Stonega archive lists men discharged from Stonega Co. mines in the summer of 1933. See Canty, "Derby Goes Union," p. 23, at HLA.

"so they'd catch you, you see.": Joe Blanton 1974 interview, cited in Canty, "Derby Goes Union," p. 23, at HLA.

"But the tipper sheet didn't show dirt in my coal.": George Sanders, quoted in *Crawford's Weekly,* Aug. 30, 1933, p. 4, cited in Canty, "Derby Goes Union," p. 23, at HLA.

Page 159

"in the midst of our colliery villages for several months.": Report of Genl. Counsel J. L. Camblos in Stonega Annual Report, 1933, p. 162, cited in Canty, "Derby Goes Union," p. 23, at HLA.

for all Virginia miners: *Big Stone Gap (Va.) Post,* Aug. 3, 1933, p. 1, cited in Canty, "Derby Goes Union," p. 23, at HLA.

as any mule driver with two arms: Author's interview with Don Givens, Sept. 14, 2002.

and wasn't about to move: Author's interviews with Don Givens and Hagy Barnett, Big Stone Gap, Va., Oct. 27, 2001.

"and he was really against the union.": Canty interview with Madge Trigg, Feb. 18, 1974, cited in Canty, "Derby Goes Union," p. 24, at HLA.

Page 161

"half of us went on.": Canty, interview with Fred Sloan, July 1, 1973, cited in Canty, "Derby Goes Union," p. 27, at HLA.

"Everything we done we had to play it safe.": Canty interview with Joe Blanton, Feb. 21, 1974, cited in Canty, "Derby Goes Union," p. 27, at HLA.

effectively preventing all motor traffic from reaching the mines: *Big Stone Gap (Va.) Post,* Sept. 28, 1933, p. 1, cited in Canty, "Derby Goes Union," p. 27, at HLA.

were working in the mines: Sarah Blanton interview, 1974, cited in Canty, "Derby Goes Union," p. 27, at HLA.

fined each guard $10: *Big Stone Gap (Va.) Post,* Sept. 21, 1933, p. 1, cited in Canty, p. 27, at HLA.

"if they thought you was goin' to work.": Blanton interview, Feb. 21, 1974, cited in Canty, "Derby Goes Union," p. 28, at HLA.

Page 162

though only on a 50 percent production basis: Stonega annual report, 1933, p. 4, cited in Canty, "Derby Goes Union," p. 28, at HLA.

"to the locals to which they belong.": Stonega annual report, 1933, p. 5, cited in Canty, "Derby Goes Union," p. 28, at HLA.

Page 163

the largest number of workers covered by a single labor agreement in any industry in American history: DiCiccio, *Coal and Coke,* p. 186.

almost on par with oil and gas workers: DiCiccio, *Coal and Coke,* p. 187.

to confirm the weight of the coal they dug: *Coalfield Progress,* Sept. 21, 1933, p. 1, cited in Canty, "Derby Goes Union," p. 30, at HLA.

in an argument over the Coal Code: Annual report, 1933, pp. 163–164, cited in Canty, "Derby Goes Union," p. 30, at HLA.

to stop all rail traffic: *Crawford's Weekly,* Oct. 4, 1933, p. 1, cited in Canty, "Derby Goes Union," p. 30, at HLA.

Page 164

The company falsely blamed the explosion on a union miner's cigarette: H. B. Humphrey, U.S. Bureau of mines, *Coal Mine Explosions,* 137; *Crawford's Weekly,* Aug. 24, 1934, p. 4, cited in Canty, "Derby Goes Union," p. 35, at HLA.

after the strike was twenty-one: *Crawford's Weekly,* Aug. 17, 1934, p. 4, and Aug. 31, 1934, p. 4, cited in Canty, "Derby Goes Union," p. 35, at HLA.

"and Franklin D. Roosevelt is trying to free us all.": *Crawford's Weekly,* Dec. 20, 1933, p. 4, cited in Canty, "Derby Goes Union," pp. 33–34, at HLA.

Chapter 12

Page 165

he could produce fifteen tons or more: DiCiccio, *Coal and Coke,* p. 198; also notes of speech by John Saxton, president of UMWA District 28, May 1939, at HLA.

more than ninety percent of all miners had joined the United Mine Workers: DiCiccio, *Coal and Coke,* p. 196.

Page 166

"provided these men report to work tonight.":, John D. Rogers, VP, Stonega C&C Co., to N.H. Ingles, Superintendent, Roda, Va., at HLA, Jan. 22, 1934.

Page 167

which, of course, was to be expected: John D. Rogers in Stonega to Ralph E. Taggart in Philadelphia, April 5, 1934, at HLA.

"will certainly be an improvement.": John D. Rogers in Stonega to Ralph E. Taggart in Philadelphia, June 15, 1934, at HLA.

"Let us, therefore, live in hopes that there is a better day coming.": H. L. Good, Westmoreland Coal Co., Irwin, Pa., to John D. Rogers, VP, Stonega C&C Co., Big Stone Gap, Va., June 22, 1934, at HLA.

Page 168

"as there was no apparent effort on the part of the men to hide them.": Report by Joseph F. Davis, district engineer, U.S. Department of the Interior/ Bureau of Mines, pp. 1–24, at HLA.

for nearly forty years: Zieger, *John L. Lewis,* p. 151. The Leisenring Group did suffer a fatal methane blast at a mine in Hampton, W.Va., in the 1970s, which killed six men.

Page 169

"and it requires time and attention.": John D. Rogers in Stonega to Ralph E. Taggart in Philadelphia, Sept. 17, 1934, at HLA.

to nearly five million: Zieger, *John L. Lewis,* p. 75.

Page 170

from Indianapolis to Washington: Zieger, *John L. Lewis,* p. 46.

"to the level of a great crusade.": Zieger, *John L. Lewis,* p. 97.

"for those to work who desired to do so.": Stonega annual report, 1938, at HLA.

"WHAT DIRECT ACTION WILL UNION OFFICIALS TAKE IN THIS EMERGENCY?": Telegram, to John L. Lewis from Ralph H. Knode, president, Stonega Coke & Coal Co., Philadelphia, April 21, 1937, at HLA.

Page 171

"PLEASE ADVISE BY WESTERN UNION THIS OFFICE IMMEDIATELY OF THE ACTION TAKEN BY YOUR LOCAL UNION.": Telegram, John L. Lewis to John Saxton, president, District 28, UMWA, Norton, Va., April 23, 1937, at HLA.

"Needless to say, we are all delighted.": E. B. Leisenring Sr. to John D. Rogers, VP, Stonega C&C, confidential, April 24, 1937, at HLA.

"I know they will think it strange if they do not receive some remembrance from me at the time of the funeral.": D. B. Wentz Jr., VP, Philadelphia, to E. J. Prescott, VP, Big Stone Gap, May 2, 1938, at HLA.

he was seventy-six by then and long since finished as a miner: Author's interview with Don Givens, Sept. 12, 2002.

Page 172

"needing credit at our commissaries.": J. D. Rogers, VP, Stonega, to all superintendents, April 13, 1939, at HLA.

restore full employment for the first time since 1929: Mark Helprin in *Wall Street Journal,* Sept. 16, 2002.

Page 173

"but they will want to work every day possible.": E. B. Leisenring Sr. to John Saxton, President District 28, UMWA, March 21, 1942, at HLA.

"John Lewis, damn your coal-black soul.": Zieger, *John L. Lewis*, p. 141.

Page 174

Young Leisenring was permanently smitten: Author's interview with Ted Leisenring, April 22, 1998.

Page 175

about a mile from the mine: I am indebted to Rod Brewster, a former resident of Amonate, for background on the town. See also the Tazewell County website at http://www.rootsweb.com/~vatazewe/CoalMining.htm.

His widow and children were left to subsist on their Social Security survivors benefits: Author's interview with Clara Givens, Nov. 14, 2002.

and both subsequently worked their way up to jobs as supervisors: Author's interview with Don Givens, Nov. 14, 2002.

On the spot he bought a farm near Sparta and soon settled his wife, his daughter, and his daughter's four children there: Author's interview with Gladys Givens Belcher, Nov. 7, 2002.

Page 176

where it could still compete with oil and gas: DiCiccio, *Coal and Coke*, pp. 197–98.

Page 177

and he was right: Author's interview with Ted Leisenring, Oct. 28, 2001.

And virtually no coal operator's son had ever joined a miner's union anywhere: Ted Leisenring's recollection. His friends and contemporaries, Herbert Jones of Amherst Coal Co. and Morgan Massey of A.T. Massey, also worked in the mines for their families' companies.

Page 178

"a good one to get on with.": Author's interview with Don Givens, Sept. 14, 2002.

Page 179

"Your daddy sure must hate the shit out of you to git you a lousy job like this.": Author's interview with Ted Leisenring, April 24, 1998.

Page 180

and Julie's first baby was delivered at the hospital in Kingsport: Author's interview with Julia Leisenring, Nov. 30, 1998.

Page 181

only to die soon after, in 1951, of black lung disease contracted in the course of his many years in the mines: Author's interview with Albert's granddaughter Gladys Givens Belcher, Nov. 7, 2002.

as a result of a mule's momentary panic in 1910: Don Givens, in an interview with the author Nov. 7, 2002, said Sam retired while Don was away in Korea, i.e., 1952–1954.

"It was better in the mines," he replied: Author's interview with Don Givens, Oct. 27, 2001.

Page 182

the company backed off and restored the men's jobs: Author's interview with Ted Leisenring, April 24, 1998.

Page 183

death-benefit checks of $1,000 each sent to ninety-nine families: : Finley, *Corrupt Kingdom,* p. 180.

and dissolve its medical department by 1957: *A Brief History of the Associated Companies,* p. 129, at HLA.

even as droves of miners were retiring and expecting pensions: Zieger, *John L. Lewis,* p. 171.

Page 184

"But what do you do with the fifty-year-old coal miner who loses his job in McDowell County, West Virginia?": Author's interview with Joseph Brennan, Oct. 6, 1998.

in a tiny apartment in the slums: Author's interview with Ted Leisenring, May 6, 1998.

both to investigate and to dramatize his cause: Finley, *Corrupt Kingdom,* p. 206.

Page 186

"I think that you will look back on this episode as though it were a mosquito trying to bite the ass of an elephant.": Author's interview with Ted Leisenring, April 24, 1998.

Chapter 13

Page 187

who had organized Harlan County for the UMW in the 1920s and '30s: Armbrister, *Act of Vengeance,* p. 76; also pp. 82–83.

Page 188

a District 19 official once remarked to Ted Leisenring: Author's interview with Ted Leisenring, April 24, 1998.

"or to close the mine until some other solution present itself.": Ted Leisenring to W. C. Schott, March 19, 1959, at HLA.

Page 189

The smell of death pervaded the industry even as the rest of the U.S. economy grew by 50 percent in the fifteen years after World War II: DiCiccio, *Coal and Coke,* p. 196.

its last Pennsylvania leasehold was sold to the Pittsburgh-Consolidation Coal Company in January of 1957: *A Brief History of the Associated Companies*, p. 9, at HLA.

the total exceeded two hundred years: Author's interview with Don Givens, Sept. 14, 2002.

Page 190

"everything just went right.": Author's interview with Don Givens, Oct. 27, 2001.

He was not yet forty years old: Author's interview with Don Givens, Nov. 5, 2002.

and built a house next door to his for his father, Saylor: Author's interview with Don Givens, Nov. 8 and Nov. 12, 2002.

As early as 1947 a survey found only three-fifths of American mining families still living in company-owned housing: DiCiccio, *Coal and Coke,* p. 196.

Page 191

Stonega set a book value of $230 per dwelling: Shiflett, *Coal Towns*, p. 210.

"Now the big companies give national leadership to the industry side.": Finley, *Corrupt Kingdom*, p. 174.

Page 192

"since May 1957, we do not seem to get the required cooperation from all the men.": Undated memo (1957 or later) from W. C. Scott, VP/general manager, Stonega Coke & Coal Co., at HLA.

"but if one should result, it would be more desirable now than after working the mine a longer period.": E. B. Leisenring to W. C. Scott, March 19, 1959, at HLA.

Page 194

and the Stonega name vanished from public use: See *A Brief History of the Associated Companies,,* pp. 9–11, at HLA; Samuel Ballam speech, Oct. 31, 1973, at HLA; DiCiccio, *Coal and Coke,* p. 197 for assets of largest coal companies. Similar naming situations have occurred in other mergers. In 1973, for example, the investment house I.W. Burnham acquired Drexel Firestone but named the merged firm Drexel Burnham, in deference to the prestige of the Drexel name. Similarly, Chase Manhattan Bank acquired J. P. Morgan & Co. in 2000 and named the merged firm J. P. Morgan Chase.

identified her as one of the hundred most influential Americans of the twentieth century: Moxell, "Environmental Movement at 40."

Page 196

And the experience offered just a taste of the sharply steeper inflation that lay ahead: Blume, Siegel, and Rottenberg, *Revolution on Wall Street*, pp. 55–57.

Page 198

"Things were getting too easy for us.": Ted Leisenring to Dwight L. Allison Jr., vice president of C. H. Sprague & Son, Sept. 3, 1968, at HLA.

Page 199

that was unlikely to threaten Ted's control: For Westmoreland's acquisition of C. H. Sprague & Son, see Baer, *Guide*, p. 172, at HLA; sale agreement dated Sept. 17, 1968, in HLA, Series 1765, Box 749; Dun & Bradstreet report, Sept. 25, 1968, at HLA; *Forbes*, Sept. 1, 1971; also author's interview with Ted Leisenring, April 24, 1998.

Page 200

which led Congress to pass the federal Coal Mine Health & Safety Act of December 1969: Baer, *Guide*, p. 173, at HLA.

Page 201

so that only delegates whose expenses were paid by union headquarters could afford to attend: Armbrister, *Act of Vengeance*, p. 44.

"Don't support them!": Armbrister, *Act of Vengeance*, p. 51.

carried twenty-eight photos of Boyle in sixteen pages: Armbrister, *Act of Vengeance*, p. 57.

"and I've got to put up with them.": Impressions of Miss Richards are from author's interview with Ted Leisenring, April 24, 1998.

Then he returned to union headquarters in Washington: Zieger, *John L. Lewis*, pp. 182–3; also Armbrister, *Act of Vengeance*, p. 36.

Page 202

Boyle listened, nodded, and did nothing: Author's interview with Herbert Jones, Nov. 2, 1998.

and was known among officials and operators alike as the "cobra woman.": Author's interview with Ted Leisenring, April 24, 1998; also see Armbrister, *Act of Vengeance*, pp. 72–73.

He turned down the Wilson job only because the price was too low: Armbrister, *Act of Vengeance*, pp. 7–8.

But publicly he said nothing: Zieger, *John L. Lewis* p. 183.

Page 203

it was an obvious reprisal for Yablonski's political activity in the union: Finley, *Corrupt Kingdom*, pp. 258–9.

three days later he was dead at the age of eighty-nine: Zieger, *John L. Lewis*, p. 184.

Boyle answered, "Yeah," and turned away: Armbrister, *Act of Vengeance*, pp. 76–77; also *Philadelphia Inquirer*, Dec. 26, 1999.

"I'm going to go public with some of this stuff.": Author's interview with Ted Leisenring, April 24, 1998.

Page 204

that had barely been disclosed in the union's annual reports: Armbrister, *Act of Vengeance*, p. 144.

"We're going to fight this thing all the way.": Carawan and Carawan, *Voices from the Mountains*, p. 189.

four hundred yards away: Armbrister, *Act of Vengeance*, p. 3.

Page 205

led investigators into the UMW hierarchy and up to the union's very highest level: On the Yablonski murders, see Armbrister, *Act of Vengeance*; Finley, *Corrupt Kingdom*, pp. 272ff. One of the best brief summaries is Ditzen, "Deadly Union Rivalry Capped '60s."

ostensibly for organizing expenses: Ditzen, "Deadly Union Rivalry."

Chapter 14

Page 210

only the second time in U.S. history that a first-degree nurder verdict was rendered without clear evidence that the alleged victim was dead: Armbrister, *Act of Vengeance*, pp. 248–249.

Page 211

Miller's qualifications, or his lack of them, mattered less than Boyle's growing vulnerability: Clark, *Miners' Fight for Democracy*, p. 27.

and pledged full cooperation with authorities: Armbrister, *Act of Vengeance*, p. 179.

her suggested $100,000 reward in half: Armbrister, *Act of Vengeance*, pp. 325–326.

"That's perfectly obvious.": Armbrister, *Act of Vengeance*, p. 252.

Page 212

massive vote fraud and financial manipulation had been committed: Armbrister, *Act of Vengeance*, pp. 287–288.

on the embezzlement charges: Armbrister, *Act of Vengeance*, pp. 294, 296.

Arnold Miller had 70,373 votes to 56,334 for Boyle: Armbrister, *Act of Vengeance*, pp. 296–297.

"Nothing like it had ever happened in the labor movement before.": Clark, *Miners' Fight for Democracy*, pp. 30–31.

"I do not expect any more arrests.": Ditzen "Deadly Union Rivalry."

Page 213

even though no such authorization existed in the union's constitution: Clark, *Miners' Fight for Democracy*, p. 45.

Page 214

of U.S. business corporations: E. B. Leisenring Jr., *Report of U.S. Coal Industry Delegation*, p. 6, at HLA. Also author's interview with E. B. Leisenring, May 6, 1998.

Page 216

"but what do you do when you get it out?": Author's interview with Joseph Brennan, Oct. 6, 1998.

"We knew nothing about surface mining in the West.": *Brief History of the Associated Companies.*, pp. 12ff; "Westmoreland Has What It Takes," *Coal Mining and Processing*, December 1978; *Forbes*, Sept. 1, 1971; author's interview with E. B. Leisenring Jr., April 24, 1998.

Page 217

a totally different business in a strange and distant land: Author's interview with John Shober, Oct. 1, 1998.

Even a coal-rich state like West Virginia might buy its coal from Montana, Wyoming, and Colorado: See, for example, Ted Leisenring's speech to an energy symposium sponsored by Union Commerce Bank of Cleveland, June 5, 1974, p. 6, at HLA.

far exceeded the company's annual operating profits of between $4 million and $5 million: Ted Leisenring's speech to an energy symposium sponsored by Union Commerce Bank of Cleveland, June 5, 1974, p. 6, at HLA, refers to "an interest cost on borrowed money well in excess of our profit" over "the past three years." Drexel Burnham & Co. report, July 1976: "The Coal Report: Paradise Found," by Lawrence J. Goldstein and Andrew J. Melnick, at HLA, lists Westmoreland's net profit as $4.1 million in 1971, $5.1 million in 1972, and $4.7 million in 1973.

Page 218

and they were said to have gone eighty years in one stretch without having killed a white man: Toole, *Rape of the Great Plains*, pp. 33–39.

and a building housing the local offices of the Bureau of Indian Affairs: Crittenden, "Coal: Last Chance for the Crow," at HLA.

Page 219

"We don't want to see this place ripped off anymore.": Crittenden, "Coal," at HLA.

"For every four points of view, there are five factions.": Crittenden, "Coal," at HLA.

"Well, they were trespassing, you see.": Toole, *Rape of the Great Plains*, p. 119.

Page 220

wrote Toole in his book, *The Rape of the Great Plains*: Toole, *Rape of the Great Plains*, p. 9.

in exchange for a royalty of 17.5 cents per ton of coal: Crittenden, "Coal," at HLA.

beginning July 1, 1974: Westmoreland Resources, answer to Crow Tribe petition to Secretary of Interior to declare coal leases invalid, p. 12, at HLA.

to power plants in Minnesota and Illinois: Toole, *Rape of the Great Plains*, pp. 44–45.

Page 221

"than any rail line in history.": Hutchinson, "So You Think You Want to Mine on Indian Land," pp. 3–4, at HLA.

what the new regulations might contain: Hutchinson, "So You Think You Want to Mine on Indian Land," pp. 5–6, at HLA.

Page 222

"Much of this risk could be reduced through tax incentives.": E. B. Leisenring testimony before Congress, March 1973, p. 3813, at HLA.

Page 223

removal of all controls on the price of coal: Nixon's order occurred on March 27, 1974. See Edward B. Leisenring speech to an energy symposium sponsored by Union Commerce Bank of Cleveland, June 5, 1974, at HLA.

suddenly began to generate returns of 15 percent and higher: Ted Leisenring's speech to an energy symposium sponsored by Union Commerce Bank of Cleveland, June 5, 1974, p. 6, at HLA.

but there was no missing his meaning: Author's interview with John Shober, Oct. 1, 1998.

Page 224

and then to $60.2 million in 1975: Goldstein and Melnick, "Coal Report," at HLA.

and its earnings rose fifteen times: Goldstein and Melnick, "Coal Report," at HLA.

he told the *New York Times*: Crittenden, "Coal," at HLA.

"the most formidable negotiating team I have ever encountered.": Hutchinson, "So You Think You Want to Mine on Indian Land," p. 7, at HLA.

Page 225

But it still required approval by the full tribal council: Westmoreland Resources, answer to Crow Tribe petition to Secretary of Interior to declare coal leases invalid, p. 14, at HLA.

"Wow, it's a long way to Chestnut Hill.": Author's interview with Pemberton Hutchinson, Sept. 16, 1998.

Page 226

the company dryly noted in a subsequent court filing: Westmoreland Resources, answer to Crow Tribe petition to Secretary of Interior to declare coal leases invalid, p. 15, at HLA. Also Hutchinson, "So You Think You Want to Mine on Indian Land," p. 7, at HLA.

Westmoreland would proceed under the original royalty agreement: Toole, *Rape of the Great Plains*, pp. 44–45.

"to avoid ending up in a closet.": Hutchinson, "So You Think You Want to Mine on Indian Land," pp. 8–9, at HLA.

and an advance royalty payment of $628, 000: Pemberton Hutchinson, "So You Think You Want to Mine on Indian Land," p. 9, at HLA.

the Crow tribe annually received about \$200 per capita: Crittenden, "Coal," at HLA.
to invalidate the Shell Oil and AMAX leases altogether: Crittenden, "Coal," at HLA.

Page 227

he mumbled as he was led through a crowd of reporters: Ambrister, *Act of Vengeance*,
 pp. 317–318.
to begin serving his term: Ambrister, *Act of Vengeance*, pp. 320–321.
to relatives and friends he spotted: Ambrister, *Act of Vengeance*, p. 321.

Page 229

But he never followed up on it: Author interview with Ted Leisenring, April 24, 1998.
 Boyle's polygraph test was taken at the Chicago Polygraph Center on March 14,
 1974; a copy with the opinion of the polygrapher, James D. Walls Jr., is dated
 April 2, 1974. Both are contained with a letter from Ethel Boyle to Ted Leisenring,
 dated July 30, 1974, made available to the author by Ted Leisenring.

Chapter 15

Page 231

the operators had the interests of both parties at heart: Author's interview with
 Joseph Brennan, Oct. 6, 1998; with Ted Leisenring, June 6, 1998.

Page 232

as great as at any time over the previous quarter-century: Clark, *Miners' Fight for Democracy*, pp. 45 and 51.
by 1975 the number of lost days had risen to 2.8 million: Author's interview with
 Joseph Brennan, Oct. 6, 1998.
"a CEO who lived in the real world.": Author's interview with Joseph Brennan, Oct. 6,
 1998.

Page 233

or hadn't voted at all: Clark, *Miners' Fight for Democracy*, pp. 52–54.

Page 234

suggested the following year: Melnick, "Coal Report," at HLA.
"someone else can step in and do almost as well.": "Westmoreland Has What It Takes,"
 Coal Mining and Processing, Dec. 1978, at HLA.
that most coal companies had never explored the idea: The estimated cost is taken
 from Westmoreland's 1974 reply to the Crow Tribe petition to declare its Montana leases invalid, at HLA.

Page 235

"You fight for what you inherit.": Author's interview with Ted Leisenring, May 6, 1998;
 also see "Keep or Sell," p. 32.

snowballed into a strike that lasted seven weeks: Westmoreland memo on the March 1977 wildcat strike at Bullitt, at HLA.

than for the same period twelve months earlier: Ted Leisenring speech at West Virginia Tech commencement, May 14, 1977, pp. 2–3.

Page 236

"along with a return to higher productivity.": Ted Leisenring speech at West Virginia Tech commencement, May 14, 1977, p. 4.

in a speech in Bluefield, West Virginia, in September: E. B. Leisenring speech, "There Is One Way Out," to Bluefield Coal Show annual dinner, Bluefield, W.Va., Sept. 16, 1977.

a sharp drop from 76 percent a decade earlier: E. B. Leisenring speech, "There Is One Way Out," to Bluefield Coal Show Annual dinner, Bluefield, W.Va., Sept. 16, 1977.

and 180,000 mine workers across the country walked off their jobs: Clark, *Miners' Fight for Democracy,* pp. 117, 121–122.

Page 237

"then your membership is going to suffer with us.": Author's interview with Ted Leisenring, April 24, 1998.

Page 238

and succeeded only in undermining Ted's committee's credibility with Miller: Author's interview with Ted Leisenring, April 24, 1998.

"subjecting us to monologues about what he wanted to do.": Author's interview with Herbert Jones, Nov. 2, 1998.

for the friendlier confines of the Heart O'Town Motel in Charleston, West Virginia: Clark, *Miners' Fight for Democracy,* p. 123.

with a group of small coal operators: Clark, *Miners' Fight for Democracy,* pp. 125–126.

"What will you give us?": Author's interview with Joseph Brennan, Oct. 6, 1998.

Page 239

not to kill but as a warning: Author's interview with John Shober, Oct. 1, 1998.

In that basic decency lay the mystical appeal of coal to miners and managers alike: Author's interview with John Shober, Oct. 1, 1998.

Page 241

whom the negotiators found no more effective: Author's interview with Ted Leisenring, April 24, 1998.

or he could call for binding arbitration: Clark, *Miners' Fight for Democracy,* p. 127.

or, at least, not since Truman had nationalized the mines in 1946: "A Coal Emergency" *Newsweek,* Feb. 27, 1978, p. 18.

and one by the UMW's general membership: Clark, *Miners' Fight for Democracy,* p. 131.

"I suspect some of his remarks are for rank-and-file consumption.": *New York Times,* March 17, 1978, with Ted Leisenring's note, at HLA.

"or don't want to talk about it.": Author's interview with Joseph Brennan, Oct. 6, 1998.

Page 242

remained intact: Clark, *Miners' Fight for Democracy*, p. 129–30.

"and erupt into wildcat strikes.": Ted Leisenring speech to Coal Mine Health, Safety & Research Institute, Va. Polytechnic Institute, Aug. 29, 1979, at HLA.

their lowest level in ten years: Clark, *Miners' Fight for Democracy*, p. 158.

Page 243

and the union wouldn't forget that: Author's interview with Joseph Brennan, Oct. 6, 1998.

"The president of this union has got to have more control.": *New York Times*, Sept. 2, 1979.

he resigned the International presidency: Clark, *Miners' Fight for Democracy*, p. 136.

Ted said hopefully in a speech in 1979: Ted Leisenring speech to Coal Mine Health, Safety & Research Institute, Virginia Polytechnic Institute, Aug. 29, 1979, at HLA.

actually lost $9.8 million: *Philadelphia Inquirer*, Feb. 18, 1979.

at its unionized mines in Virginia, West Virginia, and Pennsylvania: Ted Leisenring speech to Clinch Valley College of University of Virginia, Wise, Va., May 25, 1980, p. 3.

"and calmly to the ups and downs of the business.": "Westmoreland Has What It Takes," *Coal Mining and Processing*, December 1978, at HLA.

has declared as a policy that 'coal is dirty,' and therefore unacceptable: Ted Leisenring commencement address to Mountain Empire Community College, Big Stone Gap, Va., June 15, 1979, pp. 3–4.

Chapter 16

Page 246

a resolution he kept the fourteen remaining years of his life: Author's interview with Don Givens, Nov. 8, 2002.

Page 247

by calling out Tim Belcher's name: Ringolsby, "Late Bloomer," p. 57.

was about to embark on a career inn Major League Baseball: Tim Belcher was born Oct. 19, 1961; graduated from Sparta-Highland High School 1980.

"really enjoy what they're doing.": Joe DiPietro, quoted in *Delaware County (Pa.) Daily Times*, June 5, 1980, at HLA.

Page 248

"it does not have to kill people.": Ted Leisenring speech to Coal Mine Health, Safety & Research Institute, Virginia Polytechnic Institute, Aug. 29, 1979, p. 2.

as well as their teenage brother: Author's interviews with Don Givens, Oct. 27, 2001 and Nov. 5, 2002.

was fined $300,00 by the West Virginia Department of Mines: *Philadelphia Inquirer,* Feb. 20, 1984.

Page 250

who could freely negotiate with teams of their choice: Flood's obituary, *Philadelphia Inquirer,* Jan. 21, 1997, page F1.

Page 251

had been transported from the British Isles two centuries earlier: Cooper, "Tim Belcher," p. 37.

"within three years.": Cooper, "Tim Belcher," p. 37.

Page 252

proclaimed Ted Leisenring and Pem Hutchinson America's leading coal executives: "TWST names Leisenring and Hutchinson best chief executives, coal industry," *Wall Street Transcript,* Jan. 16, 1984.

sold the company to Gulf Oil for $325 million: *New York Times,* March 17, 1981, p. D1; *Wall Street Journal,* March 17, 1981, p. 2.

"and I don't want to take the risks.": Author's interview with John Shober, Oct. 1, 1998.

"No other coal company has been doing anything like that for its employees.": Associated Press article in *Richmond (Va.) Times-Dispatch,* May 31, 1981.

Page 253

whose coal was no longer needed by America's shrunken steel industry: *Philadelphia Inquirer,* Jan. 25, 1988.

Page 254

it was the only wise thing for his son to do: Author's interview with Ted Leisenring, May 6, 1998.

Page 255

"you can do good for people.": Associated Press report in *Kentucky Post,* Cincinnati, Ohio, Oct. 9, 1992, p. K6.

with the prospect of exponential rises in the years to come: Ted Leisenring, speech to Virginia Coal Council, Roanoke, Oct. 29, 1986.

who proved equally helpless to reverse Westmoreland's fortunes: *Philadelphia Inquirer,* May 9, 1988.

"I couldn't believe the company would come back.": Author's interview with Ted Leisenring, Dec. 11, 2002.

most of them owed to the UMW's medical fund: *Philadelphia Inquirer,* April 19, 1994 and Nov. 9, 1994; also author's interview with Ted Leisenring, May 6, 1998.

Epilogue

Page 259

and by 1999 overall coal production from surface mines in the West had eclipsed that from the East: *New York Times,* June 16, 2001.

"It's a different industry.": Author's interview with Ted Leisenring, May 6, 1998.

Page 260

it was a source of amusement: Author's interview with Ted Leisenring, May 6, 1998.

had been cut almost in half, to 75,000: *New York Times,* Aug. 12, 2001.

its membership had declined to 30,000: *New York Times,* June 25, 2001.

plus benefits worth an additional $6 an hour: *New York Times,* Aug. 12, 2001.

and more than half of all miners had at least some college education: Author's interview with Joseph Brennan, Oct. 6, 1998.

Page 262

but on the Virginia side most of the homes are still occupied: I am indebted for this information to Rod Brewster, who grew up in Amonate from 1929 to 1945 and visited there in 2001.

Page 263

Belcher responded by ordering a pizza to be delivered to McCoy in the press box: Shannon, *Tales from The Dugout,* p. 147.

"All I need is Olive Oyl and some spinach.": Shannon, *Tales from The Dugout,* p. 12.

of the game just ended : Shannon, *Tales from The Dugout,* pp. 11–12.

Page 264

they simply spelled her name as they had always pronounced the word: Author's interview with Hagy Barnett, Oct. 27, 2001.

Page 265

"Not many people do that.": Author's interviews with Hagy Barnett, Don Givens, Beecher Powers, Hyson Barnett, Big Stone Gap, Va., Oct. 27, 2001.

for decades to come: *New York Times Magazine,* July 22, 2001.

in the East and Midwest: "Panel Approves Huge Rail Plan to Move Coal," *New York Times,* Jan. 31, 2002.

declared in May of 2001: *Wall Street Journal,* May 10, 2001, p. 1.

Page 266

"We now do expect to be profitable going forward," Seglem announced in the fall of 2001: *Coal Age,* Nov. 29, 2001.

Bibliography

HLA: Archives of the Hagley Museum and Library, Wilmington, Del.

Books

Adams, Graham Jr. *The Age of Industrial Violence, 1910–1915* (New York, 1966).

Addington, Luther F. *History of Wise County, Virginia* (Wise County, Va.: Bicentennial Committee of Wise County, 1976).

Armbrister, Trevor. *Act of Vengeance: The Yablonski Murders and Their Solution* (New York: Saturday Review Press/E. F. Dutton, 1975).

Aurand, Harold. *From the Molly Maguires to the United Mine Workers: The Social Ecology of an Industrial Union, 1869–1897* (Philadelphia: Temple University Press, 1971).

Baer, Christopher T. *A Guide to the History of Penn Virginia Corporation and Westmoreland Coal Co.* Manuscript. (Wilmington, Del.: Hagley Library, 1984).

Baltzell, E. Digby. *The Protestant Establishment: Aristocracy and Caste in America* (New York: Vintage Books, 1964).

Blume, Marshal, Jeremy Siegel and Dan Rottenberg. *Revolution on Wall Street: The Rise and Decline of the New York Stock Exchange* (New York: Norton, 1993).

Burt, Nathaniel, *The Perennial Philadelphians* (Boston: Little, Brown, 1963).

Carawan, Guy and Candice Carawan. *Voices From the Mountains* (Chicago: University of Illinois Press, 1975).

Carnes, Cecil. *John L. Lewis: Leader of Labor* (New York: Robert Speller Publishing, 1936).

Carson, Rachel. *Silent Spring* (Boston: Houghton, Mifflin, 1962).

Clark, Paul. *The Miners' Fight for Democracy: Arnold Miller and the Reform of the UMW* (Ithaca, N.Y.: Cornell University Press, 1981).

Collier, Peter, and David Horowitz. *The Rockefellers* (New York: Holt, Rinehart & Winston, 1976).

Corbin, David. *Life, Work and Rebellion in the Coal Fields: The Southern West Virginia Miners, 1880–1922* (Urbana: University of Illinois Press, 1981).

DiCiccio, Carmen. *Coal and Coke in Pennsylvania* (Harrisburg: Penna. Historical and Museum Commission, 1996).

Dix, Keith. *What's a Coal Miner to Do?: The Mechanization of Coal Mining* (Pittsburgh: University of Pittsburgh Press, 1988).

Dubofsky, Melvyn. *John L. Lewis: A Biography* (New York: Quadrangle/New York Times Books, 1977).

Finley, Joseph E. *The Corrupt Kingdom: The Rise and Fall of the UMW* (New York: Simon & Schuster, 1972).

Fox, John Jr. *The Trail of the Lonesome Pine* (New York: Grosset & Dunlap, 1908).

Franklin, Benjamin. *The Autobiography* (Fort Worth: Harcourt Brace Jovanovich, 1959).

Groner, Alex. *History of American Business and Industry* (New York: American Heritage Press, 1972).

Hagner, Charles V. *Early History of the Falls of Schuylkill* (Philadelphia: Claxton, Remsen & Haffelfinger, 1869).

Hall, Frank. *Mines and Mining in Pennsylvania* (Harrisburg: Pennsylvania Department of Mines, 1919).

Harris, Joseph S. *Autobiography of Joseph Smith Harris (1835–1910),* manuscript (HLA).

Harvey, George B. *Henry Clay Frick, the Man* (New York: Scribner's, 1928).

Henson, Edward L. *General Imboden and the Economic Development of Wise County* (Wise, Va.: Historical Society of Southwest Virginia, 1965), at HLA.

Herrin, Dean Andrew. *From Cabin to Camp: Southern Mountaineers and the Coal Town of Stonega, Virginia* (Master's thesis, University of Delaware, August 1984).

History of Fayette County, Pennsylvania (Philadelphia: L. H. Everts, 1882).

Holbrook, Stewart H. *Age of the Moguls* (New York: Doubleday, 1953).

Hume, Brit. *Death and the Mines: Rebellion and Murder in the UMW* (New York: Grossman, 1971).

Humphrey, H. B. *Historical Summary of Coal Mine Explosions in the U.S., 1810–1958* (Washington, D.C.: Government Printing Office, 1960).

Josephson, Matthew. *The Robber Barons* (New York: Harcourt, Brace, N.Y., 1934).

Korson, George. *Coal Dust on the Fiddle: Songs and Stories of the Bituminous Industry* (Hatboro, Pa.: Folklore Associates Inc., 1965).

Laslett, John H. M. Ed., *United Mine Workers of America: A Model of Industrial Solidarity?* (University Park: Pensylvania State University Press, 1996.)

Levy, Elizabeth. *Struggle and Lose, Struggle and Win: The United Mine Workers* (New York: Four Winds Press, 1977).

Lewis, Arthur. *Lament for the Molly Maguires* (New York: Harcourt, Brace & World, 1964).

McDonald, David J. *Coal and Unionism: A History of the American Coal Miners' Unions* (Silver Spring, Md.; Indianapolis, Ind.: Cornelius, 1939).

McGovern, George S. and Leonard F. Guttridge. *The Great Coalfield War* (Boston: Houghton, Mifflin, 1972).

Montagu, Ashley, and Edward Darling, *The Ignorance of Certainty* (New York: Harper & Row, 1970).

Morton, Eleanor. *Josiah White: Prince of Pioneers* (New York: Stephen Daye Press, 1946).

Nicholls, C. S., ed., *Encyclopedia of Biography*. (New York: St. Martin's Press, 1996).

Richardson, Edgar P. "The Athens of America." In *Philadelphia: A 300-Year History*, ed. Russell Weigley (New York: Norton, 1982), pp. 208–257.

Richardson, Richard. *Memoir of Josiah White*. (Philadelphia: Lippincott, 1873).

Sanger, Martha Frick Symington. *Henry Clay Frick: An Intimate Portrait* (New York: Abbeville Press, 1998).

Schreiner, Samuel. *Henry Clay Frick: The Gospel of Greed* (New York: St. Martin's, 1995).

Shannon, Mike. *Tales from The Dugout* (Lincolnwood, Ill.: Contemporary Books, 1997).

Shiflett, Crandall A. *Coal Towns: Life, Work and Culture in Company Towns of Southern Appalachia, 1880–1960* (Knoxville: University of Tennessee Press, 1991).

Simpson, Henry. *The Lives of Eminent Philadelphians, Now Deceased, Collected from Original and Authentic Sources* (Philadelphia: William Brotherhead, 1859).

Thorndike, Joseph J., Jr. *The Very Rich: A History of Wealth* (New York: American Heritage Publishing Co., 1976).

Toole, K. Ross. *The Rape of the Great Plains: Northwest America, Cattle and Coal* (Boston: Atlantic-Little Brown, 1976).

Toothman, Fred R. *Great Coal Leaders of West Virginia* (Huntington, W.Va.: Vandalia, 1988).

Triolo, Melvin. *The Black Debacle: From a Thundering Voice to a Confused Whimper* (Parsons, W.Va.: McLain, 1991).

Wainwright, Nicholas B. "The Age of Nicholas Biddle." In *Philadelphia: A 300-Year History*, ed. Russell Weigley (New York: Norton, 1982), pp. 258–306.

Warner, Sam Bass. *The Private City* (Philadelphia: University of Pennsylvania Press, 1968).

Warren, Kenneth. *Triumphant Capitalism: Henry Clay Frick and the Industrial Transformation of America* (Pittsburgh: University of Pittsburgh Press, 1996).

Weigley, Russell, ed. *Philadelphia: A 300-Year History* (New York: Norton, 1982).

Whitford, Noble E. *Whitford's History Of New York Canals*. Vol. 2 (Albany: Brandow Printing Company, 1906).

Zieger, Robert H. *John L. Lewis: Labor Leader* (Boston: Twayne, 1988).

Selected Articles

Adams, Sean. "Different Charters, Different Paths: Corporations and Coal in Antebellum Pennsylvania and Virginia." *Business and Economic History 27* (Fall 1998), pp. 78–90.

Boyer, Peter J. "Rescue at Quecreek." *The New Yorker,* Nov. 18, 2002, pp. 57–73.

Canty, William S., III. *Derby Goes Union: The Decline of Welfare Capitalism and the Rise of Labor Unionism in a Southern Appalachian Coal Town, 1922–33.* Senior thesis, Yale University, October 1974.

Cooper, Craig. "Tim Belcher: Baseball through the back door." *The Sporting News,* May 14, 1984, p. 37.

Crittenden, Ann. "Coal: Last chance for the Crow." *New York Times,* Jan. 8, 1978, Sec. 3, p. 1.

Ditzen, L. Stuart, "Deadly Union Rivalry Capped '60s." *Philadelphia Inquirer,* Dec. 26, 1999.

Frost, William Goodell. "Our Contemporary Ancestors in the Southern Mountains," *Atlantic Monthly,* March 1899, pp. 311–319.

Geier, Ione. "Energy Pulsed Through Downtown Pottsville." *Pottsville Republican & Evening Herald,* June 26, 1993.

Goldstein, Lawrence J., and Andrew J. Melnick. "The Coal Report: Paradise Found." Drexel Burnham & Co. report, July 1976, at HLA.

Holt, Sharon Ann. "The Life and Labor of Coxe Miners," *Pennsylvania Legacies,* Winter 2001, pp. 7ff.

Hutchinson, Pemberton. "So You Think You Want to Mine on Indian Lands." Undated manuscript at HLA.

Irvine, Mrs. R. Tate. "Wise County in War Time: A Community History." Available in "New River Notes" (www.ls.net/~newriver/va/wiseww1.htm).

Judis, John B. "King Coal." *The American Prospect,* Dec. 16, 2002, pp. 8–9.

Lynch, Clay F. Lecture on Henry C. Frick to Westmoreland-Fayette Historical Society, June 18, 1949, at HLA.

Moxell, Hilary. "Environmental Movement at 40: Is the Earth Healthier?" *National Geographic News,* April 19, 2002.

Prescott, E. J. "History of the Virginia Coal & Iron Co." Manuscript, 1948, at HLA.

Ringolsby, Tracy. "Late Bloomer No. 1 in Draft." *The Sporting News,* June 20, 1983, p. 57.

Rottenberg, Dan. "Keep or Sell: One Family's Burden." *Family Business* magazine, August 2000, p. 32.

Shober, John A. H. "John L. Lewis: Power Baron." Senior thesis, Yale University, 1956.

Stevens, William S. "1802 Philadelphia." *Philadelphia Lawyer,* Winter 2002, pp. 44–48.

Wardley, C. W. "The Early Development of the H.C. Frick Coke Co." Address to Westmoreland-Fayette Historical Society, June 18, 1949. Available in Westmoreland Coal file at Hagley Library, Wilmington, Del.

Zagofsky, Al. "Lehigh River and Canal" (Series of Internet articles at www.enter.net/~lvcc/river.html, 1997).

"A Coal Emergency." *Newsweek,* Feb. 27, 1978, p. 18.

"TWST Names Leisenring and Hutchinson Best Chief Executives, coal industry." *Wall Street Transcript,* Jan. 16, 1984.

"Westmoreland Has What It Takes." *Coal Mining and Processing Magazine,* December 1978 (HLA).

Acknowledgments

From conception to completion, this book took five years. I shudder to imagine how long it might have required, or whether it could have been undertaken at all, without the generous help of many people whose knowledge of coal far exceeds mine.

First and foremost among these is E. B. (Ted) Leisenring Jr. It was Ted who originally suggested the book and directed me to the awesome trove of company and family records at the Hagley Library in Wilmington, Delaware. (As chairman of Westmoreland Coal, Ted had donated these papers to the Hagley and was eager to see someone put them to use.) Once the project was under way, Ted made himself available for countless hours of interviews, phone calls, and lunches, accompanied me on several trips, and introduced me to many useful people. Although Ted is one of the principal characters in this story and provided financial support for the endeavor, I can personally attest that his sole interest in this venture was intellectual. He neither sought nor received any control over the final manuscript, and readers should understand that any views expressed in this book are mine alone.

The papers of the Leisenring family's various coal companies on file at the Hagley Library, wonderful as they are, would have overwhelmed a solitary researcher like me had not an expert archivist first sifted through them and organized them. I am therefore immensely grateful to Christopher T. Baer, curator of the Hagley's Westmoreland collection, who organized the papers in 1984 and whose manuscript "Guide to the History of Penn Virginia Corporation and the Westmoreland Coal Company" provided a road map for my research. It would be an understatement to say that I could not have researched this subject without both his manuscript guide and his personal insights during the year or so that I spent at the Hagley. I am equally grateful to the Hagley's archives librarians, Marge McNinch and Katie Newell, for their unflaggingly cheerful and capable responses to my requests.

Almost without exception, the coal operators, miners, union officials, and other members of the coal industry's "family" whom I approached generously submitted to interviews and/or frequent phone calls. I'm especially grateful to Pemberton Hutchinson, John A. H. Shober, Joseph Brennan, Herbert E. Jones Jr., Julie Leisenring, Edward W. Leisenring, Don Givens, Hagy Barnett, Gladys Belcher, Clara Givens, Diane Jones of Westmoreland Coal Co., and Beecher Powers, a retired miner who is now curator of the Harry Meador Jr. Coal Museum in Big Stone Gap, Virginia. During my visit to Big Stone Gap I also benefited from the hospitality and guidance of Harry Meador III and his mother, Virginia Meador. Rod Brewster, a former resident of the coal town of Amonate, West Virginia, was very helpful with background information about that community.

For their assistance in procuring photos, I am grateful to Barbara Hall and Jon Williams of the Hagley Library in Wilmington, Delaware, Joseph Binford of the Free Library of Philadelphia, Thelma Blount of the United Mine Workers of America, David Stanhope of the Jimmy Carter Library in Atlanta, Lynda DeLoach of the George Meany Center for Labor Studies in Silver Spring, Maryland, and Burnell Yow of Raven's Wing Studio in Philadelphia.

Finally, this book could not have come to pass without the enthusiasm and critical eye of Karen Wolny, publishing director of Routledge, as well as her Routledge colleagues, most notably Nicole Ellis, who helped shepherd it through the production process; Norma McLemore, who performed an unusually detailed and astute copyedit; and Jaclyn Bergeron, who tied together many loose ends. For matching me with such a suitable editor and publishing house I am grateful to my agent, Louise Quayle, formerly of the Ellen Levine Literary Agency in New York.

Index

1. fractious -- inclined to make trouble
   peevish, unruly; cranky

2. trove - treasure trove

3. deprecate - disapproval of, belittle

4. aggrandizement - to make worse or more
   troublesome, exasperate or provoke

5. provincial - limited in perspective,
   narrow, self-centered

6. cooperage - making wooden barrels or tubs

7. alacrity - cheerful, willingness, eagerness
   speed or quickness

8. inculcate - to teach or impress by
   frequent instruction

9. in toto - totally, altogether

10. caprices - wild escapades
    playful leap or hop

11. putative - generally regarded as such

12. reticence - restrained or reserved in style
    inclined to keep one's personal
    affairs to oneself

13. nascent - coming into existence